Studies in palaeoeconomy and environment in South West England

edited by

N. D. Balaam, B. Levitan
and V. Straker

BAR British Series 181
1987

B.A.R.

122 Banbury Rd, Oxford, OX2 7BP, UK

BAR 181, 1987 : Studies in palaeoeconomy and environment in South West England.

ISBN 9780860545019 paperback
ISBN 9781407317991 e-book
DOI https://doi.org/10.30861/9780860545019
A catalogue record for this book is available from the British Library

This book is available at www.barpublishing.com

CONTENTS

**Please note that additional material is available to download from www.barpublishing.com/additional-downloads.html
The original foldouts have been reduced in size to match the A4 format of this book, the images are therefore not as clear as the original foldouts. Please refer to the original foldouts via the download for the original content.**

LIST OF CONTRIBUTORS

Nicholas Balaam	Central Excavation Unit, HBMC(E), Fortress House, 23 Savile Row, London, W1X 2HE.
Martin Bell	Archaeology Unit, Department of Geography, St. David's University College, Lampeter, Dyfed, SA48 7ED.
Tony Clark	19 The Crossways, Onslow Village, Guildford, Surrey, GU2 5QG.
Jennie Coy	Faunal Remains Unit, Department of Archaeology, University of Southampton, Highfield, Southampton, SO9 5NH.
Keith Crabtree	Department of Geography, University of Bristol, University Road, Bristol, BS8 1SS.
Andrew David	Ancient Monuments Laboratory, HBMC(E), Fortress House, 23 Savile Row, London, W1X 2HE.
Christopher Gerrard	7, Windsor Terrace, Clifton, Bristol, BS8 4LW.
Maureen Girling	*late of* Ancient Monuments Laboratory, HBMC(E), Fortress House, 23 Savile Row, London, W1X 2HE.
Louise Heathwaite	*formerly of* Department of Geography, University of Bristol, University Road, Bristol, BS8 1SS.
Julie Jones	93, Aubrey Road, Bedminster, Bristol, BS3.
Peter Leach	Birmingham University Field Archaeology Unit, Elms Road, off Pritchatts Road, P.O. Box 363, Birmingham, B15 2TT.
Bruce Levitan	University Museum, Parks Road, Oxford, OX1 3PW.
Alison Locker	"Cobblers", Old Hill, Wherwell, nr Andover, Hampshire, SP11 7JB.
Richard Macphail	Department of Human Environment, University College Institute of Archaeology, 31-34 Gordon Square, London, WC1H 0PY.
David Maguire	Department of Geographical Sciences, Plymouth Polytechnic, Drake Circus, Plymouth, Devon, PL4 8AA.
Barbara Noddle	Department of Anatomy, University College, P.O. Box 78, Cardiff, CF1 1XL.
Julia Rand	*formerly of* Department of Geography, University of Bristol, University Road, Bristol, BS8 1SS.
Mark Robinson	University Museum, Parks Road, Oxford, OX1 3PW.
Sheila Ross	Department of Geography, University of Bristol, University Road, Bristol, BS8 1SS.
Rob. Scaife	Ancient Monuments Laboratory, HBMC(E), Fortress House, 23 Savile Row, London, W1X 2HE.
Ian Simmons	Department of Geography, Science Laboratories, University of Durham, South Road, Durham, DH1 3LE.
Vanessa Straker	Department of Geography, University of Bristol, University Road, Bristol, BS8 1SS.
Charles Thomas	Institute of Cornish Studies, Trevenson House, Pool, Redruth, Cornwall, TR15 3RE.
Dominique Vaughan	Department of Urban Archaeology, Museum of London, London Wall, London EC2Y 5HN.
Nick Watson	Top flat, 12 Meridian Place, Clifton, Bristol, BS8.
Bridget Wilkinson	15, Manor Drive, Southgate, London N14 5JH

EDITORS' NOTE

The papers in this volume were presented first at a Symposium held in Bristol on 9th February, 1985. Apart from the review of Environmental Archaeology in south west England by Martin Bell (*in* H.C.M. Keeley, 1984. *Environmental Archaeology: A Regional Review*), the publication of palaeoenvironmental work in the South West has been scattered in a variety of Journals, reflecting the diversity of the research. As a result of this, it was decided to publish this collection of papers, as a follow-up to Bell's review.

Nicholas Balaam
Bruce Levitan
Vanessa Straker

PROLOGUE

Charles Thomas

Those who live and work in south west Britain are conscious of belonging to a region with a very special character. On the debit side, it is large and the process of academic communication can be unwieldy. No one place is central, the three universities (Exeter, Bristol, Bath) are concentrated in the eastern half, and at present there is an uneven distribution of full-time archaeological units, major museums with supporting facilities, independent researchers and even of established archaeological or environmental societies. Extremes of distance are not necessarily appreciated from London. A line drawn from the Isles of Scilly or the Land's End to Cheltenham - and 'regional' meetings of, for instance, the national health administration are often held at Cheltenham - is actually *longer* than a line from Cheltenham to the Scottish border. The shape of the extended peninsula implies many hundreds of miles of coastline, with exposed sections, dunes, shore habitats, differing tidal amplitudes, estuaries and islands. So incredibly rich a bio-archaeological and environmental backdrop merits, on its own, a corresponding service of monitors and researchers; one that, however, the region's sparse provisions cannot begin to afford, thereby reducing many discoveries to chance exposure.

Similar comments might of course be made (and often are made) about Scotland, where surprisingly large tracts remain unexplored by specialists. For the South West, there is a credit side. I have used the word 'rich', and advisedly so. Can any other region of England and Wales cover such a spectrum; from old river valleys, gravels, the edge of the chalklands, past the Mendip limestone and the concave landscape of the Levels to the granite bosses with their metamorphic aureoles? Not appreciably altered, save in surface vegetation and water-levels, since the third millenium BC, so diverse a landscape has through its components profoundly controlled the precise nature of clearance, settlement and exploitation. The archaeological opportunities, here, are fully matched by the palaeo-environmental ones.

Driving from, say, Bristol to Truro one might be forgiven in thinking that modern occupation (and agriculture) has obscured too much. A flight along the same route would correct that impression. High moorland - Exmoor with the Quantocks, Dartmoor, Bodmin Moor and eventually West Penwith, the Land's End stretch - still persists, and a former notion that each possessed a broadly similar and archaeologically detectable history has now been replaced by the realisation that, environmentally, they differ considerably. How considerably is probably apparent in the cases of Bodmin Moor (almost all exploited in the remote past) and Dartmoor (by no means able universally to support settlement), and we welcome renewed attention to Exmoor. But one can adduce any number of other factors that attract research. The region, without being another Burren or Peak District, has a wealth of caves; one thinks of the Kent's Cavern and Torbryan series and as for Mendip, that splendid body, the University of Bristol Spelaeological Society, maintains a *Proceedings* that deserves a far wider circulation than it gets. Nor is it wholly irrelevant to mention matters that, unexpectedly, do have distinct environmental relevance. In early historic times two unrelated languages interacted, Late British and Primitive Anglo-Saxon, by Tudor Times represented by Cornish and Middle English. Both left abundant place-names - general, descriptive and locative - that labelled environments since vanished, and the environmentalist concerned with the more recent past can use these (cf. Maxwell, 1958). With care, they can also be adduced for quite remote phases; well before the Normans, modern Somerset contained its British speakers. We need not be surprised to find that the ridge of the Poldens overlooking the Somerset

Levels was favoured for Summer grazing, since 'Polden' has, before the Old English suffix *dun* 'hill', a British element **bouo-gelt-* 'cow pasture'; and, far outside Cornwall, those places called 'Chideock' plain and compounded merely show the British adjective for 'wooded' (Turner, 1951).

This volume offers a selection, and it is both a fair and representative selection, of reports and views about the South West's early and not so early environment - results, conclusions, problems and challenges. It gives a most lively picture and the preceding, generalist, remarks should have led the reader to suppose that only a lively picture could emerge. The authors would, I know, readily agree that their work has matched an explosion of archaeological effort throughout the region since the 1950's, the days of the Bristol-masterminded motorway campaigns, CRAAGS, the efflorescence of the Council for British Archaeology's regional Groups XII and XIII and the great age of whole-region adult education symposia. At the level of individual archaeological societies, not all of this has been chronicled, though a few 'round-up' volumes have appeared (eg *Proceedings of the Devon Archaeological Society* **37**, Jubilee volume: "Prehistoric Dartmoor in its context", 1979 and *Cornish Archaeology* **25**, Silver Jubilee issue: "Work in Cornwall 1962-87"). Many Universities, from without as well as within the region, have contributed; nevertheless one must pinpoint the services of Bristol, the senior foundation. The Department of Botany formed the base for Dr. Lewis Frost's Lizard project, whose serial publications give Cornwall a unique record of one, ecologically vital, area. The Department of Geography has played a strong supporting role for the Spelaeological Society (Shaw, 1969) and through individual members of staff like Professor I.G. Simmons has proclaimed its palaeo-environmental commitment (cf. Simmons and Tooley, 1981).

The papers in this volume give a fair reflection of work in the region in the last decade or so. The (relative) absence of major, directly excavation-derived, analyses points up, as it should do, to two things. First is the contrast between the South West and (say) East Anglia, or the Thames gravels; there are simply fewer excavations west of the Wessex-Cotswold arc, and if this draws attention to long-running campaigns like the Somerset Levels project, caimpaigns of international significance, it also draws additional attention to the environmental emphasis within the work. Second is the indication, perhaps at one remove, of current intra-regional policies at city/county unit, or county committee, level. Given the realisation that (Exeter apart) Roman Britain peters out into a native countryside beyond the Devon-Somerset border, there is no particular requirement to plan for the contingency of massive, Roman, urban excavations; given the scaled-down nature of support in both fiscal and human resources - the population of Cornwall is one-*tenth* that of Kent! - there is a reluctance now to commit professional teams and amateur societies to lengthy excavations and their aftermaths.

The post-1970 renewal of interest in field survey, analysis of all preceeeding records, and landscape archaeology, has actually brought about a climate of opinion far more favourable to palaeoeconomic and environmental research than might have been the case, had the region moved strongly towards rescue-digging teams and little else. Admittedly the M5 Motorway campaign, still among the more remarkable of its kind ever undertaken, showed what *can* be done, but since then there have been no more motorways on this scale. And, allied to this (very welcome) present realisation of South West archaeology as, primarily, total-landscape archaeology must go a parallel movement - amateur as much as professional - in the natural sciences. The steady, specialist lead of the Bristol Spelaeologicals has been mentioned, but add to that the great influences within the region of the Devon Conservation Forum, Cornwall Trust for Nature Conservation, the Department of Geography at Exeter University, the Cornish Biological Records Unit, and others, (and, concerned with remoter time, the Ussher Society, and the various Quatęrnary enthusiasts who now contribute so much). It would be an interesting socio-archaeological footnote to find out how many archaeologists and environmentalists are also active in these sister-societies; one suspects very many. There has been an increased and heightened awareness, among those working in conventional pre- and proto-history, of the entire natural world. There is, and sadly there probably long will be, a large sector of those active in the natural sciences who seem quite unable (or unwilling) to take a hard, diachronic, look at their own specialist fields; bluntly, to appreciate that in the natural world the past (anthropogenic as well as genetic) has shaped the present to a far greater degree than could be claimed in archaeology and history. As we know, this particular myopia is most marked among the agencies concerned with cultural resource management. Yet - this can be freely admitted - there was once, long ago, a not dissimilar stance in the ranks of British archaeology....

My point is that field-workers in the South West - in any region- do seem to be moving towards what one might label 'a total palaeoenvironmental' perception of their investigations of the past, on all fours with that total-landscape approach noted earlier. Is this so, or is it a pious hope? The readers of these *Studies* must be the judges.

REFERENCES

Maxwell, I.S. 1958. The former distribution of vegetation in Roseland as deduced from place-names. *Trans. Papers Inst. British Geographers*, 143 ff.

Shaw, T.R. 1969. History of the Spelaeological Society. *Proc. Univ. Bristol Spelaeological Soc.*, **12(1)**, 9-30.

Turner, A.G.C. 1951. Some Somerset Place-names containing Celtic-Elements. *Bull. Board Celtic Stud.*, **14(2)**, 113-119.

Simmons, I.G. and Tooley, M. 1981. *The Environment in British Prehistory*, London: Duckworth

RECENT MOLLUSCAN STUDIES IN THE SOUTH WEST

Martin Bell

SOURCES OF EVIDENCE (Fig. 1)

The South West, as defined by Gloucestershire, Avon, Somerset, Devon and Cornwall, is by no means a geologically ideal area for subfossil mollusc analysis associated with archaeological sites. Chalk, which has produced such a wealth of data in adjacent Wessex (Evans, 1972), is confined to a tiny area around Beer. Limestone is much more extensive, forming Mendip and the Cotswolds, with some other small areas such as the Devonian Limestone near Torquay and the Lias Limestone near Watchet, both of which have proved to be of palaeoenvironmental interest. Many facies of the Cotswold and Mendip limestone do, however, weather slowly and this sometimes gives rise to non-calcareous soils which do not preserve molluscs. Physical weathering of relatively hard limestone also produces rock rubble deposits with vacuous interstices which Evans and Jones (1973) showed supported a specialized troglophile fauna (mainly *Oxychilus*, *Discus* and *Vitrea*) reflecting a shady microenvironment within the rock rubble itself rather than more general ecological conditions. Certain facies do result in much better preservation, for instance shells were much better preserved at Condicote than Hazleton (see below). On neither site were molluscs preserved in the old land surface but they were found in a possible palaeosol at Stoke Bishop. Furthermore, rock rubble is not the only sediment type associated with limestone country. Matrix supported sediments such as colluvium have produced ecologically meaningful assemblages at Brean Down and Catsgore. Also there are the tufas of limestone valleys and escarpments where mollusc preservation is of the very best and lengthy sequences can be found, as demonstrated by Martin Willing's recent research. In the coastal areas there is also the considerable potential of calcareous sand dune systems and the vast tracts of palaeo-land surface represented by submerged forests.

Even in geologically unpromising non-alkaline areas suitable conditions for mollusc preservation may be created by human activity. Some instances are surprisingly early such as Mesolithic Westward Ho!, where the import of marine molluscs to the site created calcareous conditions, and Neolithic Hazleton where the best preservation was associated with lenses rich in limestone liberated by physical weathering of the cairn. Later examples include Medieval contexts at Wells Cathedral (Bell, 1981a) and Okehampton Castle (Bell, 1982) where calcareous conditions were engendered by the import of limestone and mortar to the sites for building purposes.

Fortunately it is not necessary to summarise all subfossil mollusc work in the South West since that has recently been done (Bell *et al*, 1984). That survey and this paper do not consider Pleistocene evidence, though its importance has recently been demonstrated by the work of Gilbertson (eg 1979) and Ellis (1983). This paper is concerned with the general sequence of environmental change and some problems emerging from work on post-glacial land molluscs.

(a) The Wildwood

Throughout most of the first half of the postglacial virtually the whole of Britain was covered by dense climax forest, precise analogues for which are probably not to be found in the relict patches of drained and managed woodland which survive. Mollusc evidence for the climatic optimum wildwood comes from Kerney's (1976a) work on tufaceous loam at Tornewton Cave and a submerged forest at Blue Anchor Bay. Another assemblage is reported in this volume from Westward Ho! midden. Martin Willing (pers comm) has recently worked on a number of important tufa sequences. Willing has noted that it cannot be conclusively shown that the tufa forming habitats were wooded because the presence of some species may be explained by conditions of high humidity rather than shade. In the case of Westward Ho! and Blue Anchor Bay the associated pollen and macrofossil evidence

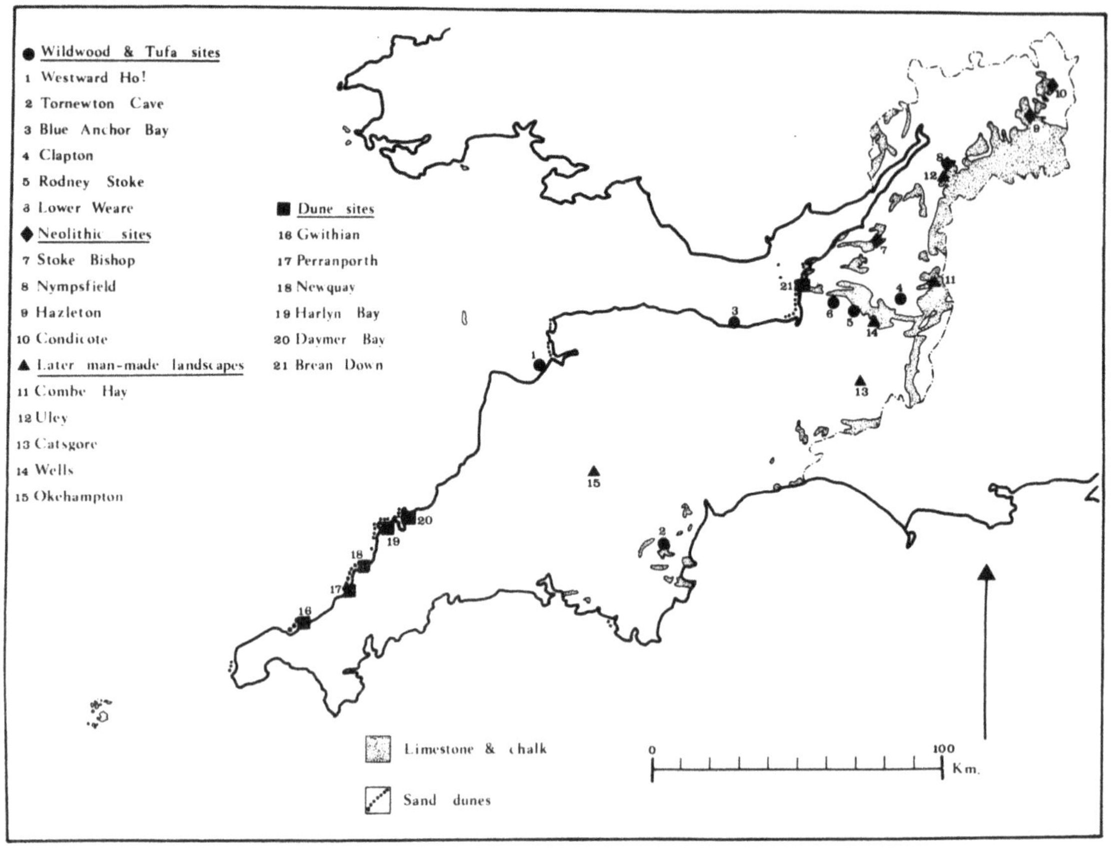

Figure 1. Map showing distribution of the main postglacial archaeological and palaeoenvironmental sites in the South West where molluscs have been studied.

(including wood) confirms that we are dealing with wooded environments and shows that tree cover came right down to the coast. This is also shown by Spencer's (1975) evidence from Newquay and calls into question the importance of the oft quoted exposure factor in preventing woodland growth in coastal areas.

A number of wildwood and tufa assemblages include species like *Vertigo pusilla* and *Spermodea lamellata* (the latter absent at Westward Ho!) which are today extinct in the South West and *Leiostyla anglica* which is today confined to a few isolated sites in the area (Kerney, 1976a). The reduction in the ranges of these and other species is probably largely the result of anthropogenic impact on woodland and marsh habitats (Kerney, 1966).

(b) Early Farmers

This evidence comes largely from Alan Saville's excavation of two Neolithic monuments on the Cotswolds. At Condicote Henge (Bell, 1983b) there was quite a large assemblage from the ditch but nothing in the very shallowly sealed palaeosol. The basal ditch levels contained a shade-loving assemblage particularly *Carychium tridentatum*, *Vitrea contracta*, *Aegopinella pura*, *Acanthinula aculeata* and *Discus rotundatus*. These shade-loving species might relate in part to a damper ditch microenvironment but the fact that they are preserved from the base of the primary fill, where molluscs are generally only found in tiny numbers, seems to suggest that the henge lay within a landscape which was not yet fully and permanently open and that following monument construction some regeneration occurred quite quickly. Associated with a peak in mollusc numbers, including woodland species, is a charcoal layer radiocarbon dated 1770±80 BC (HAR-3064). Sometime after this the ditch stabilized and a soil horizon containing Romano-British pottery formed. This seems to be associated with an hiatus in mollusc preservation because at this point there are very low numbers and a sudden ecological change. The shade-loving assemblage is replaced by a restricted group of open country species in the tertiary fill which is interpreted as the result of ploughing of the earthworks.

At Hazleton long cairn (Saville, 1984; Bell, 1983d) mollusc numbers in most contexts were low, probably because weathering of the Hampen Marley Beds did not give rise to soils which were sufficiently calcareous for mollusc preservation except in contexts where physical weathering of limestone exposed on the surface led to the incorporation of abundant small limestone pieces. Because of low numbers the mollusc diagram presented here (Fig. 2) has been prepared on the basis of absolute numbers of molluscs per 2kg dry soil. The bottom part of this diagram shows the old land surface below the cairn. Generally this light brown, limestone free soil contained no molluscs at all but at the point sampled there was a calcareous subsoil pocket at the base of the palaeosol. Even here numbers were low but the evidence they present fairly conclusive. The main species were *Carychium tridentatum*, *Oxychilus cellarius*, *Aegopinella pura*, *Vitrea contracta* and *Acanthinula aculeata* and they suggest shady woodland conditions. They are also suggested by the presence of *Acicula fusca* which is generally uncommon on Neolithic and later sites. Also significant is the presence of *Vertigo alpestris* which is today confined to small areas in North Wales, the Lake District and Scotland but was much more widespread in the mid-Flandrian wildwood. This subsoil feature is interpreted as a tree throw pit and soil development above it shows that the feature significantly predates construction of the cairn. It is unfortunate that no mollusc evidence survives from the period immediately preceding cairn construction, though fortunately there is good evidence from other palaeoenvironmental sources.

The upper part of Figure 2 shows a column of samples from the quarry close to the cairn on the north side. Three other columns including two from the south quarry were also analysed and produced a broadly similar sequence but smaller mollusc numbers. As usual molluscs were largely absent from the rapidly deposited primary fill but are relatively abundant at the base of the secondary fill. Here *Discus*, *Vitrea* and *Oxychilus* are among the most abundant species. To some extent this must relate to rock-rubble assemblages derived from the cairn, however this cannot be the whole explanation because these species are associated with many others including *Aegopinella* spp., *Carychium*, members of the *Punctum* Group (Evans, 1972: 195) and catholic species which are reportedly rare or absent in rock-rubble (Evans and Jones, 1973). Accordingly an episode of colonisation by shrubs or trees is envisaged at this time. At the top of the secondary fill is a stabilisation horizon at 76-90cm. This produced very low numbers and probably represents an hiatus in mollusc preservation as envisaged at Condicote. The subsequent tertiary fill produced a totally different assemblage much less rich in numbers and diversity and probably relating to cultivation over the monument.

One other site should be mentioned, a shallowly buried possible palaeosol beside the enigmatic Stoke Bishop megalithic tomb. A single sample produced an assemblage (Bell, 1983c) which, though not well dated, does not appear to be recent and consists largely of shade loving species. The only full mollusc sequence from a Cotswold Neolithic old land surface is Ascott-under-Wychwood, over the border in Oxfordshire (Evans, 1971), where there was good evidence for clearance and open grassland conditions prior to monument construction and for an episode of woodland regeneration in the quarry (Evans and Jones, 1973: 116).

Clearly more mollusc work needs to be done on this type of site and it will be necessary to select sites on the most favourable facies for preservation. The evidence we have so far hints that the Cotswold Neolithic landscape may not have been quite so uniformly

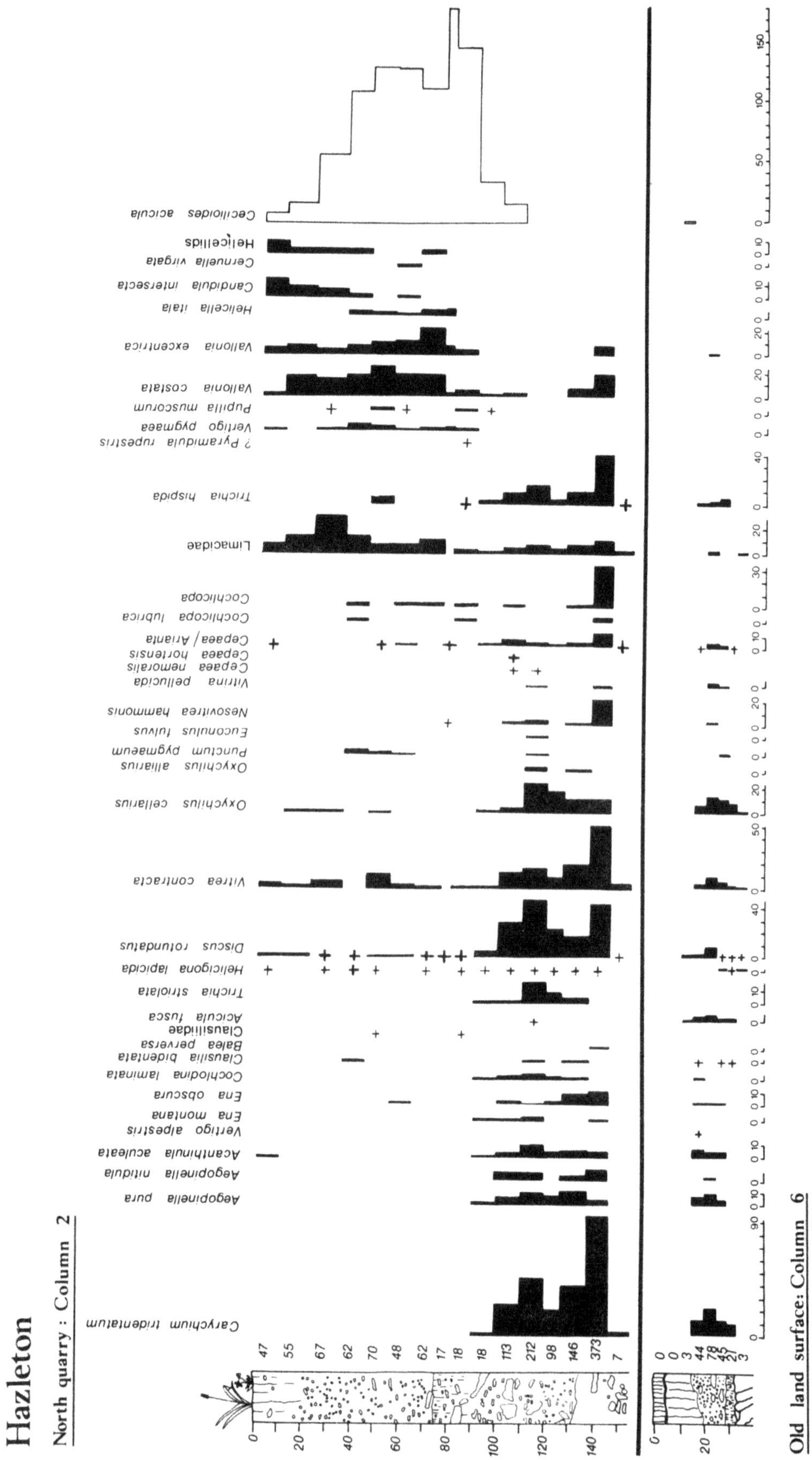

Figure 2. Mollusc diagram from Hazleton long cairn

cleared and tame as seems to have been the case in parts of Wessex. In fact spatial contrasts within the Neolithic landscape are becoming increasingly apparent. We now know for instance that most of the Neolithic monuments on the South Downs in Sussex were actually constructed in woodland (Thomas, 1982). On the Cotswolds there certainly seems to be evidence of later Neolithic scrub or woodland regeneration paralleling that for many other parts of the country (Whittle, 1978). There is also evidence for much later regeneration phases on other sites, such as Coombe Hay between the Bronze Age and Romano-British phases (Evans, 1980), and in association with certain phases of the Uley site (Collinson, 1980). All this is hardly surprising in view of the relatively extensive tree cover which is present on the Cotswolds even today.

(c) Man-made agricultural landscapes

Much woodland in prehistoric and later times would have represented not wasteland but a utilized and managed resource. Secondary and managed woodland seems to produce a more limited range of mollusc species than samples from wildwood contexts, but the real change from a molluscan perspective came with the clearance of woodland for agricultural purposes. Archaeological samples which are interpreted on sedimentary grounds as deriving from cultivated land produce a very restricted group of species chiefly *Trichia hispida*, the *Vallonias*, Limacidae and *Helicella itala*. We find this grouping in colluvium overlying a Roman road at Catsgore (Bell in Ellis, 1984), in the tertiary ditch fills at Condicote and Hazleton and in the colluvial sequences on the chalk in the South East (Bell, 1983a). The temptation is to suggest that this is a specific association of species characteristic of prehistoric to early Medieval arable environments in the same way that synecologically defined associations of species have been identified in, for example, rock-rubble (Evans and Jones, 1973) or the early stages of chalk grassland successions (Cameron and Morgan-Huws, 1975). The trouble is that it is difficult to prove this by reference to modern analogues because arable situations today tend to be dominated by Hellicelids which were apparently introduced into most parts of the country during the Medieval period. Furthermore modern arable environments are very different from those of the periods in question (Bell, 1986).

Another aspect of the man-made agricultural landscape is grassland. Though this is well represented by mollusc evidence in Wessex (Evans, 1972) and by other forms of palaeoenvironmental evidence from the South West (Bell *et al*, 1984), including the plant macrofossil evidence from Catsgore (Girling and Paradine in Ellis, 1984), it is notable that we have very little evidence of grassland conditions from Mollusca. This does not reflect environmental conditions but those of preservation; the old land surfaces examined are mostly devoid of molluscs and even stabilisation horizons in ditch sequences seem to be associated with hiatuses. This highlights a major problem for mollusc research in the South West, *ie* that preservation is often very closely related to the nature and degree of human activities. Those which give rise to the mixing of limestone into the soil (eg cultivation, monument weathering, etc) will be represented in the mollusc record whereas landuse which involves minimal bedrock disturbance will not. One solution is to use various types of palaeoenvironmental evidence in a comparative way, as was being attempted at Hazleton.

(d) Coastal Dunes

Around the coasts of the South West are extensive dunes, most containing significant proportions of sand derived from comminuted marine shells and calcareous algae. This means that land molluscs are very well preserved except in certain areas like the Scillies where there is a larger proportion of granite derived sand. Examination of a mollusc sequence from the stub end of an eroded relict dune system at Towan Head, Newquay (Spencer, 1974; 1975) demonstrated that, prior to the onset of sand blow, even this exposed coastal situation had carried woodland. Spencer also examined old land surfaces below dunes at Harlyn Bay and Perranporth and these produced open country assemblages. The dunes themselves on all these sites produced assemblages reflecting increasingly dry and open environments with some evidence for occasional stabilisation and scrub growth.

Dunes are of particular interest because they are important habitats for several species introduced to Britain during the later Post-glacial (Evans, 1979); these include *Cochlicella acuta*, *Candidula intersecta* and *Cernuella virgata*. The first of these is still mainly restricted to coastal environments in Britain but the other two species probably colonised in the Medieval period to become the major elements in the fauna of, for example, arable land. The question is, at what date did these species arrive in Britain? *Cochlicella acuta* was present at Bantham, Devon in a hearth dated 260±80 AD (HAR-5775) and 510±80 AD (HAR-5776) with which they are certainly associated because one apex was burnt. Spencer's work suggests that this species was also present in prehistory, eg in the early Bronze Age at Gwithian where there is also evidence for cf *Candidula intersecta*. At Brean Down recent work has produced *Cochlicella acuta* and cf *Candidula intersecta* from sediment directly associated with late Bronze Age bracelets (Crabtree, 1984). As regards dates of introduction, elements of uncertainty remain because some of the sites studied are coastal exposures rather than excavations and several species problems are presented by dune stratigraphies, notably slumping and winnowing effects. The latter may remove sand and leave heavy artifacts and biota of different dates erroneously associated at the same stratigraphic horizon. Current research on the major Bronze Age sequence and overlying Sub-Roman cemetery at Brean Down (ApSimon *et al*, 1961; Bell and Straker, 1984) aims to confront these problems by means of a detailed examination of artifact distributions and comparative studies of a range of associated biological evidence.

It is remarkable how many of the coastal dunes in the South West are associated with archaeological sites with a long history of continuity. This is partly because dunes preserve archaeological sequences in an almost uniquely clear way. Another factor is the special opportunities which these areas offered for what Ashbee (1974) has described as an integrated land-sea subsistence economy. One aspect of this integrated economy is the creation of man-made soils, by the introduction of shell sand, the use of shells as fertilizer (eg at Halangy Down, Ashbee, 1974: 265) and the importing of seaweed as manure (eg at Gwithian, Fowler and Thomas, 1962; Bell, 1981b).

CONCLUSION

Despite the geologically unpromising nature of the South West it has afforded a good range of opportunities for molluscan research. These are important in extending our knowledge of past mollusc species distributions into a geographical area for which, until recently, we had little information. Perhaps the chief problem is that preservation seems so often to be related to the nature and degree of human activity. Consequently specific attempts need to be made to identify sites and lithologies suitable for mollusc preservation and particularly tufa sites. Key topics which require further work include the origins of the man-made agricultural landscape, the further investigation of submerged forests and coastal dune systems. In each case the most promising approach would seem to be comparative studies using a range of palaeoenvironmental sources.

ACKNOWLEDGEMENTS

I am grateful to Professor Ian Simmons and Dr. Keith Crabtree who advised on and encouraged much of this research at Bristol and Julie Jones and Nick Watson who helped with many of the analyses described. Thanks are also due to Dr. Martin Willing for his stimulating comments on an earlier draft.

REFERENCES

ApSimon, A.M., Donovan, D.T. and Taylor, H. 1961. The stratigraphy and archaeology of the late glacial and post glacial deposits at Brean Down, Somerset, *Proc. Univ. Bristol Spelaeol. Soc.* **9**, 67-136.

Ashbee, P. 1974. *Ancient Scilly* Newton Abbot: David and Charles.

Bell, M. 1981a. Palaeoenvironmental studies at Wells Cathedral 1980. Ancient Monuments Laboratory Report No. 3338.

Bell, M. 1981b. Seaweed as a prehistoric resource *in* D. Brothwell and G. Dimbleby (eds) *Environmental aspects of coasts and islands* (Symposia of the Association for Environmental Archaeology No. 1) Oxford: British Archaeological Reports (International Series **94**), 117-126.

Bell, M. 1982. Land Molluscs, *in* R.A. Higham, J.P. Allan and F.R. Blaylock, Excavations at Okehampton Castle, Devon Part 2, The Bailey, *Proc. Devon Archaeol. Soc.* **40**, 145-146.

Bell, M. 1983a. Valley sediments as evidence of prehistoric land-use on the South Downs, *Proc. Prehist. Soc.* **49**, 119-150.

Bell, M. 1983b. The land Mollusca and other palaeo-environmental evidence, *in* A. Saville, Excavations at Condicote Henge monument, Glos. 1977, *Trans. Bristol Gloucestershire Archaeol. Soc.* **101**, 39-45.

Bell, M. 1983c. Mollusc sample from the Cromlech, Stoke Bishop, Bristol. Ancient Monuments Laboratory Report No. 3999.

Bell, M. 1983d. Land molluscs from Hazleton Neolithic long cairn. Ancient Monuments Laboratory Report No. 4000.

Bell, M. 1984. Environmental archaeology in South West England, *in* H.C.M. Keeley (ed) *Environmental Archaeology: a regional review*, London: Department of the Environment (Directorate of Ancient Monuments and Historic Buildings Occasional Paper **6**) 43-133.

Bell, M. and Straker, V. 1984. Fieldwork at Brean Down Somerset 1983-4: An interim report. Lampeter: Saint David's University College, Archaeology Unit.

Bell, M. 1986. Archaeological evidence for the date, cause and extent of soil erosion on the chalk, *J. South East England Soils Discussion Group* **3**, 72-83.

Cameron, R.A.D. and Morgan-Huws, D.I. 1975. Snail faunas in the early stages of a chalk grassland succession, *Biological J. Linnean Soc.* **7**, 215-229.

Collinson, B. 1980. Uley snail report. Unpublished manuscript.

Crabtree, K. 1984. Two gold bracelets from Brean Down, Somerset, *Antiquity* **58**, 49-52.

Ellis, C. 1983. A second British fossil occurrence of the land snail *Lyrodiscus* (family Zonitidae) from Sun Hole, Cheddar, *Proc. Univ. Bristol Spelaeol. Soc.* **16**, 191-2.

Ellis, P. 1984. *Catsgore 1979*, Bristol: Western Archaeological Trust (Excavation Monograph **6**).

Evans, J.G. 1971. Habitat change on the calcareous soils of Britain: the impact of Neolithic man *in* D.D.A. Simpson (ed) *Economy and Settlement in Neolithic and Early Bronze Age Britain and Europe*, Leicester: University Press, 27-73.

Evans, J.G. 1972. *Land snails and Archaeology*, London: Seminar Press.

Evans, J.G. 1979. The palaeoenvironment of coastal blown-sand deposits in Western and Northern Britain, *Scottish Archaeol. Forum* **9**, 16-26.

Evans, J.G. 1980. Land Molluscs *in* R. Price, and L. Watts, Rescue excavations at Combe-Hay, Somerset 1968-73, *Somerset Archaeol. Nat. Hist.* **124**, 1-42.

Evans, J. and Jones H. 1973. Subfossil and modern landsnail faunas from rock-rubble habitats, *J. Conchology* **28**, 103-29

Fowler, P.J. and Thomas, A.C. 1962. Arable fields of the pre-Norman period at Gwithian, *Cornish Archaeol.* **1**, 61-84.

Gilbertson, D.D. 1979. The Burtle sand beds of Somerset: the significance of freshwater interglacial molluscan faunas, *Somerset Archaeol. Nat. Hist.* **123**, 115-117.

Kerney, M.P. 1966. Snails and man in Britain, *J. Conchology* **26**, 3-14.

Kerney, M.P. 1976a. Two Postglacial molluscan faunas from South-West England, *J. Conchology* **29**, 71-3.

Kerney, M.P. 1976b. *Atlas of the Non-Marine Mollusca of the British Isles,* Monks Wood: Institute of Terrestrial Ecology.

Saville, A. 1984. Preliminary report on the excavation of a Cotswold-Severn tomb at Hazleton, Gloucestershire, *Antiq. J.* **64**, 10-24.

Spencer, P. 1974. Environmental change in the coastal sand dune belt of the British Isles. Unpublished BSc. dissertation, Department of Archaeology, University College, Cardiff.

Spencer, P. 1975. Habitat change in coastal sand-dune areas: the molluscan evidence *in* J.G. Evans, S. Limbrey and H. Cleere (eds) *The effect of man on the landscape: the highland zone,* London: Council for British Archaeology (Research Report **11**), 96-103.

Thomas K.D. 1982. Neolithic enclosures and woodland habitats on the South Downs in Sussex, England *in* M. Bell and S. Limbrey (eds) *Archaeological aspects of woodland ecology* (Symposia of the Association for Environmental Archaeology No. 3), Oxford: British Archaeological Reports (International Series **146**), 147-170.

Whittle, A. 1978. Resources and population in the British Neolithic, *Antiquity* **52**, 34-42.

NON-DOMESTIC FAUNAL RESOURCES IN SOUTH WEST ENGLAND

Jennie Coy

INTRODUCTION

This paper concentrates on the faunal evidence for species additional to those discussed in this volume by Noddle and Levitan: marine molluscs, fish, and wild species of bird and mammal. It emphasises those of economic importance, especially those which are likely to have been eaten.

I have referred throughout to the earlier work of Martin Bell who pinpointed in his review paper for the South West most of the sites discussed here (Bell, 1984). His work provides the regional ecological facet which is missing from the present paper and I have not attempted to duplicate this here but rather to detail recent results for each of the major relevant faunal groups, and the significance of their exploitation. At the same time some attempt is made to discuss the limitations of this evidence and the main priorities for the future for the South West.

In contrast to Bell's regional review paper this paper is therefore organised taxonomically. It does not cover all sites with bones and shells or attempt to duplicate the wide range of references given by Bell but mainly discusses recent results.

An initial attempt was made to provide tables showing recovery locations for all species. On consideration these have not been included as they provide insufficient detail to make them worthwhile. What detail there is could be misleading too, because of the wide range of treatments meted out to faunal remains from the different sites and the varying degrees of reliability of past reports. Anyone wishing to carry out a more detailed analysis for the South West when more work has been completed is welcome to contact the writer for this archive to avoid the need to repeat the literature search.

THE LIMITATIONS OF THE EVIDENCE

There are three main types of limitation that should be mentioned: preservation, the techniques of excavation, and the way in which the evidence retrieved is interpreted.

Preservation

The limitations on preservation in the South West ensure that only a patchy picture is seen for wild species. Bones and shells are not preserved in some soil types, especially the more acidic such as those described on the St Austell granite, Cornwall, by Biek (1975). There are some conditions where midden material survives; sand dune systems in Devon such as those at Braunton Burrows, (Smith *et al*, 1983) and Bantham Ham (Silvester, 1981) were productive; otherwise much of the material discussed in this paper comes from the urban centres of the South West and from the Wessex counties of Dorset and Wiltshire which are included for completeness.

The paucity of the evidence for some periods and places means that comparisons are made very difficult. But it is important to make the most of what evidence there is, and, where possible, to apply new techniques such as those for the identification of fragmentary fish vertebrae. These are discussed in the fish section below.

Excavation

All retrieval is biased but the institution of a sieving programme controls the bias to some extent, if a known proportion of each context is processed through sieves of known mesh sizes using known techniques of sample taking, sieving, and sorting. One major differential bias caused by manual retrieval is the relative representation of mammal, bird, and fish remains. The groups better represented from sieving are vertebrate micro-fauna, amphibians, reptiles, small birds, small mammals, and all fish.

Although a good picture of fishing was obtained for some of the urban sites discussed in more detail below, some of it was from manual retrieval. Sieving programmes carried out recently in Wessex (Coy and Maltby, 1987: 209) and by Levitan in the South West (this volume) have quantified the bias. Some idea of the potential for bulk sampling that lies in the South West is given by recent evaluations in Bristol (Shackleton and Douglass, 1983; Jones and Watson, this volume).

To ensure that the sampling techniques for any excavation are suited to answering the questions that faunal specialists may consider worth asking it is essential that the specialists are consulted at an early stage and visit the site during excavation. Evaluation of one or two samples during excavation may lead to a more relevant sampling programme or save time if the evidence is not preserved. Some specialists prefer to take their own samples.

Interpretation

In the past the interpretation of bones and shells from wild species has frequently been superficial and the writer admits to being guilty in this respect. Where the species were eaten today it was implied that they were eaten in the past. Where they were not currently eaten often nothing was said. There was sometimes an implicit suggestion that fish species were locally caught, by their inclusion in attempts to reconstruct the contemporary ecological picture. This is sometimes irrelevant as the trading of preserved fish must be considered, at least in the historic periods.

Both the reconstruction of past diet and the reconstruction of past ecology from this type of evidence is fraught with dangers, many of which are fully detailed elsewhere and references are given at appropriate places in this account.

Martin Bell makes a good point about generalising from small assemblages which represent specialised activity areas (Bell, 1984: 84) and most archaeozoologists in recent years have attempted to come to terms with the significance of their samples in terms of taphonomy, site-formation processes, and middle range theory generally. Such analysis now forms an essential stage in the interpretative exercise and may often give the evidence from the faunal remains as much weight in the direct interpretation of the use and history of structures as they have in the environmental or husbandry-related interpretations that may finally be arrived at. The study of the processes of deposition and attrition of these samples frequently gives unique insights into human behaviour and site formation processes at settlements, and these, together with a full understanding of the biases involved in the excavation itself, are, quite rightly, nowadays given full attention (Coy and Maltby, 1987; Maltby, 1985). Without this essential stage reliable interpretations of the archaeozoological results cannot be made.

Finally, the faunal remains discussed in this section which form important foods (notably marine molluscs, fish, and some of the larger mammals) may provide evidence of the type of exploitation strategy carried out. More detail on this is given in the individual sections.

MARINE MOLLUSCS

The main distribution of finds of shells in the South West is not surprisingly on sites on or extremely near the coast. Marine coastlines and major rivers near the sea account for the major populations of the world, so the exploitation of marine resources is not necessarily the main reason for an observed high density of habitation. On the other hand the presence or absence of Mollusca can provide additional evidence about coastline changes for some periods.

The range of molluscan species and their frequencies can tell us something about the substrate of the shore from which they were collected and results from south-western settlements do mostly reflect the nearby shores available for exploitation. Any reading of existing reports for the South West shows that limpets, Patellidae, and species of winkle, Littorinidae, predominate on sites near rocky shores whereas cockles and mussels (usually the common edible cockle, *Cerastoderma edule*, and common mussel, *Mytilus edulis*, respectively) come from sandy or muddy areas.

Further details are included in the taxonomically organised account below which examines the factors relevant to the interpretation of archaeological material. It uses the discussion of the first group - the limpets - to point to some problems and potential pitfalls in the interpretation of molluscs generally, using some discussion of modern limpets. The final mollusc section summarises the potential significance of this type of evidence to the archaeologist.

Throughout the account reference is made to published reports on the region, when relevant.

Limpets (Patellidae)

Limpets need a hard substrate to hold on to. They are mentioned both by Brooks (1974) and Ashbee (1968) as an important Cornish food source. According to work at St Austell by Miles (1975), limpets and mussels played an important part in the prehistoric diet. At the dark age site of Bantham Ham, Devon, limpets "predominated" in almost all deposits (Lees, 1981). Limpets and winkles were the commonest molluscs at prehistoric and Romano-British Mount Batten, Plymouth, (Gaskell Brown and Hugo, 1983) and Rope Lake Hole, Kimmeridge, Dorset (Coy, in press a).

Rocky shores vary considerably in their faunal detail and the biological communities themselves and the abundance of the different species has been used in definitions of "exposure" by Ballantine (1961).

Abundance indices used by Ballantine (1961) for the common limpet, *Patella vulgata* at Dale, Pembrokeshire, ranged from "over 50 per square metre or more than 50% of limpets at certain levels" down to "only a few found in 30 minutes' search", which sounds a poor return for the hungry. Stations classified on his exposure scale for Dale as "extremely exposed" had abundant but small common limpets (25mm max shell-length); very exposed, exposed, and semi-exposed stations had abundant and dominant common limpet over the midlittoral zone; and the more sheltered zones had decreasing frequencies of common limpets as fucoid seaweeds increased in density. Those common limpets present on very sheltered shores, however, commonly had a shell-length of 50mm or more.

Mention has been made only of the common limpet, but two other species may be available on shores in the South West, the china limpet, *Patella aspera*, which is confined to the lower part of rocky shores, and the black-footed limpet, *Patella intermedia*. Although these do not reach the size of larger specimens of *Patella vulgata* they may have been taken for food and therefore might be identifiable from archaeological sites. The major problem is that identification depends to some degree on body colour (eg, the black foot!), so that archaeological samples tend to have been identified to *Patella* sp or tentatively to the common limpet. However, the head scar on the shell is of diagnostic colour and coloured rays within the shell might be preserved (Terry O'Connor, pers comm).

It is important, however, that in discussions of gathering zone and limpet size we do know which species are involved (or at least which are likely) as size differences could be species differences and because the different species have different zone preferences (Graham, 1971). To suggest that a collection of uniformly small limpets is the result of over exploitation (as occurs in some earlier mollusc reports) is therefore suspect on two counts: firstly, we must know the species involved and whether it actually does grow any larger, and secondly, there is always a possibility that limpets on a particular shore might be small for their age as a result of the pressure of an extremely exposed aspect. A change in the distribution of sizes through time, although it could suggest overexploitation, might equally suggest either a different pattern of exploitation or a collection from a different location.

There are references in the archaeological literature to relative height of shell in *P. vulgata* (eg Lees, 1981). In modern limpets this is correlated with tidal zone: those higher on the shore being more pointed. In order to investigate this phenomenon and the overall size discussed above it is essential for any future work to include measurements of limpet length and height and for the shells to be examined to see whether incremental growth can be linked with at least a relative, if not an absolute, age (Rhoads and Lutz, 1980).

It would be interesting to know exactly how modern samples in the region would relate to the three species, the different zones, and the degree of exposure. Supporting modern studies are lacking for most of the species in the localities of the archaeological sites and even where these are available they are often as yet unpublished and difficult to track down. Some effort has been made to do this for Wessex by Winder (pers comm), and the South West is probably a fertile ground for a study to tie up modern sea-shore observations with archaeological investigation.

It must not be assumed, however, that such results will necessarily be wholly relevant to a different time. Thompson has shown for Bantry Bay, Eire, that limpet populations on a given area of rocky shore are by no means constant and immutable. Their most important relationships in terms of distribution are with the fucoid seaweeds or wracks, *Fucus* sp, and barnacles, Cirripedia (Thompson, 1980).

Another factor which could come into limpet gathering is ease of removal from the rock. From personal experience since childhood I have been most impressed by the steadfast way in which limpets refuse to budge once the tide has gone down, however much they are taken by surprise or by penknife; the surprise blow with the heel of the Wellington boot beloved of today's collectors (Graham, 1971: 16) being a relatively recent phenomenon. By contrast I have gathered small flat limpets from the lower regions of beaches in Greece when they are still submerged and they are easily dislodged with a finger. They are also delicious and quite tender raw.

Modern occupants of Colonsay off the West Coast of Scotland, claim to prefer limpets gathered from the low-tide area according to Mellars (1978: 389) although we are told that this is because they are more edible. We need to know far more about ease of gathering and palatability of the various species at different ages and levels of the shore. The choice of molluscs by modern gatherers such as the Australian communities studied by Meehan has been shown to be a complex process with targeting on particular species at each visit and switching to other species with deliberate, conscious behaviour (Meehan, 1975 and 1983).

Enough has been said above to warn of the danger of archaeological interpretations with a single explanation - height on shore, over-gathering... . Even more sweeping is the assumption that limpets were always eaten. Alternative theories are available. They may have been used for bait or they may have been fed to pigs as used to be the case in Brittany (Professor Giot, pers comm). Ashbee cites their use as fertiliser in the past in the Scilly Islands (Ashbee, 1974: 265). Dangerous too is the assumption in some reports that they needed to be cooked or that the incidence of burnt shells could be a measure of this.

Winkles (Littorinidae)

A number of sites near rocky shores produce both winkles and limpets as already discussed. In addition, the Roman villa at Holcombe, a mile from the sea, produced edible winkle, *Littorina littorea* (Clutton-Brock *et al*, 1974). It is difficult to know how far people would walk for such small prey. At Cleavel Point, Ower, Dorset, it was suggested that winkles and mussels would have been available in the Iron Age in the Studland Area less than an hour's walk away, nearer than today because of the subsequent accretion in the Studland area; whereas the cockles and carpet shells, *Venerupis decussata*, would have come from the mud and sand nearby. In this case, perhaps the presence of mussels and different types of marine fish (the gilthead sea bream, *Sparus aurata*, which feeds on mussels, was also found) made such trips to the coast worthwhile (Coy, in press b).

Whereas a raw limpet can be easily scraped from its shell with another shell, winkles should be dipped in boiling water to loosen them and a pin used for "winkling" them out because of the ease with which they retreat deep into the spiral shell and lie protected by the horny operculum. When they are used for bait today the shell is often smashed with a stone and this could provide an alternative technique.

Nowadays nine species of winkle, Littorinidae, are recognised in Britain. Until 1975 four was generally accepted as the likely total but two of these have now been split taxonomically - one into two species and the other into four (Heller, 1975). Not all of these would be distinguishable for archaeological material and the changes in taxonomy do not affect the edible winkle, *Littorina littorea*, the only winkle of a satisfying size (given by Graham (1971) as 25mm whereas the others are normally 10mm or less). In archaeological work a problem of identification arises with young edible winkles as they bear spiral ridges like those of the rough winkle group, *L. saxatilis*, which is now recognised as four species after Heller's work. Such young edible winkles were abundant at Cleavel Point.

Winkles are an extremely successful group commonly found on a variety of beaches, except pure sandy ones. Ballantine's scales of abundance for Dale ranged from "50 per square metre or so" down to "only one or two found in 30 minutes' search" (Ballantine, 1961). The latter order of abundance would be even meaner in meat provision than the same figure for limpets. Winder gives the average cooked meat weight of winkles from Kimmeridge Bay, Dorset, as 1.1g (Winder, 1980: 125).

The different species tend to inhabit somewhat different zones, with the edible winkle being much more common on very sheltered or extremely sheltered rocky stations at Dale. They are often quite common on muddy beaches where they may feed on detritus, being especially rich near sewage outfalls. The edible winkle is generally common throughout Britain where the shores offer some detritus, seaweed, or other form of food and anchorage, but is said not to be so in the Scillies (McMillan, 1968).

The presence of winkle shells on a settlement, especially when in small numbers, could conceivably have other causes apart from collection for eating. Because of their habit of attaching to seaweeds they could be taken along with this when it is collected for fertiliser: such use is suggested as a cause for the presence of many tiny mussel and winkle shells at Gwithian, Cornwall, (Fowler and Thomas, 1962), which is discussed along with other observations on the use of seaweed in the South West in Bell (1984: 51). Small winkles often sit on other shells, including larger winkles, and can be inadvertently collected.

Mussels (Mytilidae)

Mussels have already been mentioned in association with limpets at Cleavel Point and St Austell. At the medieval shell midden at Braunton Burrows, Devon, they are associated with cockles, Cardiidae, and the common oyster, *Ostrea edulis* (Smith *et al*, 1983), and it is suggested that cooking and distribution took place.

There are a number of species of mussel of this group likely to be found on British shores and it is not always clear that mussels from archaeological sites have been identified correctly as they are often extremely fragmented. The most likely species which would be easily accessible on beaches in the South West is the common mussel, *Mytilus edulis* (Tebble, 1966). They are common on many types of shore but grow especially well in estuaries so long as there are stones to which their byssal threads can attach. They can grow at all levels of the shore so long as such means of attachment are available and there is a constant supply of suspended detritus and plankton. Those growing high up the shore may be very small.

The densities of this highly edible and relatively large species on the rocky shores at Dale investigated by Ballantine ranged from "more than 20% cover at some levels" to "a few seen in 30 minutes search" and it was commonly present as tiny, crowded individuals on extremely exposed rocky shores; present only in cracks on very-exposed, and exposed shores; almost absent on semi-exposed, fairly sheltered and sheltered shores; and common as groups of large specimens on sheltered shores only because sheltered rocky shores there tend to deposit sand and mud in quantity (Ballantine, 1961).

The larger mussels from more sheltered locations like estuaries would be an important food source. Estuarine mussels may be in very dense populations indeed: a density of 500 - 1,000 per square metre is reported for the Menai Strait in Wales by Tebble (1966). It is conceivable that where mussels are well-preserved it might be possible to deduce something about their crowding from shell shape.

Such a rich food source then, if present, would be very significant. It is likely to be severely underestimated on the evidence from archaeological sites as it does not appear to survive as well as some other species. From the discussion above it appears that we should be prepared to consider them (as to a certain extent with winkles) as available from almost all types of shore in the South West except those of pure sand.

Cockles (Cardiacea)

Cockles are better preserved than most other marine molluscs on sites as they have much sturdier shells. They were found at Braunston Burrows and at Cleavel Point as already described, and on numerous other sites, such as the medieval deposits at Exmouth (Jones, 1980). Today they are far more indicative of sand than any other species, being so common in some places that until relatively recent times they formed the basis of the local economy. Densities of over 10,000 specimens per square metre have been recorded (Tebble, 1966). There are several species of cockle which may be encountered in the South West and shells from archaeological excavation are normally distinguishable to species unless very worn; the common edible cockle, *Cerastoderma edule*, being the most usual find. This species lives in clean sand, muddy sand, mud, or muddy gravel and can survive some way up estuaries.

They are shallow burrowers on the middle and lower shores. This, and their presence in such large numbers, is conducive to technological development in collecting techniques. Some evidence on their point of origin in an estuary might be available because of the correlation suggested between salinity and the number of radial ribs (Eisma, 1965).

Winder gives an average cooked meat weight for cockles from Burry Inlet, South Wales, as 1.73g (Winder, 1980: 125).

Oyster, Ostrea edulis

Whereas limpet/winkle or cockle/mussel assemblages might reflect a particular ecosystem, oysters more often have cultural associations and are not significantly present until Roman times. For example, they do not appear until the Romano-British phase at Rope Lake Hole (Coy, in press a).

Collections of oysters may indicate the type of oyster population exploited and the medieval collection from Okehampton, Devon, suggests a cultivated one (Backway, 1982), compared with that described for Middle Saxon Hamwic (Winder, 1980).

The detailed study of the parasites and commensals found in association with oysters seems to have some potential for pinpointing origin of supplies, for here we have shell-fish which we know from archaeological and documentary evidence to have been commonly traded (Winder, 1980 and pers comm; Backway, 1982).

Jessica Winder quotes an average wet meat weight of 7.5g for a sample of modern oysters but points out that the amount of meat varies enormously - 1.2 to 35.3g in the sample she quotes - (Winder, 1980: 121). The amount of meat provided by oysters, even so, would be far more rewarding than that from the other species discussed above.

Other Species

A wide range of other species have been remarked upon in site reports for the South West. Only a few of these provide a good quantity of meat or large exploitable populations, such as the carpet shell already mentioned for Cleavel, probably found when cockle collecting: scallop, *Pecten* sp; common whelk, *Buccinum undatum*; and razor shell, *Ensis* sp. The three last were all noted at Bantham (Lees, 1981). It is likely that the largest local molluscs, the octopus and cuttlefish, may have also played a part, but only the latter would normally leave recognizable remains on archaeological sites in the South West.

The Interpretation of Marine Molluscan Remains

Because of their static nature molluscs are often used as evidence to interpret coastal change or lack of it, eg, for Bantham, Devon, by Silvester (1981: 113).

Shell shape is sometimes used for interpretations of the method of gathering and level of the shore exploited. But the factors acting on shell shape can be very complicated. Studies of the dog-whelk, *Nucella lapillus*, have shown that the selective effects of wave action and crab predation are not always the only forces involved: in the Severn Estuary dogwhelks are unusually elongated whatever their habitat and this may be more related to gene pools and evolutionary history (Crothers, 1983).

Archaeological interpretation can gain from a consideration of mollusc data both in terms of formation processes using the taphonomic evidence available in the mollusc sample and in terms of behavioural insights. Not least, mollusc remains may provide important dietary information which is directly relevant to status.

Dietary considerations should take into account the different amounts of meat provided by the different species (Winder, 1980: 122). Oysters, the largest providers, show a great deal of variation, in size, shape, and parasitisation and for Roman and medieval deposits may yield quite complex evidence of social or behavioural variability.

The mere presence of shellfish, unless the shells are of casual origin and brought in other than by food-gathering, provides evidence of an attempt to obtain greater variety in the diet with a range of minerals and trace elements that might be lacking or low in other foods or missing at particular times of the year (Paul and Southgate, 1978). The small amounts of meat provided by most shellfish exploitation are of secondary consideration.

This paper lays particular emphasis on marine molluscs as their importance has not yet been fully appreciated in British archaeology. In the South West such marine resources were important. As for bones, the interpretation of marine molluscs may relate to more than just diet. Some attempt to discuss the archaeological rather than the zoological significance of marine molluscs, including their importance for understanding taphonomic factors, is being made currently by Jessica Winder for oysters (pers comm).

For the interpretation of marine molluscan evidence behavioural parallels with ethnographic studies on marine edge communities elsewhere must be valuable. The work of Meehan detailing strategies of shellfish-gathering peoples in Australia could provide valuable ideas for a more thoughtful discussion of the value of these resources than has taken place for the South West so far (Meehan, 1975 and 1983). Documentary and contemporary evidence for shellfish gathering in Ireland, Scotland, and Wales must also be relevant.

FISH

As for marine molluscs there would have been very little difficulty in the South West in exploiting a wide range of species of fish (both marine and freshwater) in earlier periods, even with quite simple fishing methods. In the medieval period the South West is documented as a source of fish which were traded with other areas. Wilkinson has discussed its importance for the "Southern" species - herring, pilchard, and mackerel - and the post-medieval domination of the later deep sea fishing trade with its exports to the Mediterranean, leading to Exeter's becoming a leading centre for the trade in the latter half of the seventeenth century (Wilkinson, 1979: 80). He stresses the richness of the river Exe, but other river systems in the South West would also have been rich in their individual ways.

Other parts of the South West were also active in the fish trade. For example, there are documentary records of fifteenth century imports to Southampton of herring and hake from Brixham; hake, cod, and ling specifically recorded as coming from Mount's Bay, Cornwall; and salmon from Dartmouth (Quinn, 1938: 149-153).

The three limitations discussed earlier - preservation, excavation and interpretation - play havoc with fish results. On some excavations fish bones are not preserved at all. Many published urban results are from trowelling only and will therefore overestimate the economic role of the large gadoids. Fish bone specialists stress this fact time and time again (eg Wilkinson, 1979: 74).

One important development in recent years has been the examination of large quantities of sieved material from urban archaeological deposits in other parts of Britain - from London, York, and Wessex in particular. These results have now often been compared in detail with those from normal manual recovery and show up in detail the bias that must occur in fish results from non-sieved sites. One bias which must not be overlooked is the scarcity of archaeozoologists who can identify fish bones as only this will ensure retrieval of the relevant evidence from the sieves.

Finally, there are many potential pitfalls in interpreting fish results (Wheeler, 1978: 74). Conclusions need to take into account the importance of trade, already mentioned for its link with documentation in the later periods. The ways in which fish are preserved have caused them, since at least Roman times, to be used as a trading commodity.

Marine Fishes

Despite the archaeological and documentary evidence of fishponds in the medieval period, most fish evidence from medieval sites is of common marine species, often large, which obviously formed an important item of trade and played a vital dietary role. Wheeler reveals fills of putative fish tanks of early medieval date from Priory Barn, Taunton, to contain marine fish from kitchen waste (Wheeler, 1984).

At Exeter, the multiperiod nature of the excavations allowed some comparisons to be made in fish exploitation in the different periods. Wilkinson suggests that both Roman and medieval fishing could have been mostly by hook and line whereas post-medieval finds of herring, *Clupea harengus*; mackerel, *Scomber scombrus*; and dory, *Zeus faber*, suggest the use of nets.

The medieval fishery at Exeter was dominated by hake, *Merluccius merluccius*; with conger eel, *Conger conger*; and whiting, *Merlangius merlangus*, next in order of importance. Later, haddock, *Melanogrammus aeglefinus*; ling, *Molva molva*; cod, *Gadus morhua*; and whiting became more common and hake and conger declined (Wilkinson, 1979). Results from medieval Okehampton support the picture obtained from Exeter (Wilkinson, 1982). Priory Barn, Taunton, already mentioned, produced hake, conger, whiting, and plaice, *Pleuronectes platessa* (Wheeler, 1984).

Post-medieval material reported on recently includes a large conger eel from Lydford (Jones, 1981) and a collection from two post-medieval features in Totnes which produced a wide range of marine species with whiting, herring, and flatfish dominant (Colley, 1984). The presence in this late material of so much whiting along with dory confirms the fishing patterns suggested for Exeter.

Recent work at St Katherine's Priory, Exeter, by Levitan confirms hake as the most important fish in the thirteenth century and then decreasing relative to conger and ling, although this site was not subjected to regular sieving (Levitan, forthcoming a). This dominance of hake compared with the east coast fisheries is noted in eleventh-twelfth century material in Ilchester (Wheeler, 1982: 284). The extent to which the presence of this species indicates that fishing is taking place in deep water at this time is discussed elsewhere (Wheeler, 1978: 74). Today it is heavily overfished and regarded as a deep water species but present day distributions may not be comparable with those in the medieval period. At early medieval Ilchester Wheeler remarks on the absence of ling but this species does not appear in the archaeological record at Southampton either until the fourteenth-fifteenth centuries and it has been suspected that it might come in as preserved fish (Coy, in prep a). It is also well represented in medieval Poole, Dorset (Coy, 1985).

A wide range of other species was represented at Exeter, including possible tunny, *Thunnus* sp (Wilkinson, 1979). Bones of marine fish from medieval layers may have come from the many types of preserved fish which are documented as coming into ports at this time. Records for Southampton provide an example of the wide range of preserved material, which comes into the port. I have not studied the documentation for cities in the South West but they would probably provide a similar story. The fifteenth century scale for Bristol goods entering or leaving Southampton, given in the Great Red Book of Bristol (Veale, 1933: 190-1) includes herring and other fish, and red herring as items. Herring is discussed below in connection with sieving.

Recent excavations of waterlogged waterfront sites in Bristol both had extensive sieving programmes and good recovery of fish bones (Shackleton and Douglass, 1983; Jones and Watson, this volume).

Whereas it is easy to deduce that large bones from marine fish found far inland may have been from preserved fish, and that herrings would most often have travelled in this way, it is difficult to know whether the bones of marine fish found in medieval excavations in coastal towns were from fresh or preserved fish. One clue may come from butchery and this is discussed in more detail at the end of the fish section.

Less common but important food species which are commonly found in any good fish deposit in the South West are species of gurnard, Triglidae; sea bream, Sparidae; and wrasse, Labridae. Remains of the shark family are occasionally found, and there can be little doubt that the cartilaginous fishes generally will be underrepresented from archaeological sites. Only heavily calcified remains are usually found, or well-developed dermal modifications like the "thorns" of the thornback ray or roker, *Raja clavata*, and the "spur" of the spurdog, *Squalas acanthias*. The latter was found at Exmouth (Wheeler, 1980); Exe Bridge, Exeter (Levitan, forthcoming b); and Wimborne Minster, Dorset (Graham, 1984).

Freshwater and Migratory Fish

Fish bones from prehistoric settlements are usually few, even when sieving has been carried out, but when they appear they are usually of freshwater species, such as the finds of dace, *Leuciscus leuciscus*, at Gussage-All-Saints, Dorset (Harcourt, 1979: 155).

Late Roman salmon (*Salmo salar*) bones from Gloucester led Maltby to comment on the importance of this species in the Severn area and quote relevant medieval documentation (Maltby, 1983: 244). This species is reported from many excavations, including Holcombe Roman villa (Wheeler, 1974: 155), Dark Age Bantham Ham (Coy, 1981), Roman and medieval Exeter (Wilkinson, 1979), and medieval Okehampton (Wilkinson, 1982). Wheeler makes the point that the head bones of salmon migrating up rivers are unlikely to preserve (Wheeler, 1978: 74).

Another important migratory fish is the common eel, *Anguilla anguilla*, found on most sites in the South West where sieving has taken place.

The remains of other freshwater species are found from time to time. Large specimens may be picked out in trowelling but the remains of the majority of freshwater fish such as small examples of the carp family, Cyprinidae; small pike, *Esox lucius*; and perch, *Perca fluviatilis*, often come from sieving and at the moment the main problems are still retrieval and recognition.

Size Reconstruction

Apart from a species list very little other information on fish is provided in most bone reports. Things should change gradually now that the Fish Group of the International Council for Archaeozoology (ICAZ) has made an effort to improve methodology in this subject. A series of bone measurements has now been agreed, based on an earlier manual (Morales and Rosenlund, 1979), and reconstructions of length and weight of common species are increasingly being attempted using linear regression from measured modern specimens.

The fish bones at Exeter were measured so that size distributions could be examined and size estimates made. Hake were estimated at 0.5-1m and weighing several kilograms (Wilkinson, 1979). Colley showed that whiting at Totnes had a similar size range to those at Exeter but that the majority of fish were smaller than Wilkinson's estimated 0.4-0.45m (Colley, 1984: 20).

As well as giving more information on the amount of meat that fish bones from archaeological sites actually represent, changes in size of fish represented can often be linked to changes in fishing strategies (Colley, 1983a and 1983b) and work on midden otolith material can provide detailed evidence (Wilkinson, 1981). Scope for such detailed work as this in the South West would depend upon finding preserved midden samples of a suitable type. A start was made for the material from Exeter but otoliths were few and results inconclusive (Wilkinson, 1979: 75).

Fish from Sieving

Cesspits, in particular, produce fish bones from at least Saxon times. Finds include small vertebrae of common eel, *Anguilla anguilla*, the herring family, Clupeidae, and freshwater fish. In many cases bones can be shown to have passed through the digestive system. Andrew Jones' pioneering work on this showed that if kippers were chewed and swallowed the vertebrae may survive passage through the human gut although they may be crushed or fragmented (Jones, 1984). The presence of such remains may often reinforce other evidence for the use of particular features as cesspits.

Whereas the transport of preserved herring in the medieval period is documented eels, although often smoked, may have been of more local origin. They are common in cesspit deposits and on sieved sites may be the commonest fishbone find. We need to know far more about the eels represented by the bones from archaeological sites - their exact contexts, locations, and sizes, for a start.

New Methods and Fish Bone Interpretation

Some of the problems which still hamper adequate interpretation of fish bones from the South West are soluble. Identification requires good modern skeleton collections with weighed and measured specimens. By cooperation within the European members of the ICAZ Fish Group it should be possible to solve a number of outstanding archaeo-ichthyological problems in the next decade.

Plans by the group to produce monographs for identification, measurement, and size reconstruction are already considerably advanced. Linear regresssions for sizes have already been carried out for some species by a number of authors and should be applied to all future fish material excavated in the South West (eg Morales and Rosenlund, 1979; Enghoff, 1983; Desse, Desse-Berset and Rocheteau, 1987).

One interesting area of observation linked with documentary evidence is the attempt being made to tie up the type of fish remains with the method of preservation used, by using evidence of butchery. Large fish of the cod family, Gadidae, were salted, dried, or both, and it is sometimes possible to observe butchery marks where fish were beheaded or split lengthwise. The presence of a biassed series of anatomical elements, although it can be the result of subsequent taphonomic events (even sorting by the archaeologist!), may be linked to the preparation process. At Exeter, butchery on fish bones showed that ling and pollack, *Pollachius pollachius*, may have been split for preservation (Wilkinson, 1979: 75).

A high frequency of head bones at Taunton and Totnes suggests waste from food preparation (Wheeler, 1984; Colley, 1984). Wilkinson suggests that conger at Exeter may have been dried. There are fifteenth century references to the importation of salt conger at Southampton (Quinn, 1938: 152).

The new method of X-ray identification for eroded vertebrae (Desse, 1984) will be useful for south-western material as soon as modern comparative collections have been X-rayed to provide a comparative base. Fragmentary shark material from the South West would especially benefit from this treatment and should be carefully saved until the necessary work on modern British comparative collections has been completed.

The Significance of Fish Remains

Recent work suggests that even a low habitual fish intake, equivalent of 1 or 2 fish dishes a week, may be of value in the prevention of coronary heart disease (Kromhout *et al*, 1985). Whereas such research on the different kinds of fats and their relationship with clotting factors are a modern domain, the benefits of fish for their vitamin and mineral content have been known for longer. Provision of a single dietary element such as Vitamin A could have given added significance to quite small amounts of fish in the diet, especially in times of dietary hardship, even though the reasons for its beneficial effects were not known at the time (Coy, 1982a: 114).

The association of wider dietary variety with the rich in all periods needs little explanation and the identification of collections with high dietary diversity is one way in which the bone analyst can help the archaeologist to appreciate differences in deposit types. Fish can often play a large part in this type of analysis and bulk sampling from every context is an essential if such indices are to have any worth.

Deposits of fish bone do not only result from human activity. Small fish bones on one Orkney site were reckoned to come from sprainting sites of the otter, *Lutra lutra*, exploiting nearby coastal rock pools (Colley, 1983b: 240). Fish may also have been carried about by dogs and cats or died *in situ*. The last is a particular problem when remains of very small freshwater fish are found from sieving and can only be solved by a careful examination of the archaeological context and its records.

BIRDS

As in other parts of Britain the last century has seen a severe decline or disappearance of a number of species of bird through human activities and alterations in ecosystems, the predators having been especially hard hit (Langley and Yalden, 1977). Some other species, attracted to the products of human alterations have become much more common. Yet others have altered their habits. It is important to realise that many of these changes are relatively recent phenomena. Thus the presence of a chough, *Pyrrocorax pyrrocorax*, at Exeter from fourteenth century deposits, although interesting and a painful reminder of the demise of the Cornish chough, is hardly surprising (Maltby, 1979: 73).

Whereas some finds may reflect a wider past distribution others do not. For example, Iron Age finds of kittiwake, *Rissa tridactyla*, some distance from the sea at Gussage-All-Saints, Dorset, are more likely to reflect trade (Harcourt, 1979: 155) and it has since turned up at the Iron Age hillfort of Danebury, Hampshire (Coy, 1984: 530). The list of birds represented from archaeological sites of different periods in the South West is remarkably small compared with the wide variety that must have been there so that a detailed analysis has not been attempted here. The three limitations already discussed for fish - preservation, excavation and interpretation - can usefully be discussed for birds. Preservation of bird bones is often good but many become damaged during and after excavation. Unless bird bones are separated from the big bones and, even better, boxed separately, at least after washing and marking, important information will always be lost. Once a bird bone is damaged its thin wall often allows it to break up completely.

Many bird bones are easily retrieved during trowelling, some of them are noted as coming from articulated whole or partial skeletons. Because the bones of the wing are so strongly attached for flight and because whole body carcases of birds eaten are often discarded articulated, the bones of birds may be more often found in articulation than those of the common ungulates. It is important for the interpretation of the bones that this is noted during excavation and the bones subsequently kept together. Maltby remarks on a sixteenth century deposit of duck wing bones in Exeter (Maltby, 1979: 72).

Sieving reveals many of the smaller parts of the domestic birds and certainly redresses the bias as far as the smaller bird species are concerned. A complete picture of the birds for most sites in Wessex requires washing bulk samples through at least a 4mm sieve whereas a 1mm or 600 micron sieve is necessary to gain a complete picture of fish exploitation (Coy, in prep a).

The interpretation of birds, like that of fish has been much neglected compared with that developed for the domestic ungulates. A variety of factors will have led to the deposition of bird bones in a deposit and it is important to sort these out through the detailed information on the deposit and the bone itself (Coy, 1983a).

Domestic poultry

Most bird bones found on sites from the Late Iron Age onwards in the South West, as in most other parts of England, are those of domestic fowl and to a lesser extent domestic goose. These species are of economic importance as food; and providers of eggs, down, and feathers. The treatment of their size changes, relative importance, age and sex profiles, and possibilities for unravelling taphonomic factors are discussed elsewhere (Coy, 1983a; in prep b). No further discussion of fowl and goose is given here as this would need a full analysis of the material already published for the South West and possibly some reworking of the material. Exeter remains the main published collection of fowl to have been subjected to a comparative metrical and economic analysis to date (Maltby, 1979: 67-71) but work being undertaken by Brothwell on Roman fowl at Uley Temple, Gloucestershire, will provide important detail on the fowl of that period (Levitan, 1978).

More unusual domestic poultry have frequently been sought in the South West and Maltby laments the lack of proof for peacock, *Pavo cristatus*, at Exeter (Maltby, 1979: 71) even though they are documented for the sixteenth century. Turkey, *Meleagris gallopavo*, does appear in the mid sixteenth century at Exeter (Maltby, 1979: 71) and in seventeenth-eighteenth century layers at Christchurch, Dorset (Coy, 1983b).

Other Food Species

Pigeons, *Columba* sp., are mentioned occasionally but it is not yet clear the degree to which those too small to be wood pigeon, *Columba palumbus*, might be domestic birds or could be the anatomically similar rock dove, *C. livia*, or stock dove, *C. oenas* (eg Levitan, 1982: 281, for Ilchester).

Wild species of water bird frequently turn up on sites in the South West - swan, non-domestic geese, and ducks, and the considerable problems inherent in the identification of domestic and wild forms of goose and duck should not be underestimated. Exeter produced mute swan, *Cygnus olor*; mallard, *Anas platyrhynchos*; teal, *Anas crecca*; wigeon, *Anas penelope*; and white-fronted goose *Anser albifrons*. Levitan mentions the importance of the swan in medieval Taunton (Levitan, 1984: 192). Some areas, such as the Somerset Levels, are especially suitable for water birds and some pelican remains from Glastonbury may be early medieval, although it is not always clear which of the two species - the white pelican, *Pelecanus onocrotalus*, or the Dalmatian pelican, *Pelecanus crispus*, is represented (Coy, 1977). Crane, *Grus grus*, bred in Britain until at least 1600 (British Ornithologists' Union, 1971: 82) so that finds at Iron Age Gussage (Harcourt 1979: 155) and Groundwell Farm, Wiltshire (Coy, 1982b); and medieval Okehampton (Maltby, 1982: 133) are interesting but not surprising.

Medieval Okehampton shows a typical breadth of other edible gamebirds - woodcock, *Scolopax rusticola*; common snipe, *Gallinago gallinago*; golden plover, *Pluvialis apricaria*; and partridge, *Perdix perdix*. One feature produced evidence of initial butchery of at least ten of the last (Maltby, 1982: 115). An interesting collection from Frocester Court Roman Villa, Gloucestershire, included curlew, *Numenius arquata*, and quail, *Cotur-*

nix coturnix (Bramwell, 1979).

At Exeter woodcock ranked third commonest bird, behind domestic fowl and goose. This widespread bird will overwinter in a wide variety of habitat types although it favours extensive woodland for breeding. It is likely that it bred freely in the South West in earlier times (Maltby, 1979: 72). It turns up on most sites in the South West where bird bones have been retrieved in any quantity.

A number of smaller species are often found with other food remains and usually assumed to have been eaten, one quite large passerine group often listed are the thrush family, Turdidae, as recorded for Roman Holcombe (Cowles, 1974). At Frocester Bramwell suggests that all the small birds might have been eaten (Bramwell, 1979: 61).

Medieval and post-medieval bird remains may relate to and augment information from documents. Birds of the countryside may have been widely available in the local shops and markets of the large towns so that their presence should not be used blindly to reconstruct immediate ecosystems.

Sea Birds

Apart from the finds of kittiwake on Iron Age settlements already mentioned, most of the sea bird finds are from sites on the coast or cities in river estauries. Guillemot, *Uria aalge*, was found in Iron Age deposits in Dorset at Rope Lake Hole, Kimmeridge, and Cleavel Point, Ower (Coy, in press a; in press b). The latter also produced remains of gulls, Laridae, as do a number of south-western sites.

It is difficult to know what was regarded as edible in different periods. A number of sea birds are eaten in different cultures and methods of preparation may be designed to make them more palatable. It is important to look carefully for evidence of butchery and other useage on all bird bones. Levitan points out that judging by their modern practice, herring gull, *Larus argentatus*, probably acted as scavengers in post-medieval Taunton filling the niche of raven and buzzard (Levitan, 1984).

Birds of Prey and Scavengers

Many collections of bird bones from the South West produce evidence of scavenging birds, some of which may also have killed poultry. Iron Age Gussage produced raven, *Corvus corax*; buzzard, *Buteo buteo*; and hen harrier, *Circus cyaneus*. Roman Frocester produced buzzard and red kite, *Milvus milvus*. The latter was still around in medieval Okehampton (Maltby, 1982: 134).

Osprey, *Pandion haliaetus*, was found in a very productive sixteenth century pit at Exeter (Maltby, 1979: 72). Bones of white-tailed sea-eagle, *Haliaeetus albicilla*, were found in Iron Age deposits at Cleavel Point, Dorset (Coy, in press b), at West Hill, Uley, in a 4th century context (B. Levitan, pers comm); and recently from Neolithic deposits in the Stonehenge area (M. Maltby, pers comm). This bird, as large as a golden eagle, was a breeding species in coastal districts until the nineteenth century.

Birds of prey which might have been falconer's birds occasionally turn up in medieval deposits in the South West. Sparrow hawks, *Accipiter nisus*, came from late twelfth century and sixteenth century deposits in Exeter (Maltby, 1979: 72-3). Medieval Ilchester produced sparrow hawk; goshawk, *Accipiter gentilis*; and peregrine falcon, *Falco peregrinus* (Levitan, 1982: 281).

WILD MAMMALS

The existing earlier prehistoric evidence for the exploitation of wild mammals, such as deer and the wild cattle, *Bos primigenius*, is admirably summarised by Bell (1984). More recently there have been important new finds of wild cattle, probably of Bronze Age date, from a cave near Charterhouse, Mendip (Levitan, 1986). The comments made above for the relatively recent extinction of some birds of prey apply also to a number of the mammalian carnivores (Langley and Yalden, 1977). Identification of some of the larger mammals to species and the distinction of others from their domestic counterparts is not easy for individual bones but rather depends upon building up a picture for the region as results and measurement catalogues accumulate. There has been a move towards this in the South West recently with analysis of some species in depth as a result of extensive bone reports but there is still a long way to go before the picture will reveal changes in exploitation, extinctions, and introductions.

Red, Roe, and Fallow Deer

The red deer, *Cervus elaphus*, and roe deer, *Capreolus capreolus*, are native species and traces are found on most south-western sites of all periods with a reasonable sample. Red deer records are more widely spread because of the exploitation of their antlers both as digging tools and raw material for bone working, although recent work in the Stonehenge area has produced some good evidence of roe deer exploitation (Maltby, pers comm). There is evidence of exploitation of young roe in the Middle Saxon period at Ramsbury (Coy, 1980).

Fallow deer, *Dama dama*, are on present evidence a post Norman Conquest introduction to Britain (Coy, in prep a). Some confusion of fallow bones with those of red and roe can be avoided by reference to measurement ranges which do not normally overlap for the three species in archaeological material. A series of measurement ranges for deer bones from twenty published archaeological collections is given in the above paper and measurements given for deer in the South West appear to fit within these ranges. Fallow appears at Exeter in the early twelfth century and is linked with documentary evidence of deer parks (Maltby, 1979: 60). Maltby makes the point here that in the medieval period venison was mainly for the rich. Okehampton gives a similar picture of deer and the bone report has a full treatment of the fallow remains including measurements. There appear to be more haunch remains than shoulder joint remains (Maltby, 1982: 126-7, 129). Venison is rare at medieval Bath (Grant, 1979: 68).

Wild Boar

Clutton-Brock uses measurements of teeth to show that the pigs at Roman Holcombe are likely to be domestic pigs, *Sus domesticus*, rather than the wild boar, *Sus scrofa* (Clutton-Brock, 1974). These ranges are given alongside those from Iron Age Cleavel Point and Gussage, Dorset, also said to be domestic, and the very large domestic range from Middle Saxon Southampton as a comparison (Table 1). By building up such measurement collections it should be possible to say whether there is any evidence on sites from the South West of wild boar or crossing between domestic and wild animals.

Certainly some of the Dorset measurements discussed above do seem to be at the upper end of the Hamwic range and wild boar must always be kept in mind in comparisons. Several individual long bones found in the South West have also been regarded as large but still likely to be domestic, eg at Gussage and at Cleavel.

Medium Sized Mammals

One species commonly exploited in early prehistory, the beaver, *Castor fiber*, appears to have survived until at least the Middle Saxon period. It was represented in the collections from Iron Age St Minver, Cornwall (Chaplin and Coy, 1964: 32). A whole skull with skinning marks was found at Middle Saxon Ramsbury, Wiltshire (Coy, 1980: 49). Some remains from Wirrall Park, Glastonbury (Darvill and Coy, 1985) have now been given a date of *c* 900 AD from small sample, radiocarbon accelerator dating. Other archaeological evidence of beavers has been found in the Somerset levels (Coles and Orme, 1982).

Table 1. Pig Molar Measurements from a number of sites

	Lower third molars					Upper third molars				
Site	no.	range	mean	sd	cv	no.	range	mean	sd	cv
1 Hamwic	104	25.4 - 35.4	31.1	2.0	6.3	53	25.6 - 33.0	28.8	1.9	6.7
2 Holcombe	5	30.0 - 34.0				7	26.0 - 31.3			
3 Gussage	18	30.0 - 35.0	33.0			not available				
4 Ower	9	28.0 - 34.9	31.2	2.4	7.8	24	25.4 - 32.8	29.3	2.0	7.0
5 Wild	6	36.6 - 42.6				8	32.5 - 39.9			

Key and references

sd standard deviation
cv coefficient of variation
no. number of specimens

1 Bourdillon and Coy, 1980
2 Clutton-Brock, 1974
3 Harcourt, 1979
4 Coy, in press b
5 Figures for modern wild boar quoted by Clutton-Brock, 1974.

Harcourt gives a table for Gussage showing the wild mammalian species exploited in different periods of the Iron Age there and mentions finds of hare, probably the brown hare *Lepus europaeus*; badger, *Meles meles*; pine marten, *Martes martes*; polecat, *Mustela putorius*; and fox, *Vulpes vulpes* (Harcourt, 1979: 154). In medieval Exeter fox; badger; hedgehog *Erinaceus europaeus*; stoat, *Mustela erminea*; and otter, *Lutra lutra*, were found in addition to the hare which Maltby suggests was a more regular source of meat there than the rabbit, *Oryctolagus cuniculus*. The last appears to have been introduced in the early twelfth century but is a rare find from pre-sixteenth century deposits. In post-medieval times rabbit became more popular than hare and Exeter provides evidence of skinning and butchery (Maltby, 1979: 60). This picture is mirrored by results from Okehampton although it may have been more commonly eaten there (Maltby, 1982: 129).

It is difficult to distinguish domestic cat, *Felis catus* and wild cat, *Felis sylvestris*, from bones and it is usually assumed in bone reports that big cats from Roman times onwards are domestic. Similarly the domestic form of the polecat, the ferret, *Mustela furo*, is likely to be difficult to distinguish from remains of the wild polecat unless recognisable anatomical criteria are present on a mature skull. The ferret is believed to have been introduced along with the rabbit by the Normans. A possible ferret has been recorded for thirteenth century Exeter (Levitan, forthcoming b) and one for medieval Bristol (Shackleton and Douglass, 1983).

Marine Mammals

Remains of large marine mammals are rare but a fragment from Iron Age Rope Lake Hole, Dorset, was thought by Martin Sheldrick of the British Museum (Natural History) to be the proximal end of the ulna of a large baleen whale (Coy, in press a).

CONCLUSIONS

This paper has summarised some highlights from recent archaeozoological results for the south-western peninsula and south-west Wessex relating to non-domestic mammals, birds, fish, and marine molluscs. I have made no attempt to assess the role of crustacea in the diet but the role of crabs and lobsters could have been considerable in some cases and there is slight evidence for their use on some settlements, for example, crab claws are mentioned at Roman Holcombe, (Clutton-Brock *et al*, 1974); and medieval Gwithian (Turk, 1969), Okehampton (Wilkinson, 1982), and Exmouth (Jones, 1980). The importance of eighteenth century lobster and prawn gathering is mentioned in the report for Braunton Burrows, Devon (Smith *et al*, 1983).

Conclusions for any body of archaeological material can only be as valid as the original sampling and identification techniques (Bell, 1984: 43). The problems of preservation and excavation have been discussed in each of the sections above. Many groups will rarely be preserved, such as the marine crustacea mentioned above and the cartilaginous fishes. Worse still are the areas of the South West where no bones of any kind survive. The importance of sieving for a proper understanding of the role of fish and small birds has been stressed in the relevant sections and the recent results from medieval urban sieving confirm this.

Interpretation of the available archaeozoological evidence is often of prime archaeological relevance, whether it is dietary, ecological, or economic. The dietary information can lead to a fuller discussion of the social status of the inhabitants. Shellfish and fish were important for their nutrients, including calcium from the whole bones consumed. Evidence that fish were eaten whole supports the importance of these in the diet. This sort of evidence gives insight into how earlier diets differed from today's (Eaton and Konnor, 1985).

Many specialised aspects of archaeozoological data, such as the parasitisation of oysters, can lead to discussions of which ecosystems were exploited. There is a wealth of information, as yet untapped for the South West, on the incremental growth of marine molluscs and fish which may eventually lead not only to deeper insight into which ecosystems were exploited but answer questions of seasonality. In later periods bones and shells provide evidence of trade which may tie up with documentary sources.

Finally, it is important to compare results for the South West with those for other regions. It is only in this way that larger patterns, such as the trading of saltfish in the medieval period, may be mapped.

ACKNOWLEDGEMENTS

Apart from the importance of the earlier work and continuing help of Martin Bell, I am grateful to Bruce Levitan, and Mark Maltby who have provided me with references and copies of their own work in the area and to Mark Robinson and Terry O'Connor for comments on the draft.

BIBLIOGRAPHY

Ashbee, P. 1968. Excavations at Halangy Down, St. Mary's, 1967-1968, *Cornish Archaeol.* **7**, 24-32 .

Ashbee, P. 1974. *Ancient Scilly*, Newton Abbot: David & Charles.

Backway, C. 1982. Oyster shells, *in* R.A. Higham, J.P. Allan and S.R. Blaylock, Excavations at Okehampton Castle, Devon: Part 2 the Bailey, *Proc. Devon Archaeol. Soc.* **40**, 138-144.

Ballantine, W.J. 1961. A biologically-defined exposure scale for the comparative description of rocky shores *Field Studies* **1**, 1-19.

Bell, M. 1984. Environmental Archaeology in South West England *in* H.C.M. Keeley (ed) *Environmental Archaeology: a regional review* London: Department of the Environment (Directorate of Ancient Monuments and Historic Buildings Occasional Paper No. **6**), 43-133.

Biek, L. 1975. A note on the state of preservation *in* H. Miles *op cit*, 67-71.

Bramwell, D. 1979. Bird bones *in* H.S. Gracie and E.G. Price, Frocester Court Roman Villa: second report *Trans. Bristol Gloucestershire Archaeol. Soc.* **97**, 61-62.

British Ornithologists' Union 1971. *The status of birds in Britain and Ireland* Oxford: Blackwell Scientific Publications.

Brooks, R.T. 1974. The excavation of the Rumps Cliff Castle, St Minver, Cornwall *Cornish Archaeol.* **13**, 5-50.

Chaplin, R.E. and Coy, J.P. 1964. The animal bones *in* R.T. Brooks, The Rumps, St Minver: interim report on the 1963 excavations *Cornish Archaeol.* **3**, 31-4.

Clutton-Brock, J. 1974. Animal bones *in* S. Pollard, A late Iron Age settlement and a Romano-British villa at Holcombe near Uplyme, Devon *Proc. Devon Archaeol. Soc.* **32**, 154-155.

Coles, J.M. and Orme, B.J. 1982. Beavers in the Somerset Levels: some new evidence *Somerset Levels Pap.* **8**, 67-75.

Colley, S.M. 1983a. Interpreting prehistoric fishing strategies: an Orkney case study *in* C. Grigson and J. Clutton-Brock (eds) *Animals and Archaeology: 2 Shell middens, fishes and birds* Oxford: British Archaeological Reports (International Series **183**), 157-171.

Colley, S.M. 1983b. The role of fish bone studies in economic archaeology: with special reference to the Orkney Isles. Unpublished Ph.D. thesis, University of Southampton, Department of Archaeology.

Colley, S.M. 1984. Fish Remains *in* D.M. Griffiths and F.M. Griffith, An excavation at 39 Fore Street, Totnes *Proc. Devon Archaeol. Soc.* **42**, 95-98.

Cowles, G. 1974. Bird bones *in* S. Pollard, A late Iron Age settlement and a Romano-British villa at Holcombe near Uplyme, Devon *Proc. Devon Archaeol. Soc.* **32**, 154-5.

Coy, J.P. 1977. Bird bones from Wirrall Park Farm (the Mound) Glastonbury: Ancient Monuments Laboratory Report no. 2326.

Coy, J.P. 1980. The animal bones *in* J. Haslam, A Middle Saxon iron smelting site at Ramsbury, Wiltshire *Medieval Archaeol.* **24**, 41-51.

Coy, J.P. 1981. The animal bones *in* R.J. Silvester *op cit* 106-110.

Coy, J.P. 1982a. The role of wild vetebrate fauna in urban economies in Wessex *in* A.R. Hall and H.K. Kenward (eds) *Environmental Archaeology in the Urban Context* London: Council for British Archaeology (Research Report **43**), 107-116.

Coy, J.P. 1982b. The animal bones *in* C.J. Gingell, Excavation of an Iron Age enclosure at Groundwell Farm, Blunsdon St Andrew, 1976-7 *Wiltshire Archaeol. and Natur. Hist. Mag.* **76**, 68-73.

Coy, J.P. 1983a. Birds as food in prehistoric and historic Wessex *in* C. Grigson and J. Clutton-Brock (eds) *Animals and Archaeology: 2 Shell middens, fishes and birds* Oxford: British Archaeological Reports (International Series **183**), 181-195.

Coy, J.P. 1983b. Animal bone *in* S.M. Davies, Excavations at Christchurch, Dorset, 1981 to 1983 *Proc. Dorset Natur. Hist. and Archaeol. Soc.* **105**, 43-45.

Coy, J.P. 1984. The bird bones *in* B. Cunliffe, *Danebury: an Iron Age hillfort in Hampshire* London: Council for British Archaeology (Research Report **52**), 526-531.

Coy, J.P. 1985. Animal bones from Poole, Dorset: Ancient Monuments Laboratory Report.

Coy, J.P. *in press, a.* Animal bones and marine molluscs from the 1979 excavations *in* P. Woodward, *Rope Lake Hole, Kimmeridge, Dorset*, Poole: Dorset Natural History and Archaeology Society (Monograph **6**).

Coy, J.P. *in press, b.* Animal bones and marine molluscs *in* P. Woodward, *Cleavel Point, Ower, Dorset, 1978*, Poole: Dorset Natural History and Archaeology Society (Monograph **6**).

Coy, J.P. *in preparation, a.* Wild vertebrates, ecology, and diet *in* J. Bourdillon (ed) *The environment and economy of Saxon and medieval Southampton*, Southampton: Southampton Museums.

Coy, J.P. *in preparation, b.* The provision of fowls and fish for towns *in* D. Serjeantson and T. Waldron (eds) *Diet and crafts from towns: the evidence from animal remains* Oxford: British Archaeological Reports.

Coy, J.P. and Maltby, J.M. 1987. Archaeozoology in Wessex *in* H.C.M. Keeley (ed) *Environmental Archaeology: a regional review Volume II*, London: Historic Buildings and Monuments Commission for England (Occasional Paper no. 1) 204-251.

Crothers, J.H. 1983. Variation in dog-whelk shells in relation to wave action and crab predation *Biol. J. Linnean Soc.*, **20**, 85-102.

Darvill, T.C. and Coy, J.P. 1985. Report on the faunal remains from the Mound, Glastonbury *in* J. Carr, Excavations on the Mound, Glastonbury, Somerset, 1971, *Somerset Archaeol. and Natur. Hist.*, **129**, 56-60.

Desse, G. 1984. Nouvelle contribution a la diagnose des pieces rachidiennes des poissons *in* N. Desse-Berset (ed) *2nd fish osteoarchaeology meeting* Valbonne Cedex: Centre Nationale de la Recherche Scientifique (Notes et Monographies Techniques **16**), 25-39.

Desse, J., Desse-Berset, N. and Rochetaeu M. 1987. *Contribution a l'osteometrie de la perche (*Perca fluviatilis *Linne, 1758)* Juan-les-Pins: Centre de Recherches Archeologiques du Centre Nationale de la Recherche Scientifique.

Eaton S.B. and Konner, M. 1985. Palaeolithic nutrition: a consideration of its nature and current implications *New England J. Medicine* **312(5)**, 283-289.

Eisma, D. 1965. Shell characteristics of *Cardium edule* L. as indicators of salinity *Netherlands J. Sea Res.* **2**, 493-540.

Enghoff, I.B. 1983. Size distribution of cod (*Gadus morhua* L.) and whiting (*Merlangius merlangus* (L.)) *Videnskabelige Meddelelser fra Dansk naturhistorik Forening i Kjobenhavn* **144**, 83-97.

Fowler, P.J. and Thomas, A.C. 1962. Arable fields of the pre-Norman period at Gwithian *Cornish Archaeol.* **1**, 61-84.

Gaskell Brown, C. and Hugo, T.E. 1983. Prehistoric and Romano-British finds from Mount Batten, Devon: 1979-1983 *Proc. Devon Archaeol. Soc.* **41**, 69-74.

Graham, A. 1971. *British Prosobranchs* London: Linnean Society/Academic Press (Synopses of the British Fauna No.2).

Graham, A.H. 1984. Wimborne Minster, Dorset - excavations in the town centre 1983 *Proc. Dorset Natur. Hist. Archaeol. Soc.* **106**, 77-86.

Grant, A. 1979. Animal bones *in* B. Cunliffe (ed) *Excavations in Bath 1950-1975* Bristol: Committee for Rescue Archaeology in Avon Gloucestershire and Somerset (Excavation Report 1), 60-70.

Harcourt, R. 1979. The Animal Bones *in* G.J. Wainwright, *Gussage All Saints. An Iron Age settlement in Dorset* London: H.M.S.O. (Department of the Environment Archaeological Report No.**10**), 150-160.

Heller, J. 1975. The taxonomy of some British *Littorina* species with notes on their reproduction *Zoological J. Linnean Soc.* **56**, 131-151.

Jones, A.K.G. 1984. Some effects of the mammalian digestive system on fish bones. *in* N. Desse-Berset (ed) *2nd fish osteoarchaeology meeting* Valbonne Cedex: Centre Nationale de la Recherche Scientifique (Notes et Monographies Techniques **16**), 61-65.

Jones, G. 1980. Animal bones *in* P.J. Weddell, Excavations at 3-5 Lower Fore Street, Exmouth *Proc. Devon Archaeol. Soc.* **38**, 112.

Jones, G. 1981. Animal bones *in* P.J. Weddell, Excavations at Southgate Cottages, Lydford *Proc. Devon Archaeol. Soc.* **39**, 138.

Kromhout, D., Bosschieter, E.B. and Coulander, C. de L. 1985. The inverse relation between fish consumption and 20-year mortality from coronary heart disease *New England J. Medicine* **312(19)**, 1205-1209.

Langley, P.J.W. and Yalden, D.W. 1977. The decline of the rarer carnivores in Great Britain during the nineteenth century *Mammal Review* **7**, 95-115.

Lees, D. 1981. Shellfish *in* R.J. Silvester *op cit*, 110-111.

Levitan, B. 1978. The animal bones *in* A. Ellison *Excavations at West Hill, Uley: 1977, the Romano-British Temple* Bristol: Committee for Rescue Archaeology in Avon, Gloucestershire and Somerset (Occasional Paper **3**) 41.

Levitan, B. 1982. Faunal Remains *in* P. Leach *Ilchester Volume I, Excavations, 1974-1975* Bristol: Western Archaeological Trust (Excavation Monograph **3**) 269-285.

Levitan, B. 1984. Faunal remains from Priory Barn and Benham's Garage *in* P. Leach (ed) *The archaeology of Taunton: Excavation and fieldwork to 1980* Bristol: Western Archaeological Trust (Excavation Monograph **8**) 167-193.

Levitan, B. 1986. *Bos primigenius* raised at Charterhouse *Five Arches* No.**3**, 8-9.

Levitan, B. *forthcoming, a.* The vertebrate remains from St Katherine's Priory, Exeter *Proc. Devon Archaeol. Soc*

Levitan, B. *forthcoming, b.* The vertebrate remains from Exe Bridge, Exeter *Proc. Devon Archaeol. Soc.*

McMillan, N.F. 1968. *British Shells* London: F. Warne & Company.

Maltby, M. 1979. *Faunal studies on urban sites: the animal bones from Exeter* Sheffield: University of Sheffield (Exeter Archaeological Reports Volume 2).

Maltby, M. 1982. Animal and bird bones *in* R.A. Higham, J.P. Allan and S.R. Blaylock, Excavations at Okehampton Castle, Devon: Part 2 the Bailey, *Proc. Devon Archaeol. Soc.* **40**, 114-135.

Maltby, M. 1983. The animal bones *in* C. Heighway *The East and North Gates of Gloucester* Bristol: Western Archaeological Trust (Excavation Monograph **4**) 228-245.

Maltby, M. 1985. Patterns in faunal assemblage variability *in* G. Barker and C. Gamble (eds) *Beyond Domestication in prehistoric Europe* London: Academic Press, 33-74.

Meehan, B. 1975. Shell bed to shell midden: unpublished Ph.D. thesis, Australian National University.

Meehan, B. 1983. A matter of choice? Some thoughts on shell gathering strategies in Northern Australia *in* C. Grigson and J. Clutton-Brock (eds) *Animals and Archaeology 2: Shell middens, fishes and birds* Oxford: British Archaeological Reports (International Series **183**) 3-17.

Mellars, P.A. 1978. Excavation and economic analysis of Mesolithic shell middens on the island of Oronsay (Inner Hebrides) *in* P.A. Mellars (ed.) *The early postglacial settlement of Northern Europe* London: Duckworth, 371-396.

Miles, H. 1975. Barrows on the St. Austell granite, Cornwall *Cornish Archaeol.* **14**, 5-81.

Morales, A. and Rosenlund, K. 1979. *Fish bone measurements* Copenhagen: Steenstrupia.

Paul, A.A. and Southgate, D.A.T. 1978. *McCance and Widdowson's The Composition of Foods* (4th Ed) London: HMSO/Elsevier.

Quinn, D.B. 1937. *Port Books of Southampton, Vol I, 1469-1471* Southampton: Southampton Record Society.

Quinn, D.B. 1938. *Port Books of Southampton, Vol II, 1477-1481* Southampton: Southampton Record Society.

Rhoads, D.C. and Lutz, R.A. 1980. *Skeletal growth of aquatic organisms* New York: Plenum Press

Shackleton, J. and Douglass, J. 1983. Environmental sampling in Redcliff Street, 1982-3 *Bristol Avon Archaeol.* 2, 35-6.

Silvester, R.J. 1981. An excavation on the Post-Roman site at Bantham, South Devon *Proc. Devon Archaeol. Soc.* **39**, 89-118.

Smith, P.D.E., Allan, J.P. Hamlin, A., Orme, B., and Wootton, R. 1983. The investigation of a shell midden in Braunton Burrows *Proc. Devon Archaeol. Soc.* **41**, 75-80.

Tebble, N. 1966. *British bivalve seashells* London: British Museum (Natural History).

Thompson, G.B. 1980. Distribution and population dynamics of the limpet, *Patella vulgata* L. in Bantry Bay *J. Experimental Marine Biol. Ecol.* **45**, 173-217.

Turk, F.A. 1969. Notes on Cornish mammals in prehistoric and historic times: 2 *Cornish Archaeol.* **8**, 100-104.

Veale, E.W.W. 1933. *The Great Red Book of Bristol* Bristol: Bristol Record Society.

Wheeler, A.C. 1974. *in* S. Pollard, A late Iron Age settlement and a Romano-British villa at Holcombe near Uplyme, Devon *Proc. Devon Archaeol. Soc.* **32**, 154-5.

Wheeler, A.C. 1978. Problems of identification and interpretation of archaeological fish remains *in* D.R. Brothwell, K.D. Thomas and J. Clutton-Brock (eds) *Research problems in zooarchaeology* London: Institute of Archaeology (Occasional Publication No.3), 69-75.

Wheeler, A.C. 1980. Fish bones *in* P.J. Weddell, Excavations at 3-5 Lower Fore Street, Exmouth *Proc. Devon Archaeol. Soc.* **38**, 112.

Wheeler, A.C. 1982. Medieval fish bones *in* P. Leach *Ilchester Volume I, Excavations, 1974-1975* Bristol: Western Archaeological Trust (Excavation Monograph 3) 284.

Wheeler, A.C. 1984. *in* P. Leach (ed) *The archaeology of Taunton: Excavation and fieldwork to 1980* Bristol: Western Archaeological Trust (Excavation Monograph **8**), 193-4.

Wilkinson, M. 1979. The fish remains *in* M. Maltby, 1979. *op cit*, 74-81 and 212-5.

Wilkinson, M. 1981. The study of fish remains from British archaeological sites. Unpublished Ph.D. thesis, University of Sheffield, Department of Prehistory and Archaeology

Wilkinson, M. 1982. Fish bones *in* R.A. Higham, J.P. Allan and S.R. Blaylock, Excavations at Okehampton Castle, Devon: Part 2 the Bailey, *Proc. Devon Archaeol. Soc.* **40**, 135-8.

Winder, J.M. 1980. The marine mollusca *in* P. Holdsworth *Excavations at Melbourne Street, Southampton, 1971-76* London: Council for British Archaeology (Research Report 33), 121-127.

MAMMALIAN REMAINS FROM THE COTSWOLD REGION: A SURVEY OF THE LITERATURE FROM PALAEOLITHIC TO ROMAN TIMES, AND A MORE DETAILED ACCOUNT OF THE LARGER DOMESTIC MAMMALS FROM SOME RECENT ROMANO-BRITISH EXCAVATIONS

Barbara Noddle

INTRODUCTION

The area to be considered in this survey comprises the land bordered by the rivers Cherwell in the east, Bristol Avon in the south, Severn in the west and Warwickshire Avon in the north, and thus includes the floodplains of these rivers which are covered in the north and west by the glacial gravels of the Northern Drift (Lucy, 1872). The carboniferous limestone area of Bristol is excluded. The location of the sites mentioned is shown in Figure 1.

Environmental studies in the Cotswold area have suffered because the archaeological potential was discovered long before scientific techniques were introduced. An early field outing of the Bristol and Gloucester Archaeological Society was entertained by a man actually described as a barrow opener, who demonstrated his craft! Despite the splendid example of General Pitt-Rivers' reports on his Cranbourne Chase sites, early Gloucestershire animal bone reports include such dismissive comments as "the bones and teeth of inferior animals, not a few remnants of ox, hog, dog and birds" (Nympsfield: Buckman, 1865). Cowley (Broadway, 1946) might have been justified in stating that some bones "called for no special comment" since he did make some useful comments for other sites, but it is much to be regretted that neither Hencken nor Thomas appear to have submitted even a sample of the numerous bones from Bredon Hill to an archaeozoologist.

Clifford, Cotton and O'Neil, the three great ladies of Gloucestershire archaeology, who carried out most of the excavation in the area from the mid 1920s for some forty years, usually took considerable trouble with their animal bones by the standards of the time, and submitted at least a sample to leading archaeozoologists of the period, usually Dr Wilfrid Jackson. Unfortunately publication of a quantified report was not then a normal practice, and the brief information given usually comprises a qualitative account of species frequency, perhaps a description of the size and type of animal possibly backed by a few measurements and a note of unusual or pathological specimens. Thus, King (1978) could find no Iron Age or Romano-British sites from this area which fulfilled even the limited criterion of at least 100 identified fragments for his survey. It is particularly to be regretted that the material from Bagendon received such a brief treatment, but new excavations under way there may remedy this. However, there does appear to be a genuine shortage of Iron Age bone in the area, to judge from Norbury (Levitan, 1983a), Guiting Power (Wilson, 1979) and Claydon Pike (Wilson, pers comm); at Lechlade only half the contexts from this period produced any bone at all (Jones, nd).

This paper is in two parts: the first is an account of the literature and the smaller Romano-British sites. The second comprises a more detailed study of some larger deposits, mainly evaluated by myself. Some common denominators have had to be used on the work of diverse authors. In particular, the age structure of the animals concerned is not given chronologically, but in stages of maturity: juvenile (up to eighteen months old in modern terms), immature (eighteen months to about four years), and mature (over four years). A number of archaeozoologists have most generously allowed me to include their unpublished reports in this survey, G. Jones, A. King, K. Reilly and in particular B. Levitan, whose help is most gratefully acknowledged; these reports are designated nd (no date).

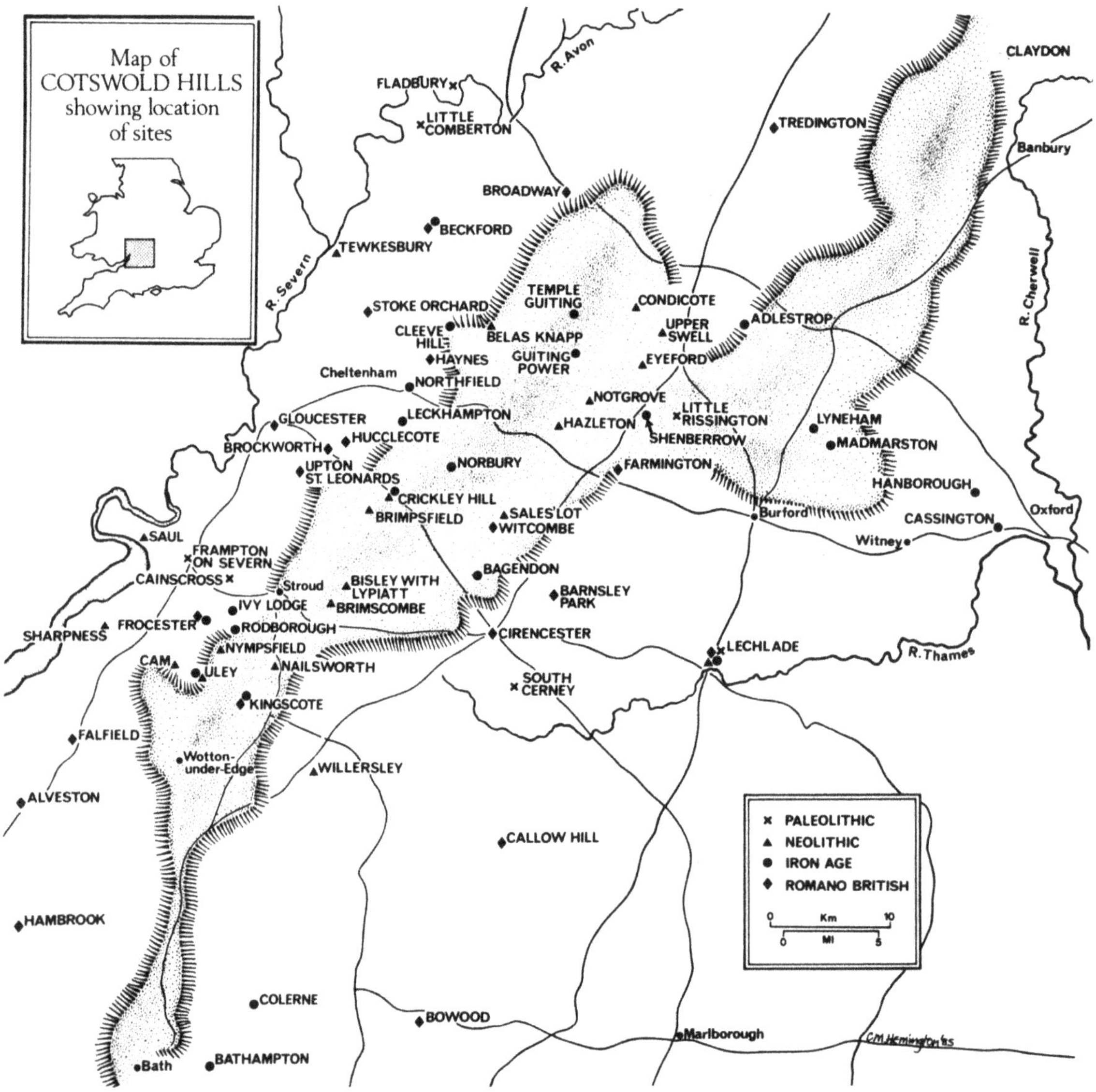
Map of
COTSWOLD HILLS
showing location
of sites
CLAYDON
FLADBURY
R. Avon
LITTLE COMBERTON
TREDINGTON
Banbury
BROADWAY
BECKFORD
TEWKESBURY
R. Severn
R. Cherwell
TEMPLE GUITING
CONDICOTE
STOKE ORCHARD
UPPER SWELL
ADLESTROP
CLEEVE HILL
BELAS KNAPP
HAYNES
GUITING POWER
EYEFORD
Cheltenham
NORTHFIELD
NOTGROVE
GLOUCESTER
LECKHAMPTON
HAZLETON
LITTLE RISSINGTON
LYNEHAM
BROCKWORTH
HUCCLECOTE
SHENBERROW
MADMARSTON
UPTON ST. LEONARDS
NORBURY
FARMINGTON
HANBOROUGH
CRICKLEY HILL
SALES LOT
Oxford
BRIMPSFIELD
WITCOMBE
Burford
CASSINGTON
SAUL
Witney
FRAMPTON ON SEVERN
BAGENDON
CAINSCROSS
Stroud
BISLEY WITH LYPIATT
IVY LODGE
BRIMSCOMBE
BARNSLEY PARK
SHARPNESS
FROCESTER
RODBOROUGH
CIRENCESTER
NYMPSFIELD
LECHLADE
CAM
ULEY
NAILSWORTH
R. Thames
SOUTH CERNEY
KINGSCOTE
FALFIELD
Wotton-under-Edge
WILLERSLEY
ALVESTON
PALEOLITHIC
NEOLITHIC
IRON AGE
ROMANO BRITISH
CALLOW HILL
Km
Mi
0
10
0
5
HAMBROOK
COLERNE
BOWOOD
Marlborough
Bath
BATHAMPTON

Figure 1.

A CHRONOLOGICAL SURVEY OF ARCHAEOZOOLOGICAL REPORTS

1. Palaeolithic

The glacial gravels deposited around the Cotswolds are sufficiently alkaline as a result of limestone outwash to preserve bone well. There is a continuous record of casual finds by quarry workers and interested amateurs from the latter part of the nineteenth century onwards. The chief areas concerned are the lower reaches of the Warwickshire Avon (eg Fladbury, Beckford, Little Comberton), the area immediately around Gloucester and west of Stroud (Barnswood, Hucclecote, Cam, Caincross, etc) and South Cerney on the eastern slope of the hills.

The most numerous finds have been mammoth teeth (described as fossilised cauliflowers by one labourer (Symonds, 1883)), but a large portion of a tusk was found more recently at Fladbury (Tomlinson, 1947). This has been dated to the last ice age (Tomlinson, 1947), and more specifically to the second glacial terrace, dated 34000 BC at Little Rissington (O'Neil and Shotton, 1972). Hippopotamus - a tooth from Little Comberton and an astragalus from Hucclecote (Lucy, 1872) - must come from the preceeding interglacial. A musk-ox skull from Frampton-on-Severn (Sandford, 1931) is a rare and important find. The only proper excavation from this period was carried out by Mrs Clifford as an adjunct to a Romano-British excavation at Barnwood (Clifford, 1930) and this is the only one listed in Bonsall's 1977 gazetteer where it is dated early Upper Palaeololithic.

Table 1. Dimensions of bones of Bos primigenius *from Gloucester and Stroud museums*

Site	Bone	Length	Proximal width	Distal width
South Cerney	Scapula			83*
Gloucester Royal Infirmary	Humerus			103
Gloucester Royal Infirmary	Radius	345	105+	108
South Cerney		330		98
Gloucester, Stroud	Metacarpal		70	
Gloucester, Stroud				73
Caincross				70
South Cerney	Tibia	440		80
South Cerney				81
Gloucester Royal Infirmary				85
Caincross	Calcaneum	190	54	

All measurements in mm, * = scapula neck width

The chance finds in the Stroud and Gloucester museums include a number of large bovid bones which appear to be *Bos primigenius.* Though these are not present in fully glacial periods, they do occur in interstadials, eg the Upton Warren (Coope, Shotton and Strachan, 1961). Smaller bovine bones are also present, particularly from South Cerney where they were collected from gravel workings. It is suggested that these might be Iron Age, as Iron Age peoples are known to have a predilection for gravel sites such as Beckford and Claydon Pike (Miles and Palmer, 1983). The aurochs bones have not been well studied, but some measurements are included in Table 1.

2. Holocene up to Bronze Age

Surface finds of mesolithic flints have been fairly numerous and widespread in the Cotswolds (Wymer, 1977), but there have been few excavations and no faunal reports from that period. The land surface beneath Hazleton Long Cairn contained a "midden" feature with concentrations of mesolithic flint, but bones from the old land surface (as yet not dated, but with radiocarbon dates forthcoming) all appear to be neolithic since the identifiable bones are all of domestic mammals (Levitan, nd). Bones from the "fine tenacious clay" of the Severn area (Lucy, 1877; Symonds, 1883) and the Nailsworth peat beds (Gardiner and Bate, 1933; Lycett, 1857) include not only red deer, beaver and other wild species, but also domestic forms. The most interesting specimen from the peat beds is an elk antler found at Nailsworth. The nineteenth century excavations for the docks at Tewkesbury and Sharpness, and the Sharpness submerged forest yielded bones mainly from wild animals: red deer, wild boar and aurochs, but sheep and horse were also identified (Symonds, 1883); it was also noted that the antlers had been removed from a red deer skull by human agency (Gardiner, 1930). Recent work on the west bank of the Severn suggests that these sediments date from the second and third millenia BC and might result from soil erosion caused by early agriculture (K. Morgan, pers comm).

A definitely dated neolithic lowland site comprising some pits was found at Cam (Westley, 1968) where the bones are mostly domestic cattle with one possible aurochs. Domestic cattle bones deposited at Gloucester museum by a water authority engineer and found during the construction of deep drainage ditches at Saul suggest that such sites might be widespread, but the silt is so deep such finds can only be made by chance. A possible settlement from which flints, rude pottery and animal bones (only wild forms identified) were recovered was buried by a landslip and revealed during construction of a reservoir on Stroud Hill (Jones, 1865; Witchell, 1895). On the other side of the Cotswolds, neolithic pits were found at Rough Ground Farm, Lechlade (Jones, nd).

Table 2. Bones of larger mammals from Neolithic and Bronze Age sites

Site	Cattle	Sheep	Pig	Horse	Dog	Other	Reference
i. Long barrows							
Belas Knapp	15	-	P	-	P	weasel	1,2,3
Bisley with Lypiatt	P	P	-	-		-	3
Eyeford	P	P			skele.		3
Hazleton	110	106	96	-	-	deer,goat	4
Notgrove	skele.	....."other bones".....					6
Nympsfield	most P	P	most P	P	P	-	7
Sales' Lot	teeth 6	-	-	-	-	fox,cat,polecat	8
Uley (Hetty Peglar)	-	-	wild	-	-	-	6
Upper Swell	P	-	-	-	-	goat	3
West Tump	1	few	-	-	-	red deer 1	9
Willersley	P	-	-	-	-	-	6
ii. other sites							
Cam	16	12	30	-	-	deer 2	10
Condicote	66	50	4	4	1	-	11
Lechlade (N)	21	1	58	-	1	red deer 16	5
(BA)	11	47	-	-	-	red deer 11	5

N - neolithic, BA - Bronze Age, P - present, skele. - skeleton

References: 1: Berry 1929, 2: Jackson 1929, 3: MacLean 1880, 4: Levitan nd, 5: Jones nd, 6: O'Neil & Grinsell 1960, 7: Bate 1938, 8: Cornwall 1966 9: Witts and Rolleston 1880, 10: Westley 1968, 11: Levitan 1983a

All other recorded neolithic bone comes from upland sites, and apart from Condicote Henge (Levitan, 1983a), it all originates from long barrows. O'Neil and Grinsell (1960) published a gazetteer of Gloucestershire barrows; of these 120 were long barrows, but only six were listed as containing animal bone. To these must be added Sales Lot, Withington (Cornwall, 1966), Hazleton (Levitan, nd) and West Tump, Brimpsfield (Witts and Rolleston, 1880). The species proportions from all these sites are set out in Table 2. Actual bone fragment counts are given where possible since percentages based on such small samples would give a spurious impression of precision. Where numbers were not stated, the authors' estimates of quantity have been used. Some of the barrows, notably Belas Knapp and Nympsfield, have been excavated on a number of occasions. Only Nympsfield and Hazleton have produced much bone, and Hazleton is unique in the quantity of deer bone (mainly antler tools) excavated. Levitan (nd) found a goat bone at Hazleton, a rare neolithic species, but another one was found at Lower Swell (MacLean, 1880). The identification of this is secure having been made by Rolleston, but the excavation context may be less certain. Of the three common domesticated species cattle are most numerous (Table 2) but at Hazleton, the most recent and reliable report, all three were in roughly equal quantities. Conversely, there was only one sheep bone at Lechlade. Animal bone in a long barrow context may well have a ritual significance, and of particular interest here are the 169 pigs' teeth found together at Nympsfield (Clifford, 1959). There were also said to be two kinds of dog at the same site. At Hazleton, limb bones of roe deer and sheep were found in the south chamber of the cairn, and two skulls of cattle plus one of pig were excavated from the forecourt from beneath cairn collapse - these might be skulls which had been displayed on the top of the forecourt wall (Levitan, nd). Witts and Rolleston (1880) describe specific bones associated with different skeletons.

Table 3. Summary of age structures from pre-Roman sites

Site	Animal	Juvenile	Immature	Mature
Neolithic				
Hazleton	Cattle	-	3	1
	Sheep	3	2	-
	Pig	8	2	1
Lechlade	Cattle	-	3	-
	Pig	-	3	3
Bronze Age				
Lechlade	Cattle	1	-	1
	Sheep	-	2	1
Iron Age				
Uley Bury	Cattle*	-	-	5
	Sheep	-	-	all
	Pig*	26%	39%	35%
Frocester	Cattle	1	2	4
	Sheep	-	1	1
Beckford	Cattle	2	2	4
	Sheep	10	8	8

* newborn/foetal also present, Beckford data calculated from Gilmour (1970)

Little information other than species proportions is available from any of the sites other than Hazleton and Lechlade. Levitan was able to age twenty individual specimens from Hazleton (Table 3). He did not record any measurements for domestic animals. Table 3 also gives the ageing evidence from Lechlade (Jones, nd). The distal humerus of a pig has width 31mm, and she recorded aurochs and wild pig. A distal portion of radius of aurochs (width 98mm) was also present from Hazleton (Levitan, nd). Westley (1968) measured a bovine distal radius of width 86mm at Cam, which may be aurochs. Neolithic bone has also been excavated from Crickley Hill and although it is still under study, a superficial examination of it by myself (courtesy Prf. Dixon) indicated a small form of cattle comparable with those of Windmill Hill (Grigson, 1965) rather than the larger forms described by Jackson (1943) from Maiden Castle, whose domestic status is not certain.

The only bone material specifically dated to the Bronze Age is a small deposit of less than 100 bones at Lechlade (Jones, nd). This is also listed in Table 2. The numbers of sheep show a marked contrast to the neolithic. Individuals that could be aged consist are summarised in Table 3. The prehistoric bones from Frocester Court, treated below as Iron Age, may well include some earlier material (Table 4c).

3. Iron Age

Iron Age sites will be considered in three groups, a) round barrows, b) hillforts, and c) settlements. There has been comparatively little bone of this period described from any source apart from the author's unpublished report on Beckford. This paucity appears to be genuine, for besides the earlier excavations mainly from the earthworks of hillforts, which might be expected to contain little food remains modern excavations such as Guiting Power (Wilson, 1979) and in particular West Hill (Levitan, 1978 and nd) have produced little bone; the contrast between Iron Age and Romano-British West Hill which produced many thousands of fragments, admittedly not all of food derivation, is very marked.

a) Round barrows.
Even less bone is recorded from round barrows than long barrows. A meagre six out of 600 odd sites listed by O'Neil and Grinsell (1960) are reported to contain bone, and in none of these are specific quantities mentioned. The information available is given in Table 4a, and sheep would appear to be the least numerous.

b) Hillforts.
In contrast, it would appear that sheep are the most numerous at hillforts (Table 4b) where rather more bone has been found. Sheep bone is more fragile and subject to greater taphonomic loss than cattle, so this difference appears to be genuine. Only amongst the small quantity of bone from the early excavations at Leckhampton (Burrow *et al*, 1925) is pig described as "most numerous" whereas at the other sites the opposite is true. However, comparisons of species frequencies from partial excavations at hillforts may be invalid because they may vary across the site. This is clearly demonstrated at Bury Wood Camp (Coy, 1967 and 1969) where numbers have exceptionally been presented as percentages to emphasise the point. Also, at Rodborough, Howard (1959) found cattle predominated, but Clifford (1937) did not.

c) Settlements.
Other Iron Age sites are grouped together in Table 4c. These include all the lowland sites and the two settlements considered by Cunliffe (1984) to be the most important in the area, Bagendon and Beckford. It is therefore unfortunate that the report on the former (Jackson, 1961b) was so brief and contained so little quantitative data. Cattle are the most numerous species other than at Lechlade, but only at Frocester Court do they really predominate. Horse is more abundant than at other Iron Age sites. The proportion found by Gilmore (1970) in her Beckford material is exceptional, and it must be supposed that the excavator selected mainly whole bones since horse bone is usually less fragmented than cattle and a false impression may be gained. However, my own Beckford material contains more horse and goat than other sites. The large number of goats at West Hill (Levitan, nd) is a phenomenon peculiar to that site, and is concerned with its ritual nature.

Table 4. Bones of larger mammals from Iron Age sites

Site	Cattle	Sheep	Pig	Horse	Dog	Other	Reference
a) Round Barrows							
Addlestrop	P	-	N	-	-	-	Cave 1938
Bevan's Quarry, Temple Guiting	P	-	P	-	-	red deer,fox, badger	Harcourt 1967
Ivy Lodge, Kings Stanley	P	P	P	-	-	-	Clifford 1950
Northfield, Cheltenham	teeth	-	-	-	-	large deer	Hill 1930
b) Hill Forts							
Bathampton	62	39	7	2	-	deer 4	Bird 1967
Bury Wood Camp*	26	63	11	4	2	cat	Coy 1967,1969
	13	83	4	-	-		
	41	55	4	-	-		
Leckhampton	P	P	most N	-	-	-	Burrow *et al* 1925
Lynham Camp	mni 1	3	-	-	1	-	Bayne 1957
Madmarston	mni 10	2	-	2	-	-	Howard 1960
Norbury	-	39	1	-	-	-	Levitan 1983b
Rodborough	P	P	P	-	-	fox,cat,badger	Clifford 1937
	predom.	P	scarce	-	1	-	Howard 1959
Shenberrow	mni 17	17	9	1	1	goat 1	Jackson 1961a
Uley Bury	29	40	28	3	-	-	Levitan 1983b
c) Settlements, etc							
West Hill, Uley	27	13	2	2	1	goat 55	Levitan nd
Bagendon	50%	P	P			goat, roe deer	Jackson 1961b
Beckford	134	115	27	156	-	-	Gilmore 1970
	1133	1326	306	268	63	deer 26,goat 57	Noddle nd
Cassington, Purwell Farm	mni 7	3	3	1	-	-	Banks 1961-2
Cleeve Hill	skull	P	N	N	1	-	Hunter & Kennard
Frocester Court	63	25	4	10	4	-	Noddle nd
Guiting Power	4	P	-	3	-	-	Houlton 1973
	78	189	6	14	3	-	Wilson 1979
Hanborough City Farm	mni 49	55	14	19	-	-	Banks 1964
Kingscote	2	-	-	-	-	-	Clifford 1963
Leckhampton	P	P	P	P	-	-	Purnell & Webb
Lechlade	18	21	4	1	-	-	Jones nd

P - present, N - numerous, mni - minimum number individuals,
* - Bury Wood Camp data are percentages

Some information is available concerning age structure and size of some of these animal populations, other than the barrows, although Harcourt (1957) does mention a six month old red deer calf from Bevan's Quarry, and Lysons (1863) specifically mentions the jaws of calves at Windmill Tump, Rodmarton. Amongst the hillfort bones there were a number of young bovine calves from Bathampton (Bird, 1967). Levitan (1983b) was able to age some of the animals from Uley Bury (Table 3). At Bagendon (Jackson 1961b) most of the cattle are described as two and a half years old, but pigs were "mostly adult". At Cleeve Hill the cattle and pigs were both "of all ages" (Hunter and Kennard, 1904). The Frocester ageing data are given in Table 3. The material from Beckford (Noddle, nd) is sufficient in quantity for a more detailed assessment to be made - this is done in the second part of the paper where it is compared with Romano-British bones. A detailed assessment of age structure is made in Figure 2. Gilmore's (1970) ageing data are summarised in Table 3.

Information is also available about the type and size of some of these animals. The cattle at Bagendon were described as being of Celtic type. Bovine measurements are given in Table 5. These data indicate animals of small stature, as are common in the Iron Age. This may be the result of poor nutrition, particularly in infancy, as well as genetic constitution. A number of bovine horncores were found at Cleeve Hill, for which the following measurements were given (mm):

basal circumference -	156,	131,	93
greatest basal diameter -	63,	43,	35
length outer curvature -	150,	120,	75.

These indicate cattle with horns of extreme Celtic Shorthorn conformation. Fewer measurements could be made on the bones of other species, but at Guiting Power there was a sheep radius 135mm in length, and a metacarpal of 117mm. At Beckford there was a sheep radius of 139mm, and at Frocester Court there were distal tibiae 22 and 24mm wide. Pig lower third molars measured 31mm (Guiting Power), 32mm (Frocester Court) and 33mm (Cleeve Hill). Also at Cleeve Hill there was a pig humerus, distal width 37mm. Again, all these measurements indicate stock small in stature, except for the pig humerus which may represent a wild specimen. The type of animal found at Becford is discussed in detail below.

Table 5. Measurements of cattle bone from smaller Iron Age sites

Site	Bone	Part measured	Measurement
Guiting Power	Radius	Length	267
Beckford			239
Bagendon	Metacarpal	Length	171
Beckford			175,192
Frocester Court		Proximal width	46
Frocester Court		Distal width	49
Frocester Court	Tibia	Distal width	48,52
Cleeve Hill	Calcaneum	Length	113,118
Bagendon	Metatarsal	Length	202
Beckford			204,210
Frocester Court		Proximal width	36
Frocester Court	Lwr 3rd molar	Length	33

all measurements in mm, Cleeve Hill coverted from imperial.

4. Romano-British

The Romano-British sites are divided into three groups: a) smaller sites, b) major urban centres (Cirencester and Gloucester), c) sites which produced large quantities of bone. (c) is dealt with in the second part of this paper.

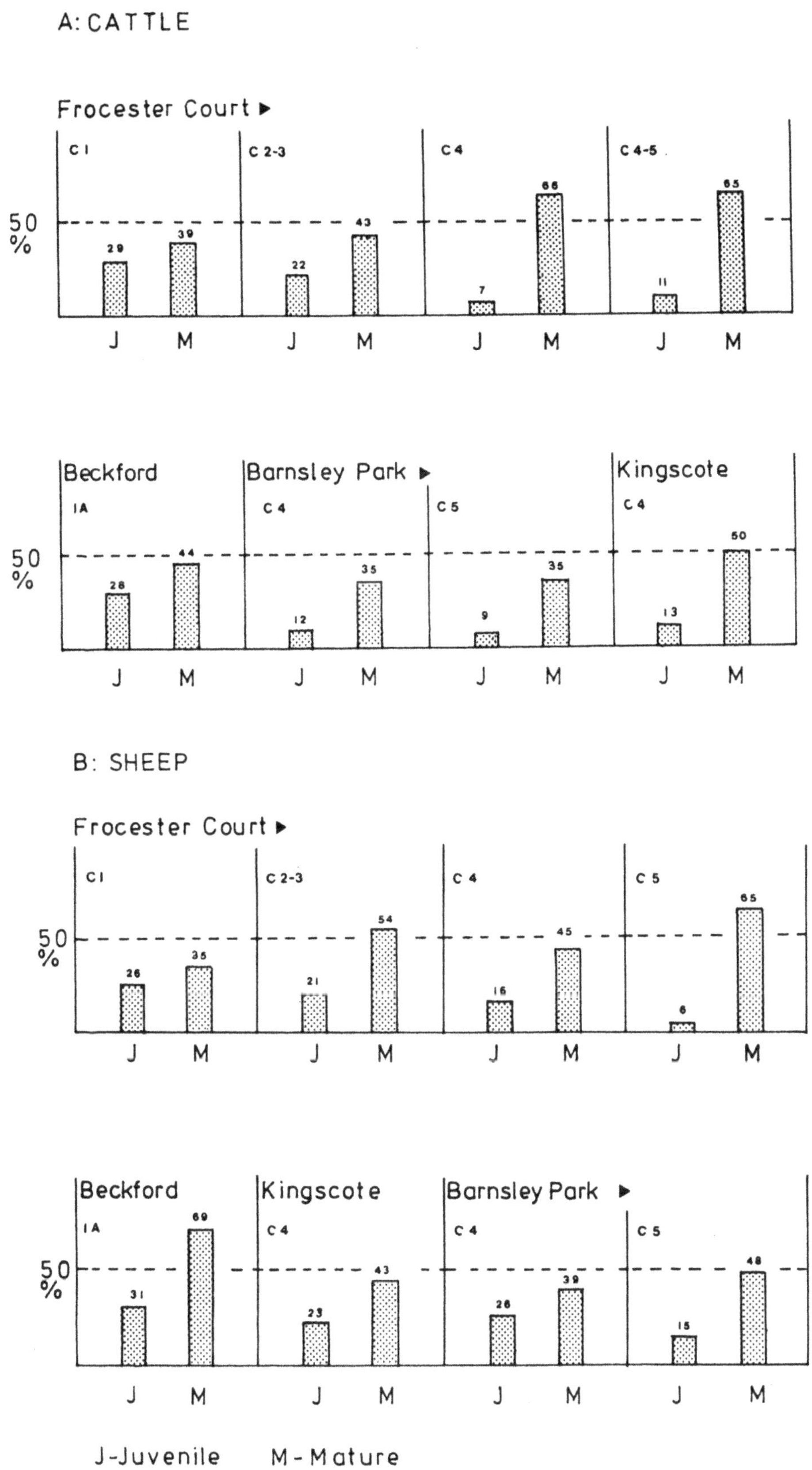

Figure 2. Age ranges of cattle and sheep from larger Romano-British sites

a) Smaller sites.

The species proportions for the smaller sites are given in Table 6a, and include a number of early sites for which proper quantification is lacking; fragment counts are used where possible. With the exception of only four sites, cattle outnumber sheep. The exceptions are Clear Cupboard Villa (Houlton, 1969) where only thirteen fragments were recorded, Lechlade (Jones, nd), and the M5 rescue sites of Heneage Court, Falfield and Lodge Farm Alveston (Miles, 1974; Everton, 1976) which were described as native in character. Non-domesticated animals were found only at the more Romanised sites. Horse is nearly always present, and in some quantity at Heneage Court (Miles, 1974), Haymes and Portway (Noddle, nd and 1984a). It is unfortunate that better quantification is not available for the apparently numerous bones from Hucclecote. Despite the small samples from Lechlade, they show the almost universal increase found in cattle numbers with the passage of time (King, 1978; Noddle, 1984b).

Table 6. Bones of larger mammals from Romano-British sites

Site	Cattle	Sheep	Pig	Horse	Dog	Other	Reference
a) Smaller sites							
Alveston Church Rd	35	8	-	2	-	-	Everton & Tratman
Alveston Lodge Farm	20	76	P	8	P	-	Everton 1976
Brockworth	P	P	P	P	-	-	Rawes 1981
Callow Hill mni	7	4	3	2	skele	red & roe deer	Jackson 1957
Clear Cupboard Villa	5	3	5	-	-	-	Houlton 1969
Hambrook Cyst	2	-	-	-	-	-	Crooke & Hinton
Haymes	307	249	41	24	21	red deer 6,goat 1	Noddle 1986
Heneage Court	37	93	3	23	-	-	Miles 1974
Hucclecote Baths	P	P	-	P	-	-	Cowley 1960
Villa	most N	P	P	mni 3	P	deer 5,badger, cat,fox,bear	Jackson 1933
Lechlade C2-3	35	62	16	1	1	fox	Jones nd
C3-4	52	50	18	3	2		
Nuthills	most N	P	P	P	-	-	Jackson 1927
Portway	181	62	4	54	11	red deer 5,goat 2	Noddle 1984a
Tredington	6	3	-	1	-	-	Everton 1976
Wadfield Villa	skull +2	-	-	-	skele	-	Rawes 1971
Witcombe Villa	most N	P	P	P	1	-	Jackson 1954
b) Urban sites:							
i) Gloucester							
Girls' High School	most N	1	1	4	-	-	Kennard 1908
Bon Marchè	71	28	19	1	-	-	Cowley 1963
Eastagte	399	158	30	16	36*	cat	Maltby 1983
Northgate	107	25	12	-	-	roe deer 2	Maltby 1983
Southgate St C1	127	112	15	10	-	-	Noddle nd
C4	51	23	12	1	1	-	
Tower St C3	2623	80	82	-	-	-	Maltby 1984
Westgate	217	30	16	4	1	red deer	Maltby 1979
ii) Cirencester							
Early occupation	391	259	142	12	1	roe deer 1,goat 1	Thawley 1982a
R-B cemeteries	1158	160	89	8	8	red deer 1	Thawley 1982b
The Beeches pre C4	186	73	38	7	1	red deer 1,hare 2 roe deer 1	Levitan 1986
C4	142	37	30	2	6	red deer 3,hare 1	Levitan 1986
C4+	176	26	19	6	2	red deer 2	Levitan 1986
C4	mni14	12	12	2	2	cat 1,goat 1,hare 1 red deer 1,badger 1	King 1986

c) major assemblages:	Cattle	Sheep	Goat	Pig	Horse	Dog	Other	Total	Reference
	%	%	%	%	%	%	%		
Beckford (Iron Age)	35	41	<1	10	8	<1	<1	3203	Noddle nd
1st century									
Frocester Court	25	48	<1	18	4	2	2	3684	Noddle 1979, nd
West Hill	24	53	19	2	2	1	-	3136	Levitan nd
Ditches Bagendon	48	34	-	15	1	1	<1	5199	Reilly nd
2nd-3rd centuries									
Frocester Court	39	37	1	12	3	4	4	5450	Noddle 1979, nd
West Hill	12	30	54	3	<1	1	<1	11592	Levitan nd
4th century									
Frocester Court	48	29	1	12	5	1	2	6016	Noddle 1979, nd
West Hill	4	22	72	2	<1	1	<1	39196	Levitan nd
Barnsley Park	30	54	1	9	4	1	1	8422	Noddle 1985
Kingscote	42	37	2	9	4	2	4	5475	Noddle nd
4th-5th centuries									
Frocester Court	54	15	2	17	5	2	5	1806	Noddle 1979, nd
Barnsley Park	32	46	2	12	5	1	1	3919	Noddle 1985

P - present, N - numerous, mni - minimum number individuals, * - dog skeleton
West Hill - calculated from Levitan's data, see text

Few of these reports give any information of the age structures of the populations, what information is available is given in Table 7.

Similarly, information on size and type of the animals at these sites is scattered. Jackson reports of two types of cattle of differing size from Hucclecote (1933), Nuthills (1930) and Witcombe (1954), which at Nuthills are described as Celtic and Longhorn. This can only be substantiated at Haymes, where there are two horncores of nearly circular cross-section, that is Romano-British type, and two of strongly elliptical cross-section, that is Iron Age. Lengths of complete bones are as follows:

radius	(Portway)	290mm
metacarpal	(Haymes)	182, 197mm
	(Portway)	180mm
	(Witcombe)	168mm
metatarsal	(Haymes)	207, 213, 222mm
	(Hucclecote)	212, 213, 237mm
	(Portway)	205mm
	(Witcombe)	216mm.

Whilst there do seem to be two sizes of bone present in these collections, the difference is probably no greater than can be accounted for by sexual and nutritional differences. Jackson describes the sheep at most of his sites as being of Soay type. The sheep at Haymes were also small (metacarpal 105mm), but larger animals occurred at Portway (metacarpal 128mm) and these differences are borne out by other bone measurements from these sites. Jackson reports a very large sheep horncore from Nuthills, and a truely gigantic specimen (basal diameters 61 and 42mm) was found at Haymes (Noddle, nd); there was also a skull with massive horn cores at fourth century Frocester Court. A solitary complete pig bone is recorded: a humerus of 162mm from Witcombe.

Table 7. Age structures from smaller Romano-British sites

Site	Animal	Juvenile	Immature	Mature
Haymes	Cattle	23		51
	Sheep	26		38
Portway	Cattle	9	23	58
	Sheep	4	18	48
Lechlade	Cattle	3	3	8
	Sheep	3	6	12

b) Urban Romano-British sites.

Urban Romano-British sites in this area are the major towns of Corinium (Cirencester), at one time second only to London, and Glevum (Gloucester). Excavations, both deliberate and incidental, have been carried out at both towns from the nineteenth century onwards. In recent years both have undergone extensive planned excavations, and large quantities of animal bone have been found. However, very little evaluation of this material has yet taken place, though Levitan is working on major assemblages from Cirencester (pers comm), and work on Gloucester is scheduled for the future. The following account is therefore both scrappy and unsatisfactory as even when proper evaluations are made, single small sites cannot give a true indication of the situation for the whole town.

Gloucester:

The species proportions from the Gloucester reports are presented in Table 6b (i). The bones are few in number - only about 1000 overall - but a fairly consistent picture is presented. With the exception of first century Southgate Street cattle comprise between half and two thirds of the identified fragments. The decrease in sheep numbers at Southgate Street between the first and fourth centuries is paralleled at Frocester Court, a conceivable supplier. Pig is consistent, but horse is variable. Horseflesh may have occasionally been consumed by Romano-British people but its popularity is unclear. There is relatively little evidence of tertiary butchery on horse bone, perhaps because horse fat was not liked, and occasionally whole skeletons are found with no evidence of butchery at all. A portion of a complete horse skeleton was found at the Kingsholme site (Hurst and Garrod, 1975). Dogs were also buried complete, and a high proportion of dog bone may represent only a few individuals. Dogs were, of course, kept as pets and non-utility "breeds" were developed (Harcourt, 1974). A solitary cat bone comes from Eastgate. A few deer bones were present at North, West and Southgates. Deer are probably under-represented in reports where considerable fragmentation of other species has taken place; however, hunting cannot be said to have made a substantial contribution to the meat supply.

Ageing data are available for only two sites: Eastgate (Maltby, 1983) and Southgate (Noddle, nd). Fifteen cattle from Eastgate included one juvenile, four immature and five mature, whilst for sheep the figures are two, five and three respectively (based on state of epiphyseal closure). Age range for pigs (from dentition of the mandible) was 6% newborn, 22% juvenile, 27% immature and 39% mature. Both sources of ageing were used for Southgate; there were eight immature and two mature cattle, ten immature and twelve mature sheep and one juvenile, two immature and one mature pigs. A population consuming large amounts of meat will affect the surrounding animal husbandry system, and once animals are being bred specifically to supply that market, there should be a predominance of immature specimens. Juveniles would not make good butchers' carcasses and are likely to have been actual or anticipated casualties, whereas mature animals would have fulfilled some other economic function such as breeding or wool production.

Bone measurements are available from Bon Marchè (Cowley, 1963) and Southgate. Distal bovine metacarpal widths from Bon Marchè (45, 47 and 49mm) are from very small animals, and an oval horncore (basal diameter 42 x 31mm) suggest native Iron Age type stock was being supplied. There was a mixture of sizes at Southgate, though most were again small. Whole bones include a metacarpal of length 172mm and metatarsals 195 and 200mm. Proximal metacarpals of 60 and 63mm were recorded which represent larger animals, possibly bulls. There are five horncores of which four are nearly circular and of similar curvature, and one is oval in cross-section, straight, and much larger: these suggest Romanised females and an Iron Age type bull (basal diameters (in mm): 250x380, 260x380, 300x390, 360x460 and 430x600). Changes in shape of bovine horncores are discussed by Armitage and Clutton-Brock (1976) and Noddle (1983). Kennard (1908) was of the opinion that the unspecified number of cattle bones he examined from the Gloucester Girls' School site came from animals of two sizes.

Cirencester:

Early excavators have little to say about animal bones from Cirencester except that they were almost always present. St Clair Baddeley (1922) mentions horse and the horncore of a cow, and Cripps (1907) found the skeleton of a dog during an excavation of the Basilica. So far, the recent Wacher and McWhirr excavations have produced little result (but see below). Thawley (1982a; 1982b) has reported on two small early groups. Levitan (1986) and King (1986) have evaluated material from the Beeches site which, though rather small in quantity, covers a long period. King also included these bones in his 1978 paper. Maltby (1984) discussed a rather larger third century deposit excavated by Darvill which relates to the much larger St Michael's Field assemblage presently being evaluated by Levitan (pers comm). Otherwise it is apparent that much work remains to be done on such bones as were retained.

The species proportions from the above are set out in Table 6b (ii). The relative proportions of cattle and sheep are much as might be expected except at Darvill's Tower Street site where sheep are very much in the minority (Maltby, 1984). Sheep numbers decline with time at the Beeches, and there is a decrease in their numbers between the two early sites discussed by Thawley. As at Gloucester, the number of wild animals suggests only a modicum of hunting, at least as a contribution to the pot.

The ageing data presented by Thawley and King are rather confused, and Levitan had very little bone to work on. However, it is clear that the majority of cattle and sheep were killed at the prime immature stage and that there were few juvenile or mature individuals. This indicates that even in the first century Corinium was affecting the local agriculture rather more than Glevum, but the relative prosperity of the two towns concerned may have had a purely local effect on the quality of meat consumed. At the early sites, the majority of pigs were also immature, but at the Beeches both King and Levitan found a surprisingly large number of mature animals. This might indicate that pigs were being reared in the immediate area. Mature pig was also common at Glevum. It should be borne in mind that meat production was not the only return on animal slaughter, and that hides were probably valuable, particularly in a military context. Furthermore, other secondary products may have been important, such as milk, wool and traction. Rearing of animals for these latter reasons would explain the presence of older animals since they would have been kept longer than those reared for meat production only.

There were few complete bones to measure from the Beeches, but Levitan's results for bone ends indicate a wide range of cattle sizes. Thawley found two bovine metacarpals 161 and 167mm in length (very small) and two metatarsals, both 198mm long. She also measured the bases of six bovine horncores which gave ratios of greatest to least diameter of 1.21, 1.25, 1.36, 1.44 and 1.89. Taking size as well as ratio into consideration, these appear to represent four Romanised females and two Celtic males, a finding similar to that of Haymes.

Large Romano-British assemblages

There remain five large bone deposits which have been recently examined and from which rather more information can be gained than from the forgoing. Such information is also much more reliable as it is less likely to have been influenced by mere chance. Among the sites to be considered is the extraordinary ritual complex at West Hill, Uley, currently under investigation by Levitan (1978 and nd) where there are many thousands of fragments, but the economic information is obscured by the number of obviously sacrificial animals present. Reilly (nd) is examining a large first century assemblage from the Ditches, Bagendon and has kindly made his initial findings available. The other three sites have been investigated by the author. They are Frocester Court, a prosperous villa of the first to the fifth centuries on a site, still being excavated, with origins in the Bronze Age, and lying at the foot of the Cotswolds, between Glevum and Uley (interim report: Noddle, 1979); Barnsley Park, a fourth to fifth century villa a few miles north of Corinium (Noddle, 1985); and Kingscote, a small township of the fourth century on top of the Cotswolds, not far from Frocester and Uley.

The species proportions for these sites, and for Iron Age Beckford, are summarised in Table 6c. Here percentages can be properly employed, and the data are set out chronologically. The century boundary chronology employed here does not fit the actual phases for West Hill, and these data have been modified to compare with the other sites. Levitan's sheep/goat count has been calculated on the basis of proportions established for those bones where the differences between the two species are clear.

The following trends can be deduced from Table 6c. There is a steady increase in cattle numbers at Frocester, at least up to the fifth century, at the expense of sheep. The same trend is observed at Barnsley, but cattle numbers are lower at this upland site than for corresponding periods at Frocester. Kingscote, a consumer as opposed to a producer site, has an intermediate number of cattle, which is less than that of the Beeches, Cirencester. There are, however, fewer cattle at first century Frocester than at Iron Age Beckford. At Beckford there were more horses than any of the Romano-British sites (see also Gilmore, 1970). There is a much higher proportion of cattle at the Ditches than at any other site. West Hill is dominated by the vast numbers of sacrificial sheep, and, in particular, goat, but Levitan attempted to dissociate these from the more utilitarian livestock, and concludes that the basic meat economy was not very different from other Roman sites (as discussed in this paper).

Though the ritual aspect of West Hill is spectacular, there are other instances of non-utilitarian bone deposits. Several burials at Frocester have included a young pig or sheep, either complete or represented by the forelimb. Another instance of this is recorded from a late Roman burial at Bagendon (Grew, 1981). Token "sacrifices" may be made using discarded head and feet of the animals; in one instance at Barnsley Park these had been carefully laid out in the position they would have occupied had the animal been trussed up and laid on its back, other odd arrangements of sheep bones were also found (Noddle, 1985).

The age ranges of the cattle and sheep from all but West Hill are presented in Figure 2. Those from West Hill are not given because the majority of animals were not slaughtered for normal consumption, and in contrast to the results in Figure 2, the sheep and goats from West Hill (all periods from first to sixth centuries) were mostly killed when mature (Levitan, nd). In Figure 2, juveniles and mature animals are shown: the difference being made up by immature animals. The long period covered by the Frocester excavations includes apparent changes in cattle husbandry: the first century age proportions are very similar to those from Iron Age Beckford, with more than a quarter dying as juveniles. Whilst this could be the result of using the cattle for milk production, it is more likely to be the result of actual or anticipated caualties. In later periods, the proportion of juveniles decreases and that of mature animals increases. The proportion of juvenile animals from Barnsley Park also decreases with time, but here the increase is in immature rather than mature stock. One would expect that Kingscote, as a consumer site, would have rather more immature specimens - that is prime meat - than a producer site such as Barnsley, but in fact there is a greater proportion of mature animals; though the excavations at Kingscote covered only a small portion of the township, they appear to have been from a prosperous part of the town with mosaic floors and the like.

In the case of sheep, juvenile proportions also decrease, suggesting an all round improvement in standards of husbandry. However, Beckford had a higher proportion of mature animals which may indicate a greater emphasis on breeding and production of wool as opposed to that of meat and skins for the Iron Age site. The pattern at Barnsley is similar to that of Frocester. Again Kingscote has a higher than expected proportion of mature animals.

Some of the earliest archaeozoological studies in the nineteenth century were carried out with a view to establishing genetic differences between livestock of different populations (Noddle, 1984b). This concept is of particular interest for the period covered by these sites, there being a strong tradition of introductions by the Romans. Noddle (1978) tried to establish criteria by which genetic differences could be discerned in the skeletal material, the collection of data being termed a type profile which includes both metrical and non metrical items. The method might also enable distinction to be made between geographical races, for example the difference between upland and lowland types. However, it should also be stated that only inference may be drawn as there was not sufficient material, in particular measurable horncores, for definite conclusions to be made. The profiles are given in Table 8.

Cattle were of similar size at Kingscote and Barnsley, and of much the same horncore type, but the material from Barnsley is very scanty. The two populations differ for location of the femoral nutrient foramen, but it is not known how strong the inheritance factor is in cattle for this character; a single sire can affect a sheep population considerably (Noddle, 1978). It is tentatively suggested that there was a Cotswold upland type of cattle since there are greater differences between Frocester and Kingscote than the above despite their closer proximity. Frocester cattle are larger - possibly a result of improved nutrition on the lowland site. The femora are similar, but the horncores differ, those at Frocester being more circular (Romanised) in cross-section than the elliptical horn core of the Iron Age Celtic ox. However, a number of smaller Romano-British sites indicate either mixed cattle populations or wide divergence in horncore sexual dimorphism - a topic on which there is very little information available.

More data are available for the differentiation of sheep types, but the results are not particularly clear. The Barnsley animals were larger than those from Kingscote with respect to astragalus measurements, but smaller for metacarpals. The proximal metacarpal ratio indicates a more primitive sheep at Kingscote, but the opposite is indicated by the scapula ratio. The inheritance of metacarpal shape is not known, but the primitive scapula conformation is difficult to breed out (Noddle, 1978). Little distinction could be made between these upland specimens and those at Frocester. Metacarpal and scapula evidence again conflict, though the femora show some difference. Thus, there would seem to be only one sheep type present, and the Frocester evidence suggests little change through time. However, this type can be contrasted with Beckford which had smaller animals with very elliptical horns, comparable with the primitive Soay. It is suggested that the Romano-British Cotswold sheep was a stabilised Iron Age - imported breed cross. This type may have been strongly horned in the male - a skull with massive horncores set at a primitive angle was found at Frocester (fourth century), and the largest sheep horncore that the writer has ever encountered in a domestic animal was found at Haymes; Jackson (1927) also reported large horns from Nuthills. The female of this type on occasion lacked horns, and some rudimentary horns were also found, but at Barnsley a

Table 8. Type profiles of cattle and sheep from larger Iron Age and Roman sites

a) Cattle

Frocester Court	1st century			2nd-3rd century			4th century			4th-5th century		
	Mean	Number	Range	Mean	Number	Range	Mean	Number	Range	Mean	Number	Range
Metacarpal length	168	1	-	179.5	4	172-184	191.2	5	180-202	187	4	182-192
Astragalus length	60	8	56-65	62.2	21	57-71	63.6	27	58-74	61.5	2	60-63
Horncore basal index	-	-	-	1.39	11	1.00-1.56	1.25	3	1.16-1.39	1.32	1	-
Femur foramen	.... 43% medial, n=7											
	Kingscote 4th century			4th century		Barnsley Park	5th century			Beckford (Iron Age)		
Metacarpal length	183.6	21	158-217	187	13	163-222	192	3	187-195	166.5	8	160-170
Astragalus length	61.2	25	54-69	61.9	23	51-71	62.5	7	57-70	58.4	14	55-66
Horncore basal index	1.34	5	1.19-1.42	1.39	2	1.38-1.41	1.41	1	-	5 of 7 elliptical		
Femur foramen	40% medial, n=8			85% medial, n=11				-------		54% medial, n=23		

b) Sheep

Frocester Court	1st century			2nd-3rd century			4th century			4th-5th century		
Metacarpal length	125	1	-	116.2	4	112-115	120	1	-		-------	
Astragalus length	25.3	8	24-28	25	7	24-26	25	3	24-26		-------	
Metacarpal prox. index	1.42	17	1.20-1.47	1.34	14	1.25-1.50	1.39	6	1.29-1.50	1.35	1	-
Scapula index	1.05	14	1.00-1.17	1.04	5	1.00-1.12	1.06	4	1.00-1.13	1.06	1	-
Horncore basal index	1.45	15	1.06-1.68	1.44	8	1.27-1.56		-------			-------	
Femur foramen	78% p, 22% d, n=9			64% p, 8% d, 18% m, n=11			60% p, 40% d, n=5				-------	
	Kingscote 4th century			4th century		Barnsley Park	5th century			Beckford (Iron Age)		
Metacarpal length	122	5	115-131	121	8	116-129	122	5	115-129	114	6	112-118
Astragalus length	25.5	11	23-27	26.4	39	24-29		-------		27	2	25-29
Metacarpal prox. index	1.42	14	1.31-1.50	1.33	25	1.29-1.50	1.35	16	1.20-1.50	1.39	16	1.31-1.50
Scapula index	1.04	6	1.00-1.08	1.09	6	1.02-1.21	1.06	3	1.06-1.07	1.10	27	0.04-1.20
Femur foramen	41% p, 59% d, n=17			46% p, 46% d, 8% m, n=13			100% p, n=1			77% p, 23% d, n=18		

p - proximal, d - distal, m - midshaft, n - number examined

complete ewe's skeleton carried quite substantial horns. Such a mixture occurs in the Soay breed and need not indicate cross breeding.

The overall impression gained from this study of the four non-ritual sites is that the Roman occupation had a considerable effect on livestock husbandry. A meat market was established, particularly in Corinium, which changed the age structure at death of populations kept by producers. Husbandry improved, but this could also be due to climatic changes as well as the dissemination of superior knowledge or even a straight cash injection from improved sales.

CONCLUSIONS

Although the Cotswold-Severn area has been farmed for several thousand years very little is known about its agriculture. The alkaline soils preserve bones well and there have been many missed opportunities during the numerous excavations that have taken place. This may be partly because of the long tradition of amateur excavation in the area: this, though otherwise often excellent, does not lend itself to the expense of environmental studies. The author has been able to work on material from a number of such excavations, but only one has been funded. Investigation of both major urban sites is now in progress, together with the large Romano-British assemblage from Claydon Pike on the eastern side of the Cotswolds (by R. Wilson and B. Levitan). However, much work remains to be done, and it is hoped that this work will give a clearer picture of domestic livestock present in the Cotswold area during the prehistoric and Romano-British eras.

BIBLIOGRAPHY

Armitage, P.L. and Clutton-Brock J. 1976. A system for classification and description of the horncores of cattle from archaeological sites *J. Archaeol. Sci.* **3**, 329-48.

Banks, J. 1961-2. Animal bone report *in* G.J. Dawson, Excavations at Purwell Farm, Cassington *Oxoniensia* **26-7**, 1-6.

Banks, J. 1964. Animal bones from City Farm, Hanborough, Oxon *Oxoniensia* **29**, 89-91.

Bate, D.M.A. 1938. Animal bone report *in* E.M. Clifford, Excavations at Nympsfield Long Barrow, Gloucestershire *Proc. Prehist. Soc.* **4**, 212-3.

Bayne, N. 1957. Excavations at Lyneham Camp, Lyneham, Oxon *Oxoniensia* **22**, 1-10.

Berry, J. 1929. Belas Knapp Long Barrow, Gloucestershire: report of the excavations of 1929 *Trans. Bristol Gloucestershire Archaeol. Soc.* **51**, 273-304.

Bird, F. 1967. Animal bone report *in* G.J. Wainwright, The excavations of an Iron Age hillfort at Bathampton Down, Somerset *Trans. Bristol Gloucestershire Archaeol. Soc.* **86**, 57.

Bonsall, C.J. 1977. *Gazeteer of Upper Palaeolithic sites in England and Wales* London: Council for British Archaeology (Geological Abstracts).

Buckman, J. 1865. Notes on an ancient British tumulus at Nympesfield, opened by the Cotteswold Club *Proc. Cotteswold Natur. Field Club* **2**, 184-189.

Burrow, E.J., Knowles, W.H., Paine, A.E.W. and Gray, J.W. 1925. Excavations on Leckhampton Hill, Cheltenham during the summer of 1925 *Trans. Bristol Gloucestershire Archaeol. Soc.* **47**, 81-112.

Cave, A.J.E. 1938. Animal bone report *in* H.E. Donovan, Excavations at Addlestrop Hill Barrow, Gloucester *Trans. Bristol Gloucestershire Archaeol. Soc.* **60**, 162-164.

Clifford, E.M. 1930. A prehistoric and Roman site at Barnwood near Gloucester *Trans. Bristol Gloucestershire Archaeol. Soc.* **52**, 201-254.

Clifford, E.M. 1937. The earthworks at Rodborough, Amberley and Minchinhampton, Gloucestershire *Trans. Bristol Gloucestershire Archaeol. Soc.* **59**, 287-308.

Clifford, E.M. 1950. The Ivy Lodge Round Barrow *Trans. Bristol Gloucestershire Archaeol. Soc.* **69**, 59-77.

Clifford, E.M. 1959. Fossil reptiles and other animals *Proc. Cotteswold Natur. Field Club* **33**, 94-9.

Clifford, E.M. 1963. Burial at Kingscote, Gloucestershire *Trans. Bristol Gloucestershire Archaeol. Soc.* **82**, 205-207.

Coope, G.R., Shotton, F.W. and Strachan, I. 1961. A late Pleistocene fauna and flora from Upton Warren, Worcestershire *Phil. Trans. Roy. Soc. London B.* **244**, 379-418.

Cornwall, I.R. 1966. Animal bone report *in* H.E. O'Neil, Sale's Lot Long Barrow, Withington, Gloucestershire 1962-1965 *Trans. Bristol Gloucestershire Archaeol. Soc.* **85**, 32-34.

Cowley, L.F. 1946. Excavation of a Romano-British site at Broadway *Trans. Worcestershire Archaeol. Soc.* **23**, 21-34.

Cowley, L.F. 1960. Human and animal bones *in* A.G. Hunter, A Romano-British bath block at Trevor Road, Hucclecote, Glos. *Trans. Bristol Gloucestershire Archaeol. Soc.* **79(2)**, 172.

Cowley, L.F. 1963. The animal bones *in* A.G. Hunter, Excavations at the Bon Marchè site, Gloucester 1958-9 *Trans. Bristol Gloucestershire Archaeol. Soc.* **82**, 62-63.

Coy, J.P. 1967. Appendix ii *in* D. Grant King, Bury Wood Camp, Colerne. Excavations in the area of the SW opening *Wiltshire Archaeol. Natur. Hist. Magazine* **62**, 13.

Coy, J.P. 1969. Appendix ii *in* D. Grant King, Bury Wood Camp: Excavations in the NE and NW areas *Wiltshire Archaeol. Natur. Hist. Magazine* **64**, 47-48.

Cripps, W.J. 1898. Notes on the Roman basilica at Cirencester lately discovered by Wilford J. Cripps Esq CB *Trans. Bristol Gloucestershire Archaeol. Soc.* **21**, 70-78.

Crooke, M.A. and Hinton, D.D. 1950. Report on the bones and pottery from the Hambrook Cyst *Trans. Bristol Gloucestershire Archaeol. Soc.* **69**, 202.

Cunliffe, B. 1984. Gloucestershire and the Iron Age of Southern Britain *Trans. Bristol Gloucestershire Archaeol. Soc.* **102**, 5-16.

Everton, R.F. 1976. Animal bone report *in* P.J. Fowler, J. Bennet and V.S. Hill (eds), Archaeology and the M5 motorway 4th report. Well at Stoke Orchard and Rudway, Lodge Farm, Alveston *Bristol Gloucestershire Archaeol. Soc. Trans* **94**, 49 and 81-83.

Everton, R.F. and Tratman, E.K. 1976. Animal bone report *in* P.J. Fowler, J. Bennet and V.S. Hill (eds), Archaeology and the M5 motorway 4th report. Church Farm, Alveston, *Bristol Gloucestershire Archaeol. Soc. Trans* **94**, 89.

Gardiner, C.I. 1930. Recent discoveries in the Stroud Valley *Proc. Cotteswold Natur. Field Soc.* **34**, 163-180.

Gardiner, C.I. and Bate, D.M.A. 1933. Occurrence of elk and other mammalia in the alluvium of the Stroud and Nailsworth valleys *Proc. Cotteswold Natur. Field Club* **35**, 77-82.

Grew, F.O. 1981. Roman Britain in 1980 *Britannia* **12**, 354.

Gilmore, F. 1970. Animal remains from Beckford *in* A. Oswald, Excavations at Beckford *Trans. Worcestershire Archaeol. Soc.* **3**, 18-28.

Grigson, C. 1965. Faunal remains: measurements of bone, horncores, antlers and teeth *in* I. Smith (ed), *Windmill Hill and Avebury: Excavations by Alexander Keiller 1925-1939* Oxford: Clarendon Press, 145-67.

Harcourt, R. 1967. Animal bone report *in* H.E. O'Neil, Bevan's Quarry Round Barrow, Temple Guiting *Trans. Bristol Gloucestershire Archaeol. Soc.* **86**, 37-38.

Harcourt, R. 1974. The dog in prehistoric and early historic Britain *J. Archaeol. Sci* **1**, 151-175.

Hill, H.C. 1930. Northfield Tumulus, Cheltenham *Trans. Bristol Gloucestershire Archaeol. Soc.* **52**, 305-308.

Houlton, J. 1969. Animal bones *in* P.E. Gasciogne, Clear Cupboard Villa, Farmington, Gloucestershire *Trans. Bristol Gloucestershire Archaeol. Soc.* **88**, 64-65.

Houlton, J. 1973. Animal bones *in* P.E. Gascoigne, An Iron Age pit at Wood House, Guiting Power *Trans. Bristol Gloucestershire Archaeol. Soc.* **92**, 207.

Howard, M.M. 1959. Animal bones *in* D.M. Renne, The excavation of an earthwork on Rodborough Common in 1954-55 *Trans. Bristol Gloucestershire Archaeol. Soc.* **78**, 43.

Howard, M.M. 1960. Animal bones *in* P.J. Fowler, Excavations at Madmarston Camp, Swalciffe 1957-8 *Oxoniensia* **25**, 46-47.

Hunter, D.D. and Kennard A.S. 1904. Evidences of ancient occupation on Cleeve Hill. Appendix: the vertebrate and molluscan remains *Proc. Cotteswold Natur. Field Club* **15**, 58-67.

Hurst, H. and Garrod, A. 1975. Excavations at Gloucester. 3rd interim report, Kingsholm 1966-75 *Antiq. J.* **55**, 267-294.

Jackson, J.W. 1927. Animal bone report *in* the Marquess of Lansdown, A Roman Villa at Nuthills, near Bowood *Wiltshire Archaeol. Natur. Hist. Magazine* **44**, 54-55.

Jackson, J.W. 1929. Animal bone report *in* W.J. Kemp, Belas Knapp Long Barrow *Trans. Bristol Gloucestershire Archaeol. Soc.* **51**, 271-272.

Jackson, J.W. 1933. The animal remains *in* E.M. Clifford, The Roman Villa, Hucclecote *Trans. Bristol Gloucestershire Archaeol. Soc.* **55**, 370-374.

Jackson, J.W. 1943. Animal bones *in* R.E.M. Wheeler, *Excavations at Maiden Castle, Dorset* London: Society of Antiquaries (Research Report no. **12**), 360-371.

Jackson, J.W. 1954. Animal bones *in* E.M. Clifford, The Roman Villa, Witcombe, Gloucestershire *Trans. Bristol Gloucestershire Archaeol. Soc.* **73**, 64-66.

Jackson, J.W. 1957. Animal bones *in* N. Thomas, Excavations at Callow Hill, Slympton and Stopefield, Oxon *Oxoniensia* **22**, 51.

Jackson, J.W. 1961a. Report on the animal bones from Shenberrow Hill Camp *in* C.I. Fell, Shenberrow Hill Camp, Stanton, Gloucestershire *Trans. Bristol Gloucestershire Archaeol. Soc.* **80**, 37-39.

Jackson, J.W. 1961b. The animal bones *in* E.M. Clifford, *Bagendon: a Belgic Oppidum* Cambridge: Heffer & Sons, 268-271.

Jones, T. 1865. On some flint instruments and the geological age of the deposit in which they were found upon Stroud Hill *Proc. Cotteswold Natur. Field Club* **2**, 97-111.

Kennard, A.S. 1908. Notes on the Roman remains at the Girls High School, Gloucester *Proc. Cotteswold Natur. Field Club* **16**, 125-126.

King, A. 1978. A comparative study of bone assemblages from Roman sites in Britain *Bull. Inst. Archaeol. Univ. London* **15**, 207-232.

King, A. 1986. Animal bones from site DE/DF *in* A. McWhirr, *Cirencester Excavations III. Houses in Roman Cirencester* Cirencester: Cirencester Excavation Committee (Monograph 3), 142-152.

Levitan, B. 1978. The animal bones *in* A. Ellison, *Excavations at West Hill, Uley: 1977 The Romano-British temple* Bristol: Committee for Rescue Archaeology in Avon Gloucestershire and Somerset (Occasional Paper No. **3**), 40.

Levitan, B. 1983a. The animal bones *in* A. Saville, Excavations at Condicote Henge monument, Gloucestershire 1977 *Trans. Bristol Gloucestershire Archaeol. Soc.* **101**, 38-39.

Levitan, B. 1983b. The animal bones *in* A. Saville, *Uley Bury and Norbury Hill Forts: Rescue excavations at two Gloucestershire Iron Age sites* Bristol: Western Archaeological Trust (Excavation Monograph No. **5**), microfiche C6-D5, E14-F6.

Levitan, B. 1986. The vertebrate remains from site CQ and CX/CY *in* A. McWhirr, *Cirencester Excavations III. Houses in Roman Cirencester* Cirencester: Cirencester Excavation Committee (Monograph 3), 133-141.

Lycett, J. 1857. *The Cotteswold Hills* Stroud: Piper Stephenson & Spence, 116-7.

Lysons, S. 1863. An account of the opening of the tumulus on his property at Rodmarton *Proc. Soc. Antiq.* (Second Series) **2**, 275-279.

MacLean, Sir J. 1880. Description of the chambered tumuli of Uley and Nympsfield *Trans. Bristol Gloucestershire Archaeol. Soc.* **5**, 86-118.

Maltby, J.M. 1979. Appendix 4; the animal bones *in* C.M. Heighway, A.P. Garrod and A.G. Vince, Excavations at 1 Westgate Street, Gloucester 1975 *Medieval Archaeol.* **23**, 182-185.

Maltby, J.M. 1983. The animal bones *in* C.M. Heighway, *The East and North Gates of Gloucester* Bristol: Western Archaeological Trust (Excavation Monograph No. **4**), 228-245.

Maltby, J.M. 1984. Animal bones and the Romano-Birtish economy *in* C. Grigson and J. Clutton-Brock (eds), *Animals in Archaeology 4: Husbandry in Europe* Oxford: British Archaeological Reports (International Series **227**), 125-138.

Miles, D. and Palmer, S. 1983. Claydon Pike *Curr. Archaeol.* **8(3)**, 88-92.

Miles, G. 1974. Animal bones from Heneage Court, Falfield *in* P.J. Fowler and J. Bennett (eds), Archaeology and the M5 Motorway 3rd report *Trans. Bristol Gloucestershire Archaeol. Soc.* **93**, 121.

Noddle, B.A. 1978. Some minor skeletal differences in sheep *in* D. Brothwell, K. Thomas and J. Clutton-Brock (eds), *Research problems in zooarchaeology* London: Institute of Archaeology (Occasional Publication No. **3**), 133-142.
Noddle, B.A. 1979. The animal bones *in* H.S. Gracie and E.G. Price, Frocester Court 2nd report 1968-77: the courtyard *Trans. Bristol Gloucestershire Archaeol. Soc.* **97**, 51-62.
Noddle, B.A. 1983. Size and shape, time and place: skeletal variations in cattle and sheep *in* M. Jones (ed), *Integrating the Subsistence Economy* (Symposia of the Association for Environmental Archaeology No. 4) Oxford: British Archaeological Reports (International Series **181**), 211-238.
Noddle, B.A. 1984a. Animal bones *in* B. Rawes, The Romano-British site on the Portway, near Gloucester *Trans. Bristol Gloucestershire Archaeol. Soc.* **102**, 68-70.
Noddle, B.A. 1984b. A comparison of the bones of cattle from ten Iron Age and Romano-British sites *in* C. Grigson and J. Clutton-Brock (eds), *Animals in Archaeology 4: Husbandry in Europe* Oxford: British Archaeological Reports (International Series **227**), 105-124.
Noddle, B.A. 1985. The animal bones *in* G. Webster, P. Fowler, B. Noddle and J.L. Smith, The excavation of a Romano-British rural establishement at Barnsley Park part 3 *Trans. Bristol Gloucestershire Archaeol. Soc.* **103**, 82-97.
Noddle, B.A. 1986. The animal bones *in* B. Rawes, The Romano-British settlement at Haymes, Cleeve Hill, near Cheltenham. *Trans. Bristol Gloucestershire Archaeol. Soc.* **104**, 92.
O'Neil, H. and Grinsell, L.V. 1960. Gloucestershire Barrows *Trans. Bristol Gloucestershire Archaeol. Soc.* **79(1)**, 5-149.
O'Neil, H.E. and Shotton, F.W. 1972. Mammoth remains from gravel pits in the North Cotswolds *Proc. Cotteswold Natur. Field Club* **36**, 196-197.
Purnell, F. and Webb, E.W. 1948. An Iron Age A site near Cheltenham *Trans. Bristol Gloucestershire Archaeol. Soc.* **69**, 197-198.
Rawes, B. 1971. An outbuilding at Wadfield Roman Villa *Trans. Bristol Gloucestershire Archaeol. Soc.* **90**, 124-128.
Rawes, B. 1981. The Romano-British site at Brockworth, Gloucestershire *Britannia* **12**, 45-77.
Sandford, K.S. 1931. The Proboscidean remains found at Barnwood near Gloucester *Trans. Bristol Gloucestershire Archaeol. Soc.* **53**, 243-249.
St Clair Baddeley 1922. The excavations at Cirencester (Corinium Dobunorum) Feb-March 1922 *Trans. Bristol Gloucestershire Archaeol. Soc.* **44**, 101-115.
Symonds, W.S. 1883. *The Severn straits: notes on glacial drifts, bone caverns and old glaciers* London: Simpkin Marshall & Company.
Thawley, C. 1982a. Animal remains *in* J. Wacher and A. McWhirr (eds), *Early Roman occupation at Cirencester* Cirencester: Cirencester Excavation Committee (Cirencester Excavations Vol. I), 211-227.
Thawley, C. 1982b. Animal bones *in* A. McWhirr, L. Viner and C. Wells (eds), *Romano-British cemeteries at Cirencester* Cirencester: Cirencester Excavation Committee (Cirencester Excavations Vol. II), microfiche 2/5 E13-F14.
Tomlinson, M.E. 1947. The Lower Morr (Fladbury) mammoth and its environment *Trans. Worcestershire Archaeol. Soc.* **24**, 55-59.
Westley, B. 1968. Animal bones *in* I.F. Smith, Report on late Neolithic pits at Cam, Gloucestershire *Trans. Bristol Gloucestershire Archaeol. Soc.* **87**, 25-26.
Wilson R. 1979. Animal bone *in* A. Saville, *Excavations at Guiting Power Iron Age site, Gloucestershire, 1974* Bristol: Committee for Rescue Archaeology in Avon Gloucestershire and Somerset (Occasional Paper No.7), 141-144.
Witchell, E. 1865. On a deposit at Stroud Hill containing flints, land and freshwater snails etc *Proc. Cotteswold Natur. Field Club* **2**, 208-211.
Witts, G.B. and Rolleston, G. 1880. Description of the Long Barrow called West Tump in the parish of Brimpsfield, Gloucestershire *Trans. Bristol Gloucestershire Archaeol. Soc.* **5**, 201-211.
Wymer, J.J. 1977. *Gazeteer of Mesolithic sites in England and Wales* London: Council for British Archaeology (Research Report No. **20**).

MEDIEVAL ANIMAL HUSBANDRY IN SOUTH WEST ENGLAND: A SELECTIVE REVIEW AND SUGGESTED APPROACH

Bruce Levitan

INTRODUCTION

Despite rich sources of evidence including field monuments and documents, relatively little work had been done on the medieval period in south west England until recently. In his general survey for Somerset, Aston (1982: 123) stated that "Medieval archaeology is a relatively new subject ... Work had been done on some major monuments ... but there had been no work on, or even recognition of, medieval settlements before the 1950s". For Gloucestershire, Leech (1984: 302) concluded "The archaeological study of medieval towns in Gloucestershire has been fairly typical of that of many other areas. Research has focussed largely on the more important towns, with often no attention at all having been paid to the smaller market centres founded in considerable numbers in the twelfth and thirteenth centuries".

It is not surprising, therefore, that until 1970 almost no work on bones from medieval sites in south west England had been published. Since then, and particularly in the wake of rescue excavations, a number of sites have yielded important bone assemblages. By 1980, sites in all the major towns in the region had been excavated, and the outstanding comparative study for Exeter published (Maltby, 1979a). Unfortunately, many assemblages have not yet been studied or published, and there is still no major study from the main towns of Gloucestershire and Avon in published form. Somerset has fared rather better, with work from Ilchester and Taunton recently published (Levitan, 1982a; 1984a), but Cornwall is another blank area, the few sites from the county being coastal, dune deposit sites where the shelly nature of the sands allows bone to be preserved.

A starting point for this paper was the regional review by Martin Bell, completed in 1981 and recently published (Bell, 1984). This provides a useful summary of published and some unpublished work, but a consideration of the evidence and its implications in terms of animal husbandry was beyond his brief. He did, however, comment on the lack of adequate analyses, and, for example, could list studies from only fourteen urban sites in the region, many of which comprised very small assemblages (Bell, 1984: 94-97).

The aims of this study are, firstly to bring Bell's review up to date for bone analysis in the region (concentrating on urban sites), secondly, to attempt a synthesis of the available information and to identify problem areas, and, finally, to highlight aspects requiring future research. The third aim will be illustrated with two examples from Exeter. It should be said at the outset that much of the information is patchy, and many samples are small, so that criticisms concerning the validity of comparing such material may be well founded, however, this is at least a starting point.

SOURCES OF EVIDENCE

Table 1 lists the sites used in the following analysis. I have included all important recent reports and as many minor sites as I could find, but the list may be incomplete. Size of assemblage is also indicated. In all cases number of identified fragments has been used for comparisons because this was felt to be the most comparable data. Minimum numbers (MNI) are not used because methods of calculation may have differed, and not all authors calculated MNI anyway. In any case MNI do not describe the size of an assemblage objectively (O'Connor, 1985: 27-28).

Exeter is clearly the city with the best samples (in terms of numbers of bones and quality of material), levels of coverage and standards of analysis. Bristol, the major city of the region, has little extant work of value, but if analysis of the two most recent sites can be completed, our understanding of the city economy will be radically

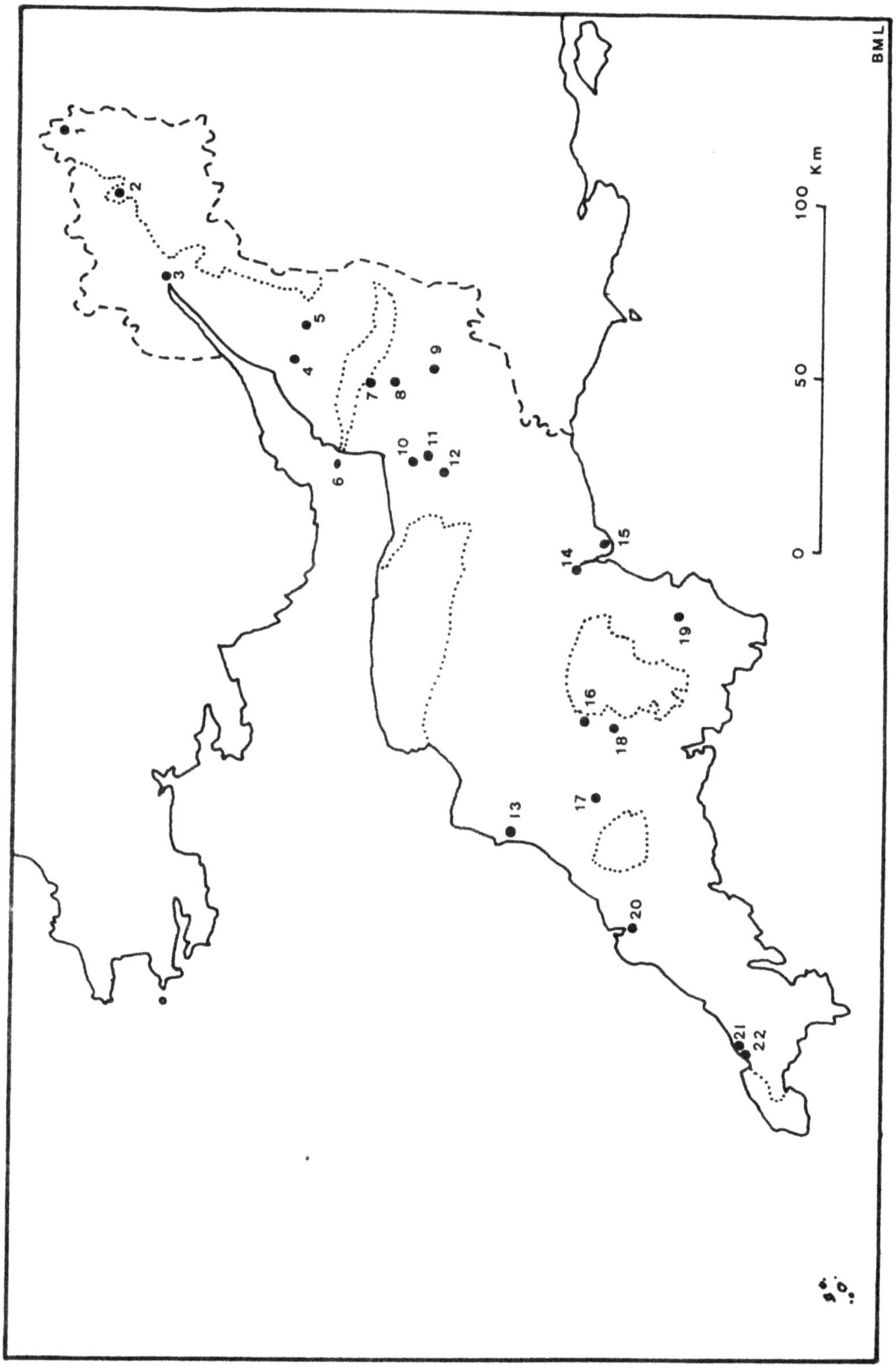

Figure 1. Location of sites cited in text and Table 1.

1: Upton, 2: Winchcombe, 3: Gloucester, 4: Bristol, 5: Bath, 6: Steep Holm, 7: Cheddar, 8: Glastonbury, 9: Ilchester, 10: North Petherton, 11: Crandon Bridge, 12: Taunton, 13: Braunton Burrows, 14: Exeter, 15: Exmouth, 16: Okehampton, 17: Launceston 18: Lydford, 19: Totnes, 20: Wadebridge, 21: Crane Godrevy, 22: Gwithian

improved (see Jones and Watson, this volume). Similarly, there is little evidence available from the main city of the northern part of the region, Gloucester, but the possibility for improving this seems remote at present. Taunton and Ilchester are the main towns of Somerset, and both have been relatively well covered recently, Ilchester in particular, and there are monographs on the archaeology of both towns now available (Leach, 1982; 1984), with a second volume on Ilchester forthcoming (Leach, in prep).

Somerset also contains the important monastic foundations of Glastonbury and Wells, but despite a series of excavations at both sites (Aston and Burrow, 1982: 119-121), little bone evidence is available, with only the bones from Silver Street, Glastonbury having any direct relationship to the abbey (Ellis, 1982).

Finally, there are a limited number of small township/village sites including deserted settlements. Recent medieval research has centred particularly on deserted settlements, but despite a reasonable number of excavations (Aston, 1982: 130; Aston and Viner, 1984: 276-9) very few rural sites have had bone assemblages studied. This seems surprising since many villages were enclosed in the fifteenth century specifically for sheep farming (Aston, 1982: 130). The detailed analysis of rural sites is also important in studying the incidences and effects of plague on rural economies (see, for example, Rahtz and Rowley, 1984; Levitan, 1984b). Furthermore, the intricate network of economic inter-relationships which is reflected in the documentary evidence provides a clear example of the complexity and importance of the rural economy (Gerrard, this volume). Clearly the level of information from bones is inadequate (Fig. 1)

RELATIVE IMPORTANCE OF THE MAJOR DOMESTICATES

Cattle, sheep (with goat) and pig form the main meat supply throughout the medieval period, but the relative importance of the species may vary. The incidence of and explanation for such variations is clearly central to an interpretation of site and regional patterns.

This paper is not the place to consider the methodological problems of using and comparing archaeological vertebrate evidence though it is realised that the validity of such comparisons may be called into question. In particular, three main problems may be identified: the representativeness of the assemblage (particulary if numbers are small), the comparability of published data, and the comparability of the material (ie the bones).

The second point is of particular relevance because the sources of evidence are variable and not necessarily wholly reliable, eg few of the assemblages are sieved, in some cases the sample size is small. Nevertheless, it is felt that they at least provide a starting point for new thought and analysis. Accordingly data from the sites have been plotted, and they are summarised in Table 1. The proportions of cattle, sheep and pig are plotted on triangular co-ordinates in Figures 2-4.

Some readers may not be familiar with triangular graphs. The three axes comprise sheep/-goat, pig and cattle. The sheep/goat base-line reads from 0% to 100% down from the apex to the right hand base, with the percentage axis lines running parallel to the cattle base-line; eg the 20% sheep/goat line runs from the 20% point on the sheep/goat base-line to the 80% point on the pig base line. Any plotted data which fall along that line represent 20% sheep/goat. The pig base-line reads from 0% to 100% right to left along the horizontal. Percentage axis lines run parallel to the sheep/goat base-line; eg the 20% pig line runs from the 20% point on the pig base-line to the 80% point on the cattle base-line. The cattle base-line reads 0% to 100% upwards on the left hand side. Percentage axis lines run horizontally (ie parallel to the pig base-line). In the previous examples, the 20% sheep/goat line and the 20% pig line should cross at the point where the 60% cattle line crosses them, this running from the 60% point on the cattle base line to the 40% point on the sheep/goat base-line.

Table 1. Site summaries

Site	Cattle	Sheep	Pig	Horse	D+C	Game	Other	Bird	Fish	Total	Reference
1.Upton											
	21	51	2	-	-	-	8	1	-	83	Noddle 1969
	452	106	23	-	-	1	109	-	-	691	Yealland & Higgs 1966
2.Winchcombe											
10-11 C	343	512	76	3	8	2	-	-	-	944	Levitan 1985
12-15 C	280	259	23	2	5	5	-	-	-	574	
3.Gloucester											
Bon Marchè	44	9	8	-	-	-	-	3	-	64	Hunter 1963
Eastgate	1219	942	283	38	81	5	-	51	2	2621	Maltby 1983
Northgate	168	100	37	1	6	1	-	11	-	75	Maltby 1983
Westgate	162	91	84	3	4	-	-	26	-	370	Maltby 1979b
4.Bristol											
Castle, pre conq.	150	86	92	3	1	3	-	14	P	349	Noddle forthcoming a
medieval	1237	972	959	30	66	93	4	1031	P	4392	
Spicers almshse	12	17	11	-	-	1	-	26	308*	375	Levitan forthcoming a
Mary-le-Port	547	491	98	4	5	-	-	17	-	1162	Noddle 1985
Greyfriars	168	166	47	6	16	81	-	?	?	484	Noddle forthcoming b
5.Bath Citizen's House											
10-11 C	143	183	51	3	-	1	-	55	7	443	Grant 1979
11-12 C	1023	1115	292	16	30	7	-	153	1	2637	
13 C	166	369	76	1	10	4	-	26	2	654	
6.Steep Holm medieval	43	106	34	4	-	262@	1	P	P	450	Levitan 1983b
7.Cheddar Palace											
11-13 C	143	183	51	3	-	1	62	120	72	506	Higgs *et al* 1979
14 C+	118	141	134	-	-	80	240	-	-	713	

Table 1. Site summaries (cont)

Site		Cattle	Sheep	Pig	Horse	D+C	Game	Other	Bird	Fish	Total	Reference
8.Glastonbury												
Silver St	10-12 C	482	806	195	16	10	3	-	81	-	1593	Levitan 1982b
	13-15 C	122	158	51	5	19	6	-	29	4	394	
Tor	early med	1	11	-	-	-	-	-	120	72	202	Harcourt 1970
	late med	1	"a few"	-	-	-	-	-	-	-	?	
Wirral Park	11-12 C#	175	236	180	158	1	29	9	53	-	841	Darvill & Coy 1977
		-	-	-	-	-	-	-	39	-	39	Coy 1977
Beckery Chapel												Rahtz & Hirst 1974
9.Ilchester												
Kingshams		1483	1614	250	6	2	1	-	373	53	3782	Levitan 1982a
Limington Rd Sth		226	136	28	3	7	5	-	16	1	422	Levitan forthcoming c
Church Street		14	18	6	2	3	-	-	-	-	43	Levitan forthcoming c
Manor House (A)		97	46	12	1	4	1	2	4	2	169	Levitan forthcoming c
Manor House (B)		99	53	21	1	-	-	-	6	-	180	
10.North Petherton												
	10-13 C	11	8	1	-	-	-	-	2	-	22	Adcock 1977
	14 C	6	11	9	-	-	-	-	2	-	26	
	14 C+	52	34	10	-	-	4	-	-	-	100	
11.Crandon Bridge		735	555	82	34	12	5	-	-	-	1423	Noddle forthcoming c
12.Taunton												
Priory Barn	12-13 C	199	367	60	1	1	-	-	-	52	628	Levitan 1984
	14 C	12	16	1	-	-	-	-	-	-	29	
	15 C	56	21	12	-	3	-	-	-	-	92	
	16 C	21	22	2	1	1	-	-	-	-	47	
Benham's G	10-11 C	374	242	20	3	4	1	-	8	-	652	Levitan 1984
	12-13 C	1346	1316	125	16	20	1	-	1	2	2827	
	14 C	154	120	6	12	2	-	-	34	-	328	
Taunton Priory	18 C+	................"with a few medieval exceptions"................										Langley 1978

Table 1. Site summaries (cont)

Site		Cattle	Sheep	Pig	Horse	D+C	Game	Other	Bird	Fish	Total	Reference
13.Braunton Burrows (auger)	medieval	-	P	-	-	P	-	-	P	P	?	Smith *et al* 1983
14.Exeter												
G S I-II	11-12 C	1233	1549	443	12	48	31	13	364	1479	5175	Maltby 1979a
	13 C	425	510	103	5	16	9	4	146	183	1422	
	14 C	253	196	51	7	10	5	-	45	37	604	
	15 C	77	87	19	1	4	4	-	19	8	219	
	16 C	610	662	121	8	34	37	-	120	70	1662	
G S III	11-12 C	1535	1605	637	36	112	28	1	363	482	4799	Maltby 1979a
	13 C	1165	1355	526	17	98	40	1	428	796	4417	
	14 C	33	83	35	1	4	-	-	26	27	209	
	15 C	1	1	-	-	-	-	-	-	1	3	
	16 C	552	867	237	1	499	175	27	927	1330	4615	
Trickhay St	11-12 C	821	683	295	9	28	12	-	230	181	2259	Maltby 1979a
	13 C	283	325	88	4	17	6	-	102	80	905	
	14 C	131	137	38	2	7	5	-	82	15	417	
	15 C	35	45	17	2	6	2	-	38	112	257	
	16 C	17	13	5	-	1	1	-	14	40	91	
High St	11-12 C	105	57	42	4	1	6	-	14	2	231	Maltby 1979a
	13 C	18	26	8	2	-	-	-	7	1	62	
St Kath Pr	13 C	342	546	111	6	-	-	5	112	12	1124	Levitan forthcoming b
	14 C	334	333	87	-	4	-	8	133	27	926	
	15 C	313	247	45	1	-	-	6	42	1	655	
	16 C	3849	1334	405	7	6	-	24	125	44	5814	
Exe Bridge	13 C	2760	2458	170	15	23	4	18	92	77	5617	Levitan forthcoming b
	14 C	451	438	45	1	4	10	2	103	51	965	
	15 C	511	571	114	4	5	21	3	198	310	1737	
	16 C	612	597	54	3	11	22	4	150	92	1545	

Table 1. Site summaries (cont)

Site	Period	Cattle	Sheep	Pig	Horse	D+C	Game	Other	Bird	Fish	Total	Reference
15.Exmouth	13 C	15	9	6	-	-	-	-	1	3	34	Jones 1980
16.Okehampton												
	pre 13 C	382	422	538	4	27	220	8	528	P	2129	Maltby 1982
	14 C	264	271	214	2	1	383	29	670	P	1563	
	15-16 C	489	674	185	5	29	1399	296	663	P	3740	
17.Launceston		....		not	seen	....						Jones 1975
18.Lydford	12 C	32	14	6	2	-	-	-	-	1	55	Jones 1981
	13-14 C	P	P	P	P	P	P	-	P	P	194	Harcourt 1980
19.Totnes	16 C	79	169	21	-	-	P	P	P	2443**	>2712	Bovey, Colley 1984
20.Wadebridge		....		not	seen	....						Turk 1965
21.Crane Godrevy	12 C	P	P	P	P	P	P	-	P	-	?	Turk 1970
22.Gwithian												
Hockin's Pit	10-12 C	few	sparse	1	few	1	2	-	P	P	?	Turk 1969
Sandy Lane	11-13 C	P	P	few	-	-	-	-	-	-	?	Turk 1969

P - present, G S - Goldsmith Street, St Kath Pr - St Katherine's Priory, Benham's G - Benham's Garage
#- with contamination from earlier periods, * - 79 bones from sieving, ** - 2426 fish bones from sieving,
Game = deer+lagomorpha, D+C = dog & cat, @- mostly rabbit (= intrusive ?),
Site numbers refer to location map: Figure 1

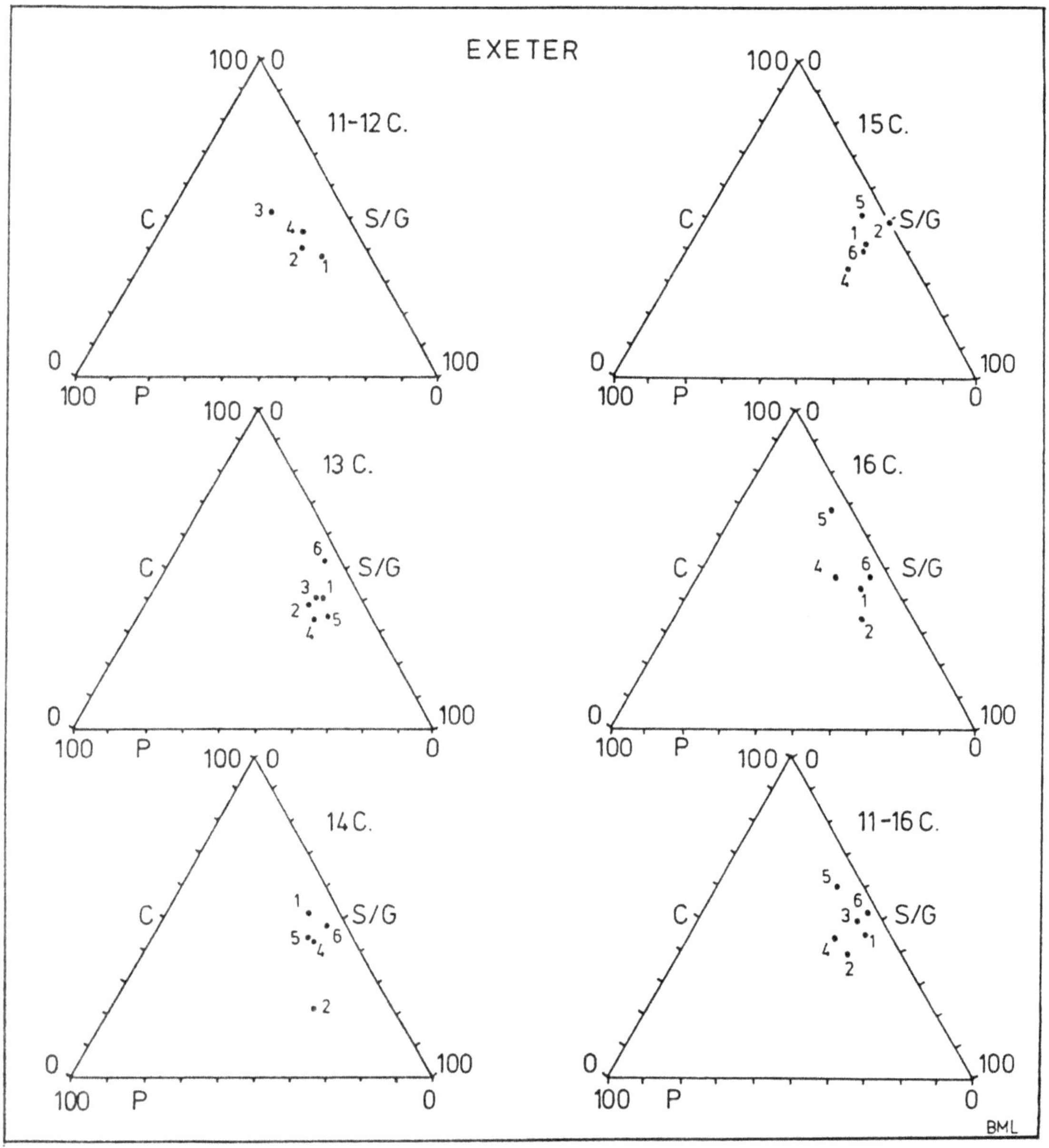

Figure 2. Proportions of cattle, sheep and pig at six sites from Exeter.

1: Goldsmith Street I-II, 2: Goldsmith Street III, 3: Trickhay Street, 4: High Street, 5: St Katherine's Priory, 6: Exe Bridge.
C = cattle, S/G = sheep/goat, P = pig

Figures 2-4 subdivide the sites by period, but some sites can only be divided into very broad periods, and others consist of single assemblages only. This makes assessment of temporal and lateral variations difficult because not all sites can be included. The most detailed breakdown is possible for Exeter (Fig. 2), where six assemblages can be compared for five periods spanning early/middle medieval to early post medieval. The general impression is of little overall change over the half millenium, with cattle and sheep predominating at all sites and all periods. This is somewhat of an oversimplification, and not only do individual assemblages vary through time (Goldsmith St. III in particular), but there is quite a lot of variation for any one period: for example in the eleventh to twelfth centuries cattle proportions range from under 40% to over 50% , and sheep from *c* 30-50% . The variation is not too great to invalidate an "Exeter average" for each period, but such an average should not be taken to represent more than a crude approximation of the importance of the three species. If such variation is apparent for Exeter, no doubt it also occurs in other towns, but there is not enough information here to assess the other sites.

Figure 3a illustrates the temporal evidence for those sites where a period breakdown is possible. In general, three "groups" can be defined: sites with over 20% pig ("pig" group), those with over 45% sheep ("sheep" group) and those with over 50% cattle ("cattle" group). Although these groups overlap in principle (eg 25% pig, 52% cattle, 23% sheep = pig/cattle groups), in fact no sites fall into the overlap areas.

Table 2 summarises the results, and with a few exceptions most sites are in the cattle group, though some are always outside it (eg Okehampton, Glastonbury) and others are outside for some periods (eg Exeter thirteenth century, Winchcombe pre-thirteenth century).

These results show that cattle were the most important group, generally exceeding 50% of the proportions. Okehampton is an interesting exception with cattle still important, but pig much better represented during the medieval period than at the other sites. Similarly, sheep are more important at Bath and Glastonbury than the other sites, though Winchcombe, Taunton and Exeter all have fluctuating sheep:cattle proportions. It is impossible to say to what extent these data are representative of the animal exploitation patterns of the time, but there is evidence of an important general pattern with individual variations which may reflect local conditions. Such "local conditions" may sometimes be fairly obvious (eg Okehampton Castle - see below), or more complicated, such as interactions between urban sites (? different activities) and between towns and their hinterlands.

On the last point, one can attempt to investigate the variations by comparing the temporal sequence of several sites from a single town; the example of Exeter has been given above (Fig. 2), but unfortunately it is the only town in the region for which this can be attempted at present (see below).

At a more general level, sites can be compared over a broader temporal grouping, and Figure 4 illustrates this for the medieval period (plus sixteenth century) for five towns. Perhaps significantly, the cattle/sheep/pig template given for Figure 3a does not work for all sites in Figure 4. To some extent this undermines the validity of the above results, but although the template was chosen somewhat arbitrarily, it does at least serve to identify variations.

Table 3 summarises the groupings for the sites in Figure 4, with a fourth "intermediate" group indicating those sites which fall between these groups. These results confirm the impression given above that cattle are the most important group, and all the towns except Bristol have at least one site in the >50% cattle group. They also indicate quite a lot of variation as no one town has all sites in one group, though Exeter shows the most uniformity. These results also underline the problems of summarising data, particularly in the case of Ilchester. In order to include the other sites, where there were

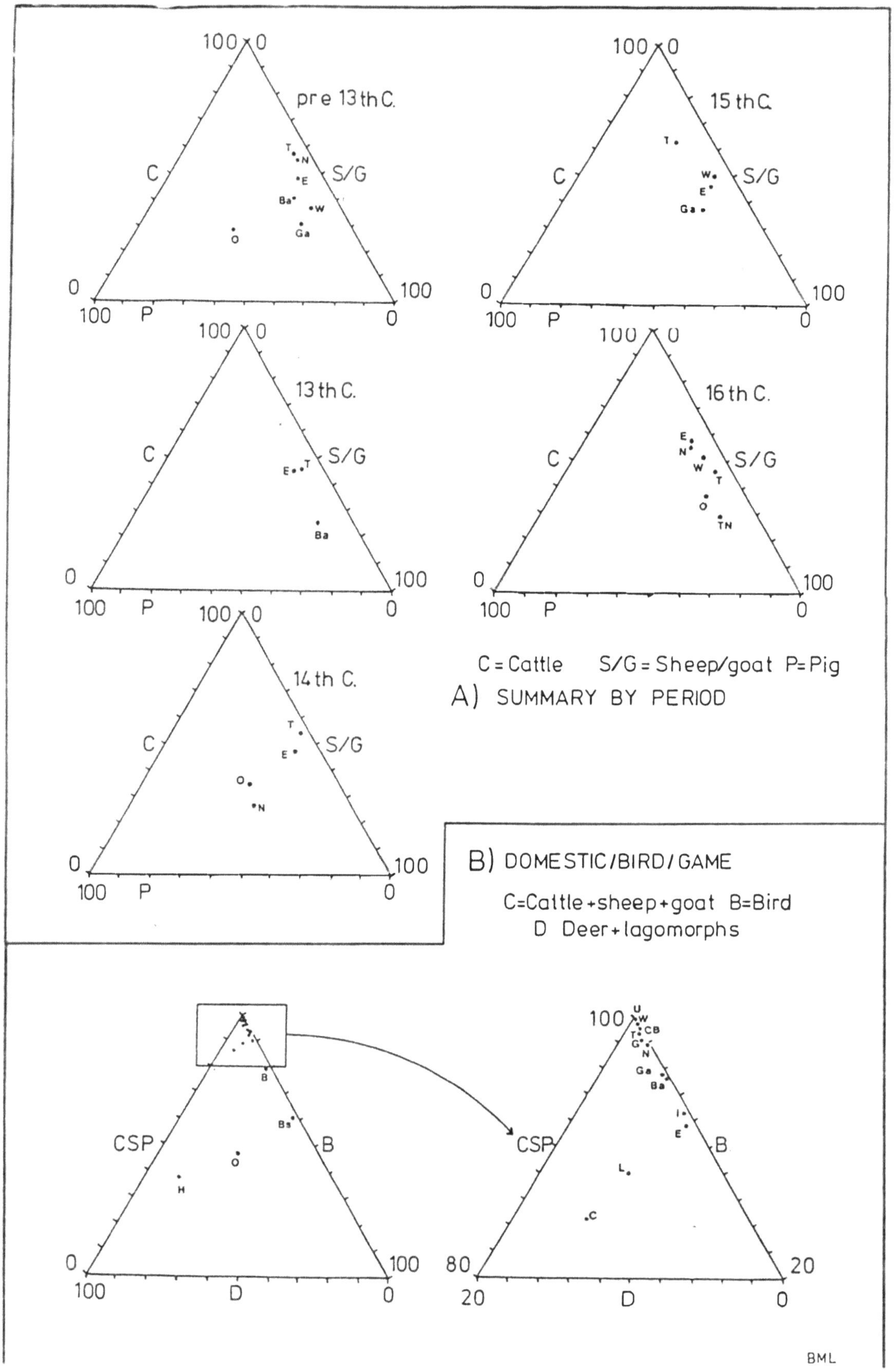

Figure 3. (a) Proportions of cattle, sheep and pig: summary by period. (b) Proportions of domestic, game and bird meat sources.

B: Bristol, Ba: Bath, Bs: Spicer's Almshouse, Bristol, C: Cheddar, CB: Crandon Bridge, E: Exeter, G: Gloucester, Ga: Glastonbury, H: Steep Holm, I: Ilchester, L: Launceston, N: North Petherton, O: Okehampton, T: Taunton, U: Upton, W: Winchcombe.

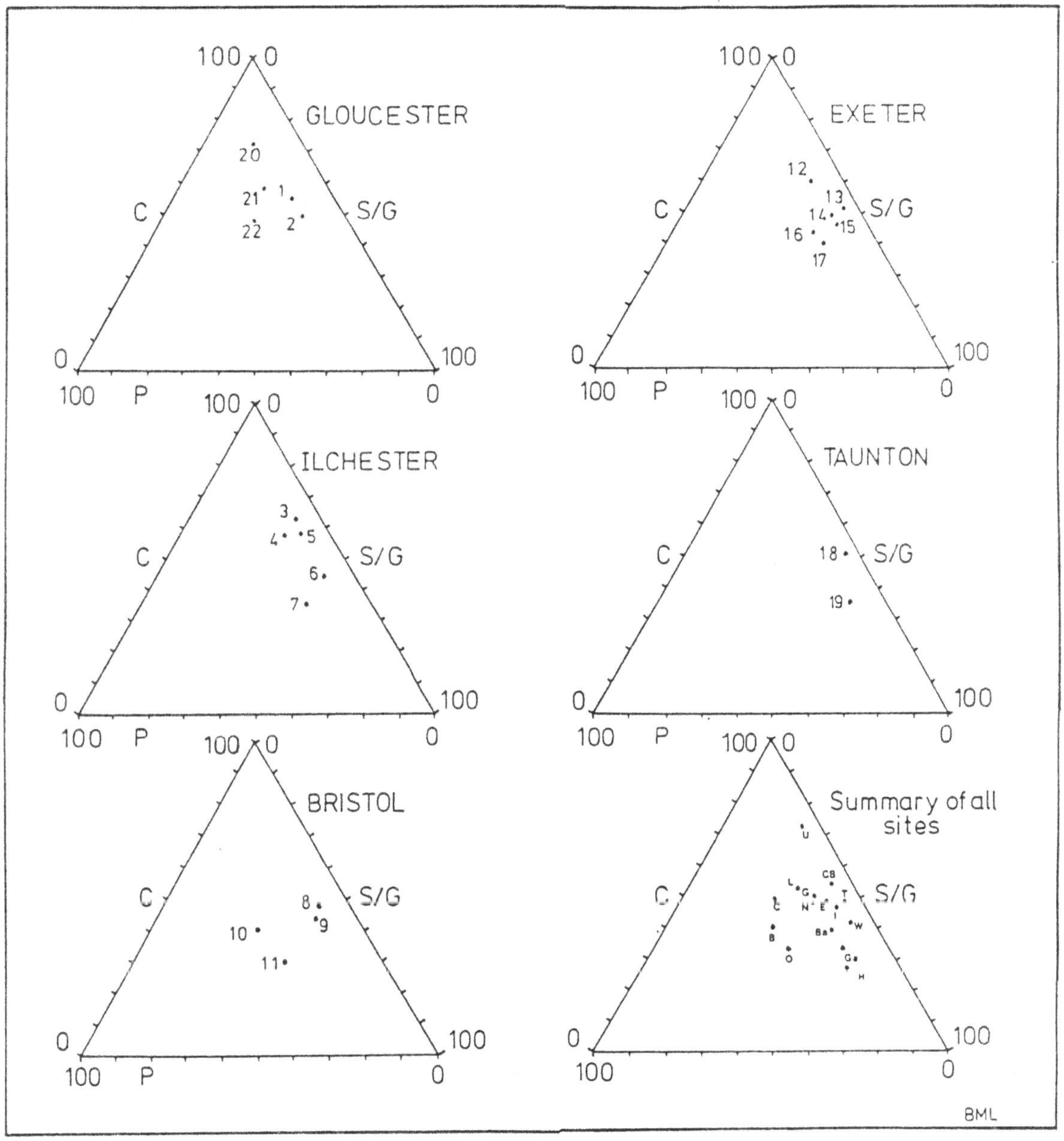

Figure 4. Proportions of cattle, sheep and pig for the medieval and early post medieval period.

1:North Gate, 2: East Gate, 3: Manor House A, 4: Manor House B, 5: Limington Road South, 6: Kingshams, 7: Church Street, 8: St Mary-le-Port, 9: Greyfriars, 10: Bristol Castle, 11: Spicer's Almshouse, 12: St Katherine's Priory, 13: Exe Bridge, 14: Trickhay Street, 15: Goldsmith Street I-II, 16: High Street, 17: Golsmith Street III, 18: Benham's Garage, 19: Priory Barn, 20: West Gate, 21: Bon Marchè, 22: Southgate
Key to letter codes: see Figure 3

Table 2. Summary of sites by period

Period	>20% pig group	>50% cattle group	>45% sheep group
pre 13th century	Okehampton	Bristol North Petherton Taunton Exeter Lydford	Winchcombe Bath Glastonbury
13th century		Exmouth	Bath Taunton Exeter
14th century	North Petherton Okehampton	Taunton Exeter	
15th century		Winchcombe Taunton Exeter	Glastonbury
16th century		Winchcombe North Petherton Exeter	Taunton Okehampton Totnes

only single excavations, the "All sites" group in Table 3 is given. Here the sites in the first part of the table are combined for each town concerned to give a town "average". The effect of mis-representation in summarised data is well demonstrated in the case of Ilchester. Combining the five sites, all of which fall into either the "sheep" or "cattle" groups, gives an overall "intermediate" determination.

OTHER SOURCES OF MEAT

A second important source of meat in medieval economy is from "minor" sources such as birds (domestic and wild) and game mammals. Coy describes and discusses the evidence from wild animals in detail in her review paper (this volume), so the present discussion is more superficial and is meant to complement her paper. The contribution of fish is not discussed here as this is also dealt with by Coy. The major problem with summarising bird and fish bone data is the paucity of the evidence. This can be attributed mainly to lack of sieving - an aspect which still receives too little attention, a point which also, aptly, is made by Coy (this volume).

Figure 5 illustrates the relative contributions (by bone counts) of lagomorpha (rabbits and hares) and deer (red deer, roe deer, and some fallow deer). This is not meant to show the relative importance of the two groups since deer provide a great deal more meat *per capita* than the lagomorpha. It does illustrate and reiterate the variation that occurs, ranging from deer only (though to what extent this is a result of sampling bias resulting from poor recovery of the smaller lagomorph bones is unknown), to mainly rabbit and hare. It should also be noted that lagomorph and bird bones will occur in

Table 3. Summary of intra-site variation

Town	>20% pig	>50% cattle	>45% sheep	overlap group
Gloucester		North Gate		East Gate
Ilchester		Manor Hse A Manor Hse B Limington Rd	Kingshams Church St	
Bristol	Bristol Castle Spicers Almshse			Mary-le-Port Greyfriars
Exeter		St Kath Priory Exe Bridge Trickhay St		Goldsmith I-II Goldsmith III High St
Taunton		Benhams Gge	Priory Barn	
All sites	Bristol Okehampton Cheddar	Crandon Bri Gloucester Upton Lydford Exmouth	Taunton Bath Winchcombe Glastonbury Totnes	North Petherton Exeter Ilchester

different types of deposit to slaughteryard waste which would include deer: the former are much more likely to appear in "kitchen" waste (O'Connor, pers comm) - thus some of the differences can be accounted for by the types of deposit and not by patterns of exploitation.

The sites fall into three groups: mostly or only deer (Taunton (Benham's Garage), Crandon Bridge, North Petherton, Cheddar and Okehampton); deer in the majority, but lagomorpha a large minority (Ilchester, Gloucester and Winchcombe); and rabbit/hare in the majority (Glastonbury, Bristol, Exeter and Upton). A similar comparison of birds has not been attempted because they are more difficult to group and sampling biases are likely to have been even greater.

Figure 3b illustrates the comparison of bird/game/cattle+sheep+pig proportions. Game (deer+lagomorpha) are present in very small proportions, generally less than 2.5% at most of the sites, and birds are mostly less than 10%. Exceptions are Okehampton, Bristol and Cheddar.

Okehampton stands out with large proportions of game (principally deer), an aspect which is discussed in detail in the site report by Maltby, but which may be summarised here as being related to the castle economy with game exploitation and sport important pursuits (Maltby, 1982).

The Cheddar Palace evidence - with game at around 15% - is somewhat similar, and there are, of course, the obvious parallels in type of site. In contrast are two other sites of similar type but much lower percentages of game, Bristol Castle and Taunton Benham's Garage (which lies just outside the castle walls). Taunton Castle was, in fact, one of the few important castles in Somerset, and related to holdings of the Bishop of Winchester (Aston, 1982: 125). Both Taunton and Bristol castles are urban sites, and the hunting/sporting aspects may have been less important, though it seems equally likely that game remains were deposited elsewhere (see, for example, Levitan, 1984b).

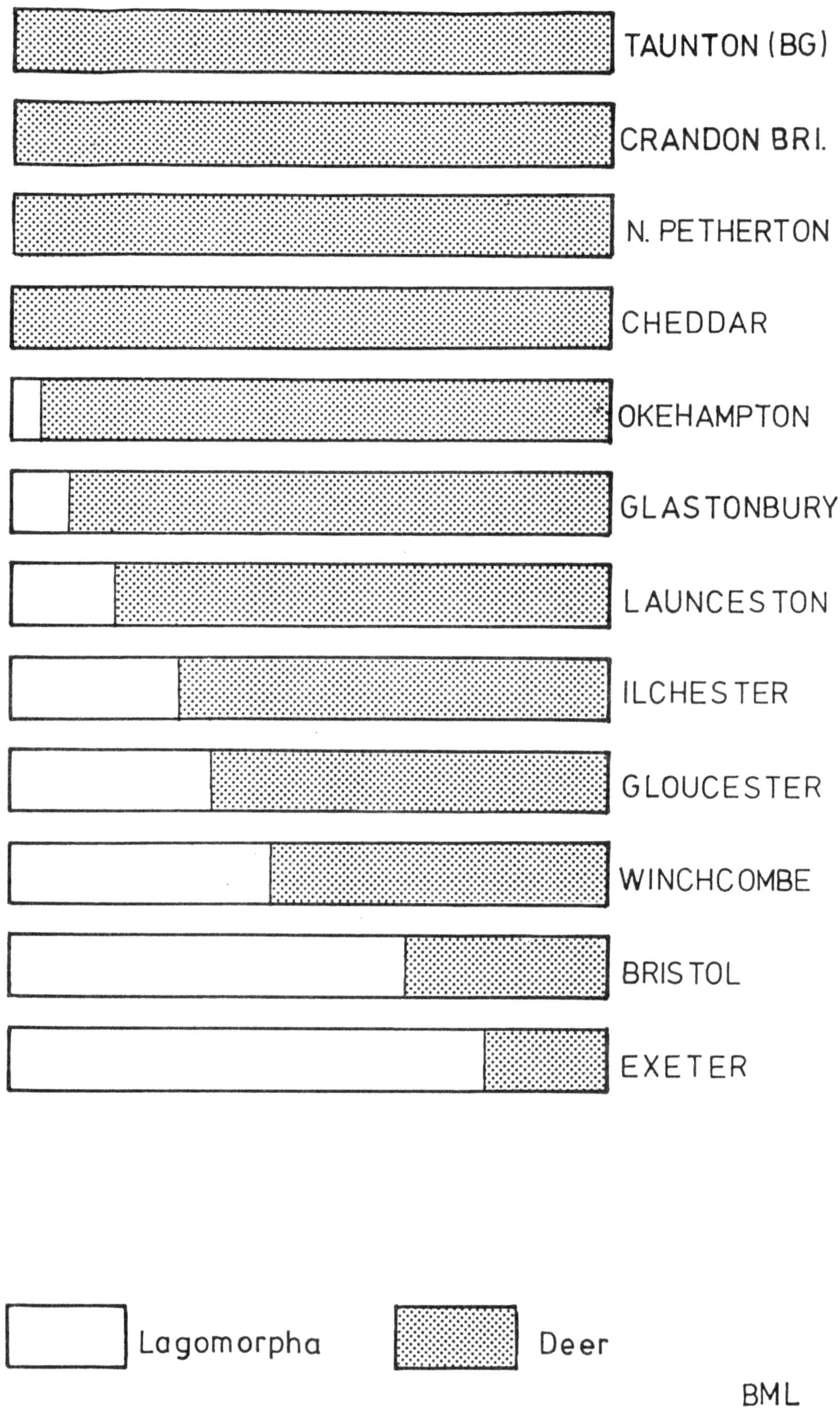

Figure 5. Proportions of lagomorpha and deer.

Spicer's Almshouse, Bristol is given separately from the rest of the Bristol sites because it is so different. Large and notable components being domestic fowl and pig, and it is argued that homegrown food supplies were important as the relatively poor nature of the site precluded much buying in of meat (Levitan, forthcoming a). The other sites have less than 2.5% game but birds range from less than 1% to about 10%, with four sites, Glastonbury, Bath, Ilchester and Exeter having over 5% birds.

This rather simple approach to inter- and intra-site comparison clearly shows that the variation present within towns is great enough to throw suspicion onto attempts to make generalised comments about the economy and husbandry without first making a series of detailed comparisons at least on the scale of Maltby's study for Exeter (Maltby, 1979a). It does, however, provide a useful exercise in highlighting points of similarity and contrast between the sites/towns which may then bear further investigation. Careful examination of the results given above indicates some evidence of patterning either within one town or between towns. In general, however, the interpretations are hampered by a lack of enough sites. Maltby has shown that detailed examinations of several sites within a city provides the basis for an assessment of specialised activities ranging from city-rural interactions through marketing and processing to domestic activities and rubbish.

TWO EXETER SITES

The final part of this paper provides a selective detailed comparison between two further sites from Exeter (as yet unpublished - Levitan, forthcoming b). These have been chosen because they illustrate the above point concerning selectivity and comparison for sites from a town, and they complement Maltby's interpretations. They specifically illustrate the kind of results which can be achieved and underline the point that too little such information is available to form an adequate record for a successful regional survey.

The two sites to be considered are St. Katherine's Priory, directed by John Allan; and Exe Bridge, directed by Stuart Brown. The excavation and analysis of both sites was funded by the HBMC, and they were excavated by the Exeter Archaeological Field Unit. They stand outside the medieval city walls and span the thirteenth to the seventeenth centuries. They make good sites for comparison due to their similarity as extra-mural sites and because they are otherwise dissimilar, Exe Bridge being a site where riverside reclamation was succeeded by tenement buildings (in much the same way as at Dundas Wharf, Bristol - see Jones and Watson, this volume), and St. Katherine's Priory being a Benedictine nunnery.

Three aspects of the assemblages have been chosen for comparison here: i) proportions of cattle, sheep and pig, ii) body part representation for these species, iii) butchery evidence.

Relative importance of cattle, sheep and pig

Calculating species representation

The method of calculating species frequencies and body part representation is different to the more usual methods such as MNI, fragment counts, etc. It is based on the weighting proceedure used by O'Connor for York (O'Connor, 1984), but also takes into account fragmentation of the bones (Levitan, forthcoming d). Essentially, an index is calculated which converts all bone elements into a value which compensates for the degree of fragmentation of the bones. For example, four humeri, one of which is a quarter complete, one half complete, and two complete have an index of 2.75. This value is further standardised for skeletal frequency (eg there are two humeri, but eight first phalanges in a cattle skeleton, so the former indices are standardised by dividing by two and the latter by dividing by eight). The final value in the example, therefore, is 1.38.

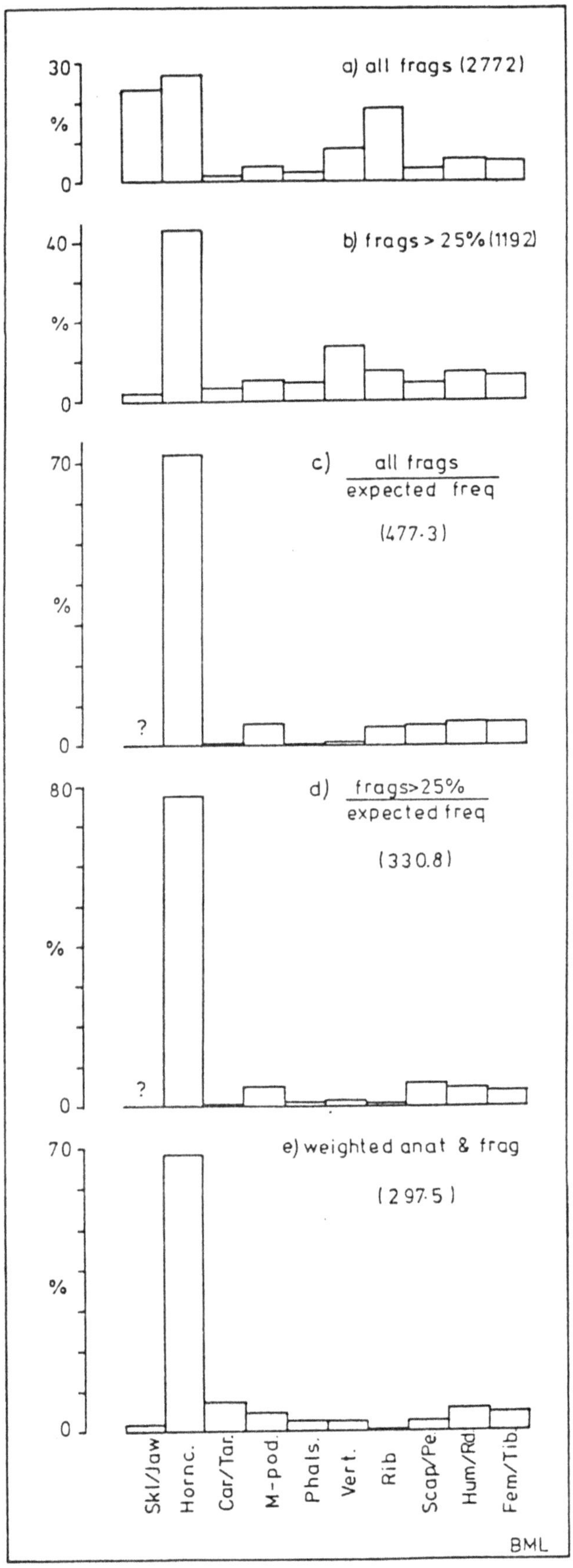

Figure 6. Comparison of anatomical representation methods

How do results calculated in this way compare with fragment count analysis and with O'Connor's method? An example from Exe Bridge (Fig. 6) illustrates the same data (body part representation in cattle) calculated in five ways. The top histogram (Fig. 6a) is a simple count of all 2772 fragments of cattle from thirteenth century contexts. It may be argued that this is swamped by bones such as the skull which break up into many more fragments than more compact bones such as phalanges, and indeed there are apparently higher frequencies of skull (Skl) than phalanges (Phals).

One can compensate by only counting fragments more than 25% complete as in the second diagram (Fig. 6b), and one can see the difference in skull and rib in particular.

Both these methods are sometimes used, and the next two histograms give the results if each count is then standardised for skeletal part along the lines suggested by O'Connor (1984), although the Exeter data used here are fragment counts, not zone counts as employed by O'Connor. Differences from the histograms described above are immediately apparent, these differences being common to both (Fig. 6c and 6d). Horncores appear to be much more dominant than in the two top histograms, and all other elements relatively less frequent.

The last histogram (Fig. 6e) uses the fragmentation formula described above which provides a very conservative estimate, but takes O'Connor's method one step further (ie it standardises the data for fragmentation as well as for skeletal element). It should be noted that where zone counts are used (Watson, 1979), fragmentation standardisation is not required. Zone counts, however, may disregard part of the sample, whereas the fragmentation count method attempts to use all the sample. It is felt that the last method removes many of the biases inherent particularly in urban material where fragmentation can be very variable and might be the result of a number of different causes (eg butchery, post-butchery uses such as glue making and bone working, scavenging, trampling, etc) and is recommended for large samples. It is not necessarily appropriate in all cases, especially those where the causes of fragmentation are known or where the fragmentation patterns appear to be similar throughout the sample. In the present case it does appear to be appropriate, and the following analysis is based on the use of this method. The relative abundance of the species may also be calculated using this method, the individual anatomical indices simply being summed.

Relative importance of the major species

The relative abundance of cattle, sheep and pig is shown in Figure 7. (Note that the proportions of the three species in the preceeding part of this review were calculated using fragment counts). The results may be presented as percentages based on independent totals as is the normal case for bone reports (Fig. 7i). Here the percentages of the species are calculated separately for each period, and the total percentage for each period is thus 100% The use of the fragmentation index method means that species counts are directly comparable between phases, so counts based on an aggregate total may be used. In this case the species counts are based on the total for all phases, and shows the relative abundance of the species from phase to phase as well as their relative importance between the phases. Thus, Figure 7ii not only shows the relationship between the species, but the temporal changes in the sample. At Exe Bridge, for example, there is an overall decrease in sample size indicated, and at St. Katherine's Priory the reverse trend is apparent. Such results cannot be shown by the percentages based on independent totals.

The apparent differences in the two diagrams (Fig. 7) arise from the use of percentages and not from any real differences. The aggregate method is useful because it provides a direct view of how the species inter-relate within and across periods. Thus, Figure 7i shows that sheep percentages from St. Katherine's Priory decrease after the fifteenth century, and this is set against an increase in cattle. Figure 7ii, however, shows an increase of sheep, because in fact the sample size in the sixteenth century is

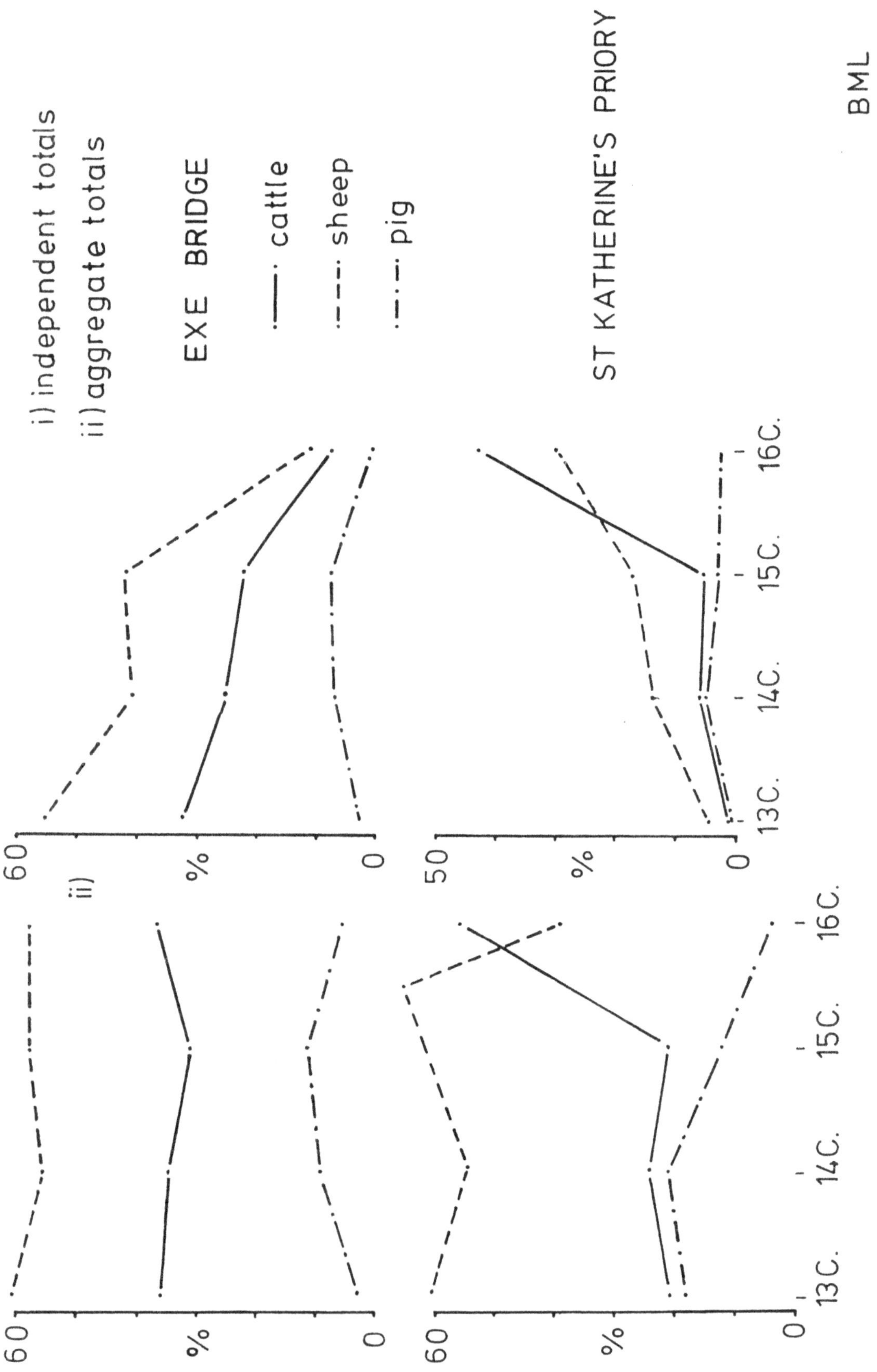

Figure 7. Proportions of cattle, sheep and pig at Exe Bridge and St Katherine's Priory, Exeter.

considerably larger than in the fifteenth century.

Cattle increase in importance relative to sheep at St. Katherine's Priory, eventually becoming the more numerous in the sixteenth century; at Exe Bridge cattle are never more numerous than sheep, though perhaps they gain a little in importance relative to sheep in the sixteenth century. Pigs at first rise in importance, and then decrease, but in any case do not appear to react to cattle and/or sheep changes, so possibly market forces acted on pigs in a different way to market forces on cattle and sheep. The pattern whereby pig proportions are apparently unrelated to cattle and sheep has been found at other sites, and O'Connor (1982: 11) observed this for Flaxengate, Lincoln and concluded that pigs may have been marketed by specialist butchers, or simply reared at the household level rather than farmed on a larger scale. This would certainly fit some of the documentary evidence, Bristol for example (see Jones and Watson, this volume).

The interesting result from comparing proportions using aggregate totals is that cattle and sheep rise in importance at St. Katherine's Priory relative to Exe Bridge. It would be facile to suggest that there is some direct inter-relationship here, but it makes the point that several such comparisons across the city would be worthwhile, building up a kind of species contour map of the city. In the present case, the results may be related to the activities at the sites, with the major bone deposit events at Exe Bridge being early, and the major deposits at St. Katherine's Priory later.

This exercise also underlines the point that single sites cannot be taken to represent towns or cities, and here neither pattern should be taken to represent medieval Exeter. Despite this truism, the idea that one or just a few sites from a town can be taken as a basis for general comments about the town's economy persists in some quarters. The underlying point is that far from concentrating too much on urban sites, not enough has been done yet. This is not to suggest that ever more urban sites should be dug, but that key towns and cities should be identified, and resources concentrated there.

Regional centres are of obvious importance, and Maltby found at Exeter, for example, that size of cattle was smaller than contemporary material from elsewhere in the country (Maltby, 1979a: 36-38). Furthermore, each new site from the city provides new information: the point of replication has yet to be reached even though thirteen sites have been analysed so far. At present we have no more than a number of keyhole pictures of the city. Each site can make intrinsic sense - such as the horncore dumps at Exe Bridge, or the enclosed nunnery deposits at St. Katherine's Priory - but we are far from a complete understanding of the city economy and its variation. In such cases it would not be necessary to analyse all the data from particular sites, but to concentrate on specific topics which add to our knowlegde of the town; the data which merely replicate information already available need not be considered. A good example of this approach is the recent report on two sites from York by O'Connor (1984). This topic is explored further by the author elsewhere (Levitan, forthcoming e).

Anatomical representation

Anatomical representation has been calculated and summarised as in Figures 8-10. The summary is on the basis of related groups of bones in terms of animal processing, and is similar to the groups employed by O'Connor (O'Connor, 1984: 14). For cattle there is a major difference between the two sites which remains more or less consistent throughout the four centuries. It is characterised by extremely high proportions of horncores at Exe Bridge, particularly in the thirteenth century, and high proportions of limb extremities, whereas St. Katherine's Priory has a virtual absence of horncores, and high proportions of girdles and upper limb bones (Fig. 8). The evidence for sheep displays the same pattern, but even more sharply defined (Fig. 9). However, in the case of pigs, the pattern is different (Fig. 10), and the two sites can be said to be essentially similar, with upper limb bones predominant.

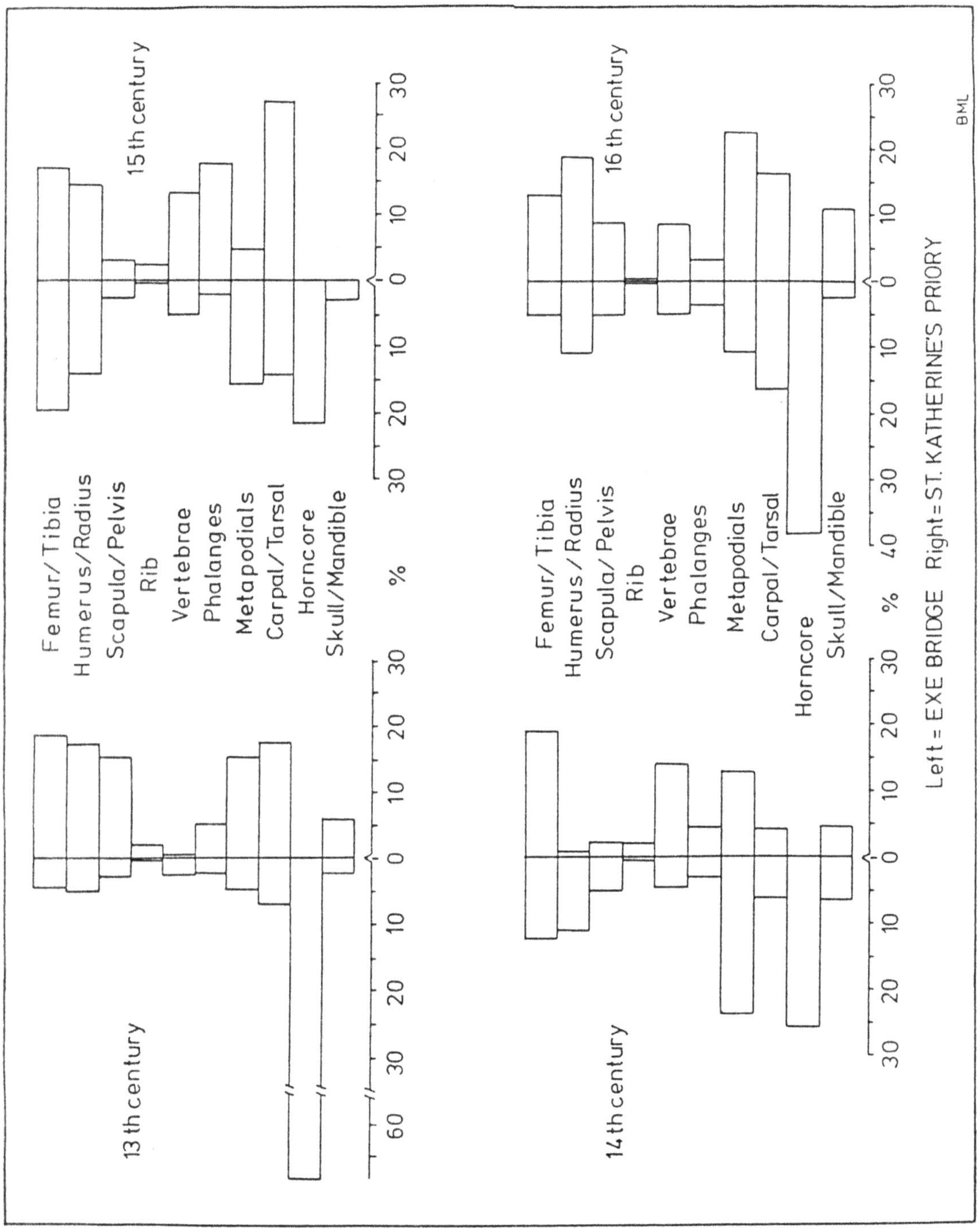

Figure 8. ***Anatomical representation for cattle at Exe Bridge and St Katherine's Priory, Exeter.***

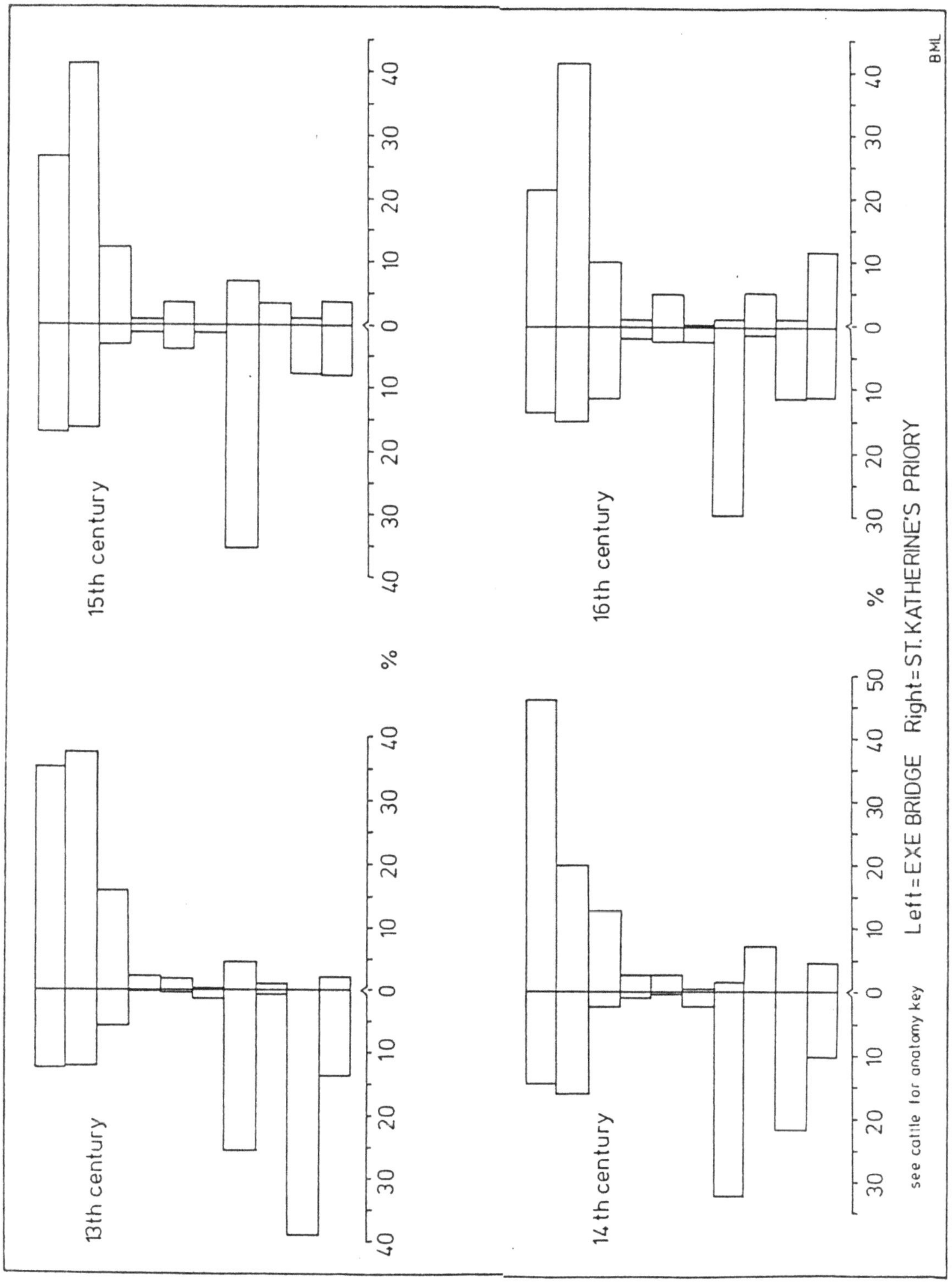

Figure 9. Anatomical representation for sheep at Exe Bridge and St Katherine's Priory, Exeter.

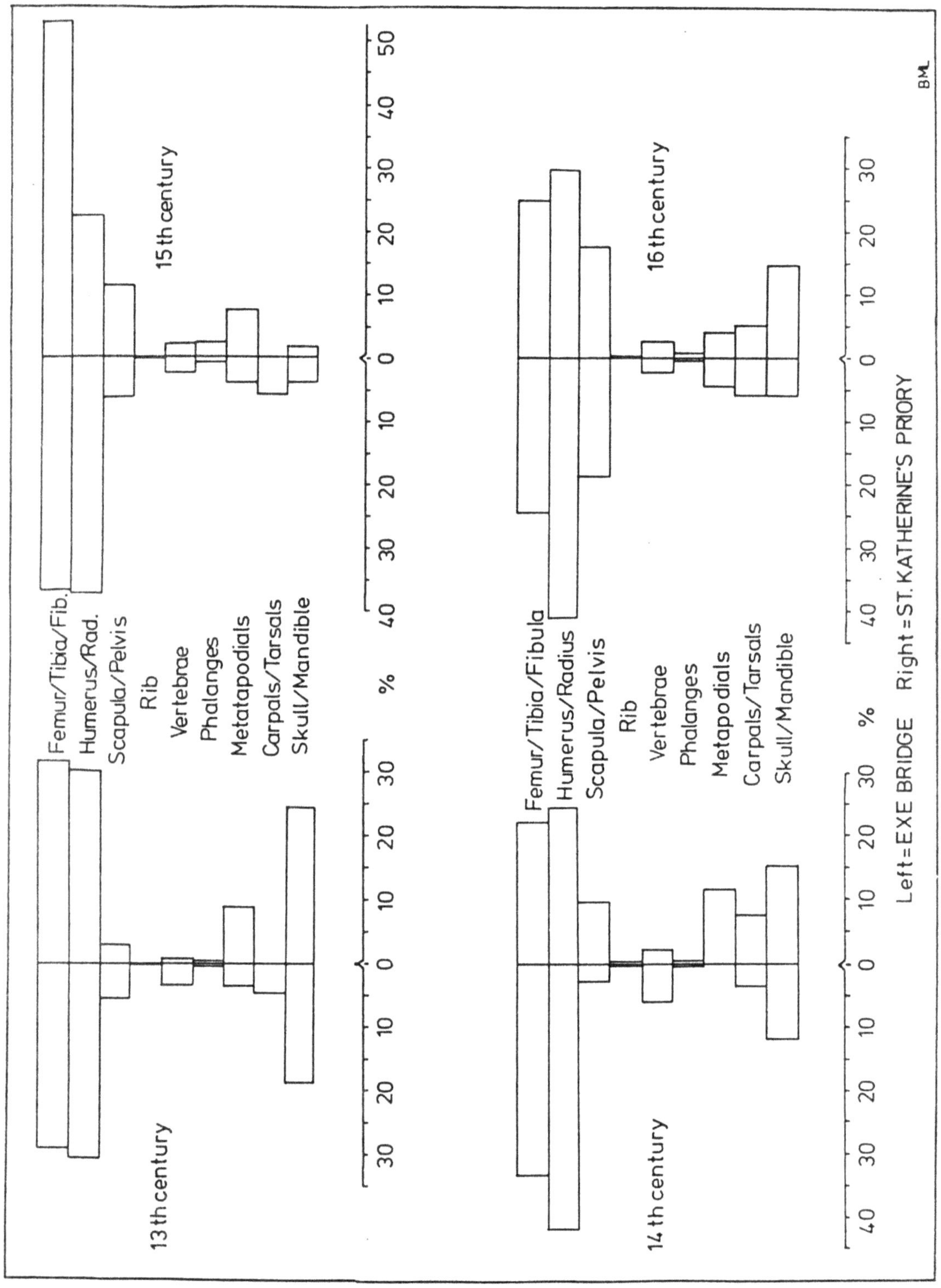

Figure 10. Anatomical representation for pig at Exe Bridge and St Katherine's Priory, Exeter.

This can be interpreted as representing, at Exe Bridge, a series of deposits which are dominated by horncores, probably being the waste from tanning or horn working. Superimposed on this pattern are deposits which are more typical of food processing, and, in the case of pigs, probably representing a mixture of primary and secondary butchery, with only a small domestic waste constituent. The cattle remains echo this with higher proportions of limb bones and skull+mandibles than of ribs and vertebrae, the latter being more typical of domestic waste, while skull and mandibles would be removed at the primary butchery stage, and long bones and girdles at the secondary butchery stage (ie jointing and boning out in preparation for commercial retail).

The sheep have the same pattern of anatomical representation as cattle, but here long bone joints may have been left bone in, so a higher proportion of domestic waste is, perhaps, indicated.

St. Katherine's Priory can be seen as entirely the result of food processing since the main anatomical groups are typical of butchery for meat. Generally the deposits seem to represent secondary and domestic waste rather than primary butchery - but there are puzzling exceptions where high proportions of metapodials are present, for example fifteenth and sixteenth century cattle, so some primary butchery waste is present.

Perhaps, in the case of pigs, - particularly in the thirteenth to fourteenth centuries - some were actually slaughtered at the site, and we cannot rule out their being reared there as well, but the sheep all appear to have been bought in as prepared joints. The cattle also were probably bought as prepared joints, but the presence of metapodials might bear witness to bone working as well.

A second major point is that the main deposits at Exe Bridge date from the thirteenth century, with deposit size decreasing considerably in the fourteenth century, and thereafter remaining fairly constant. At St. Katherine's Priory the reverse is the case, with the largest group of bones concentrated in the sixteenth century. This pattern is shown up well by the aggregate counts plot in Figure 7. The case of Exe Bridge can easily be related to the site history. The major bone deposits in fact represent dumping on the river bank related to reclamation of the bank which eventually became part of a tenement with a yard and buildings on it, and the dumping of waste was reduced.

At St. Katherine's Priory, the picture is more complicated and too detailed to explore here (see Levitan, forthcoming b). In essence the deposits are related to their location within the Priory. The concentration of bones in the sixteenth century owes its existence to a deposit outside the kitchens which represents the infilling of a drain and need not be seen as representing greater activity than in the previous periods.

Thus, the concentrations of bones can mean different things, and should be interpreted with caution: in neither case do the deposits here necessarily indicate greater acivity overall, but represent specific concentrated dumping for different reasons. This refers back to the earlier point concerning different species and anatomies occurring in different types of deposit; it may be that primary slaughter waste was present elsewhere at St. Katherine's Priory.

Butchery evidence

Evidence for butchery here is confined to cut and chop marks on the bones, although in practice ancient fragmentation patterns are also important (Levitan, 1984a: 183-188). The butchery can be conveniently summarised diagrammatically (Fig. 11). These illustrations show that butchery processes were similar for both sites, though the removal of horncores is a particular feature of the Exe Bridge remains. The much heavier butchery on cattle is an indication that a lot more preparation occurred, with bones often being filleted out. Many bones were also smashed open, possibly for extraction of marrow. Butchery of sheep (Fig. 12) and pigs, by contrast, is mainly at the joints, and indicates division into basic units which are left bone in. Pigs are not illustrated, because the samples are small and the butchery evidence limited.

Interpretation

In summary it can be seen that these two sites are characterised by points of similarity which indicate an aspect of uniformity, for example in the preference for cattle and the essentially similar manner of butchery - further analysis shows that age structures are also similar (Levitan, forthcoming b). Perhaps it may be concluded that the sources of meat supply were similar, and that the Exeter butchers followed essentially similar patterns of joint preparation (compare these with the butchery from Dundas Wharf (Jones and Watson, this volume) where some points of dissimilarity may be seen).

There are also a number of important differences between the sites, particularly in anatomical representation. We have seen that there are mixtures of food and industrial remains at Exe Bridge, and of mainly food remains at St. Katherine's Priory. It has been suggested that the horncore concentrations at Exe Bridge represent waste from industrial processes (presumably hornworking rather than tanning as there are relatively few limb extremities which might be expected if the latter obtained). These deposits can be related to the site development and can be interpreted as representing dumping on the banks of the River Exe leading to reclamation and construction of tenement buildings. The "industrial" nature of the deposits suggests a hornworking industry nearby, and the fact that horncores remain predominant in later periods suggests both the continuation of the industry and possibly its pursuance at this site.

In the case of St. Katherine's Priory it is possible to assume that the bones relate to actual activities at the Priory since it is unlikely that waste from outside would be disposed there. We can see that the bones are mainly remains of food, and the presence of secondary butchery waste indicates that perhaps the meat was bought wholesale after primary butchery, so that the secondary butchery occurred at the Priory itself, which thus functioned as its own butchers - and perhaps some pigs were reared and butchered at the site itself. The large sixteenth century deposits might be seen in the context of a large rubbish accumulation, perhaps relating directly to the neighbouring kitchen ranges (Levitan, forthcoming b).

At Exe Bridge the chronological sequence indicates little change in the *status quo*, though cattle perhaps become a little more important in the sixteenth century. In the case of St. Katherine's Priory the nature and change in the deposits is rather different: the deposits are mostly food remains and cattle were the most important source of meat. In contrast to Exe Bridge, cattle increase in importance in the sixteenth century to become the most numerous of the three species, so their contribution to the diet must have been enormous.

CONCLUSIONS

The two sites from Exeter emphasise the problems of dealing with complex sites such as found in urban contexts. We cannot take single sites to represent urban activities in general, and studies such as the one by Mark Maltby (1979a) on Exeter are much nearer our ideal than single site reports. Furthermore, there is the need, as Maltby himself pointed out (1979a: 94) to relate urban economies to their hinterlands.

My intention initially was to attempt a detailed regional summary and interpretation. The survey that has been presented, however, has merely emphasised the lack of evidence and the shortfall in the quality of the record. The lesson is valuable: we must concentrate our resources into particular lines of enquiry if we are to find out anything worthwhile about such economies. I have worked in this region for nine years and have analysed some twenty sites during that time - yet the combined data from these sites and from other workers such as Barabara Noddle and Mark Maltby is still insufficient - we

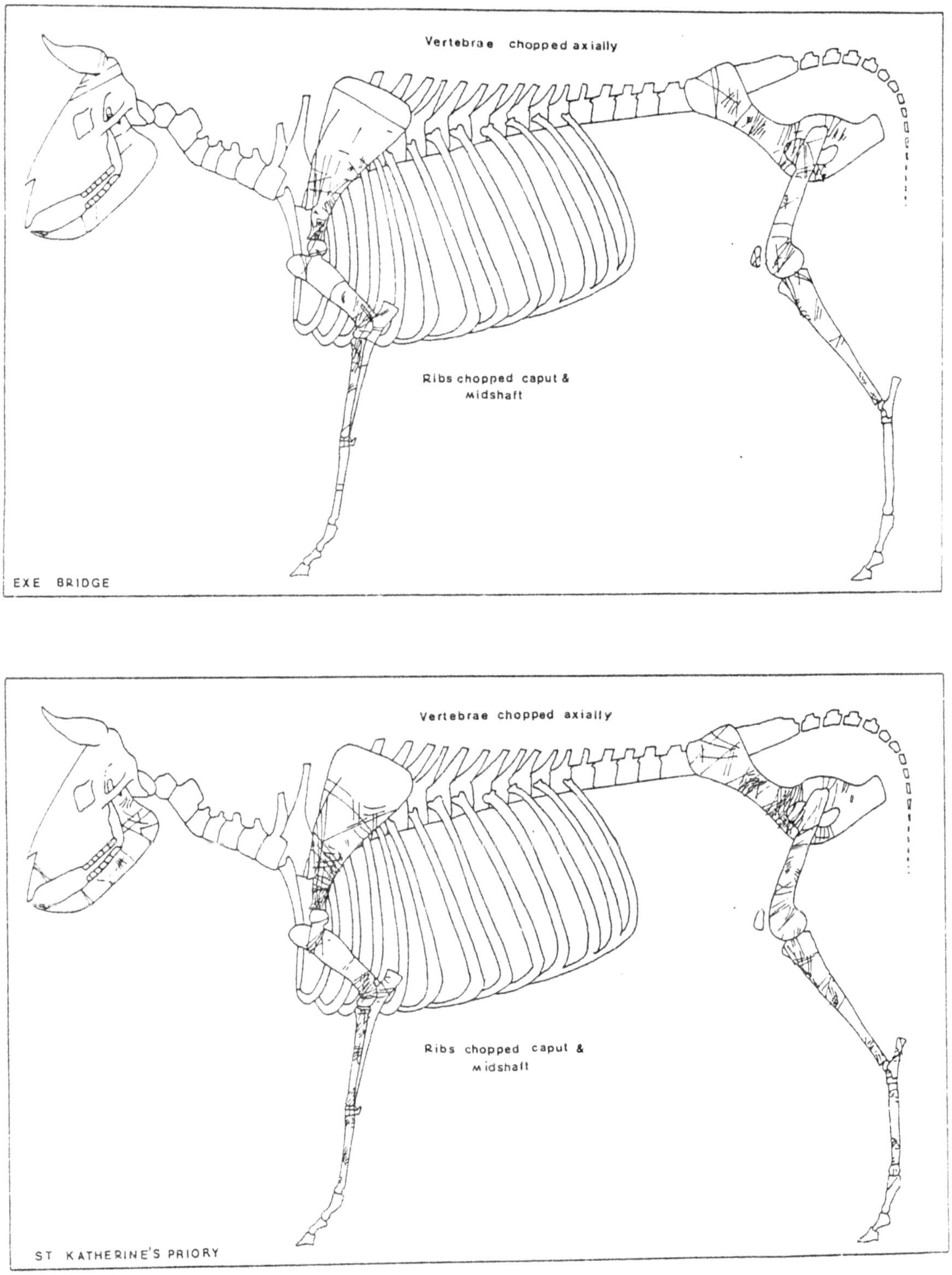

Figure 11. Summary of butchery on cattle bones from Exe Bridge and St Katherine's Priory, Exeter

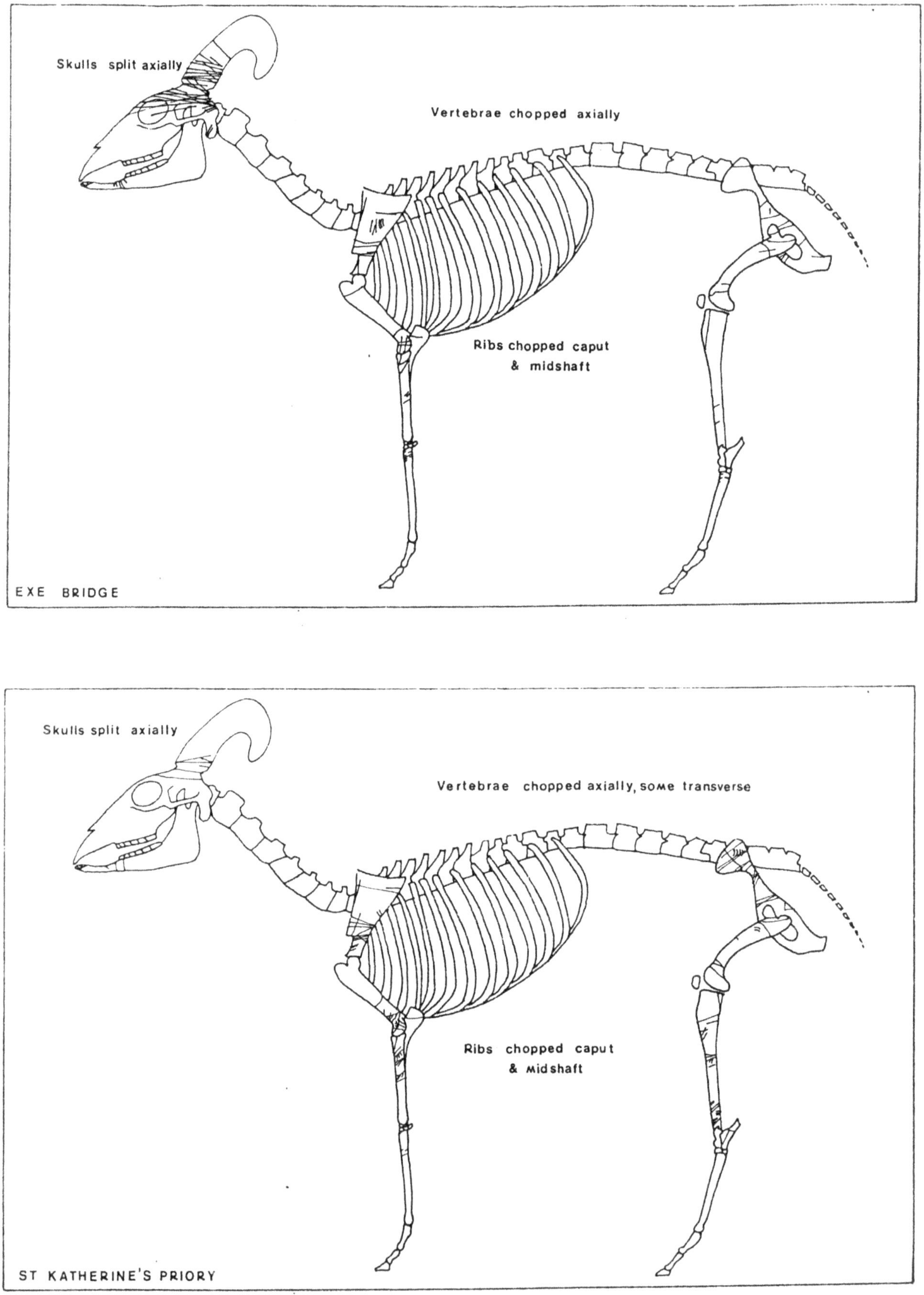

Figure 12. Summary of butchery on sheep bones from Exe Bridge and St Katherine's Priory, Exeter.

have been spreading the work too thin. Instead, the aim should be to concentrate resources into a project which would take in a large number of sites within a city, and have related work in the hinterland. The South West region is varied and is an ideal area for a study which could concentrate on selected regional centres (Gloucester, Bristol, Exeter) and on a number of key townships and rural sites in their hinterlands.

By so defining and limiting our aims we might have a chance to obtain some useful and worthwhile information. The willy-nilly approach solves nothing and clearly is a waste of resources - bones from a site should not be analysed simply because they exist, but should be considered only when the excavation strategy has clearly defined aims to which a bone analysis could contribute. Furthermore, the problems aired here are important enough to warrant, in some cases, the faunal element being the primary reason for excavation.

Taking Gloucester and its county as an example, there are indications that the medieval development of Gloucester was slower and more conservative than Southampton. Both sites had similar origins and functions, yet at Gloucester there appears to have been little long distance trade through its port and there is the continuation of spelt wheat agriculture into the late Saxon period in contrast to Hamwic (Southampton) where free threshing wheats had been introduced (Leech, 1984: 295-6). Clearly such issues raise more questions, some of which relate directly to faunal evidence. The twelfth and thirteenth centuries witnessed an increase in urbanisation in Gloucestershire with planned medieval towns such as Cheltenham being created (Leech, 1984: 299). The reasons for this and the impact on other towns and on rural economies should be considered. There is also the importance of the woollen industry and trade, which influenced so much settlement in Gloucestershire, but towards which little faunal analysis has been directed.

Research should also be directed towards rural sites, in particular to consider the relationships between farms, hamlets, villages, townships and regional centres. Desertion of hamlets and villages should be more carefully considered. Not all were due to plague, and economic changes such as the change from arable to pasture for sheep and cattle may have been responsible for many cases (Aston and Viner, 1984: 287), whilst the attraction of urban centres with prospects of higher paid work may have been an incentive to leave the countryside then, just as now. The change from an earlier pattern of dispersed hamlets to one of fewer but larger nucleated villages, as at Frocester for example (Aston and Viner, 1984: 289) also should be considered.

Such studies would call for excavation planning on a regional basis with individual sites being keyed into an overall research design, one in which the study of faunal remains and other forms of economic and environmental evidence are central aspects.

Finally, a plea must also be made for some form of standardisation since the ability to carry out local and regional reviews must rest on comparability of data. The present review has been necessarily superficial because the bone reports are not directly comparable - even reports by the same author differ from site to site. To some extent this is desirable, and the present author would not wish to be confined to a single inflexible system of analysis. However, there must be some common ground, and this need not preclude the addition of new and different methods as the individual site dictates. The standardisation should be at a high enough level to allow flexibility of approach in recording and analysing the sample, and so that the recording process itself does not allow the data to become fossilised. So long as the basis for the analysis is clearly defined, it matters little whether the bones were recorded using computer codes, hand written record sheets, or notched wood. It is the standardisation in terms of information exchange that is required, not of the raw data itself.

ACKNOWLEDGEMENTS

I am grateful to Barbara Noddle for allowing me to use information from unpublished reports. Thanks also are due to Terry O'Connor and Elspeth Levitan for reading and commenting upon an earlier version of this paper.

REFERENCES

Adcock, A. 1977. The animal bones *in* P. Leach, Excavations at North Petherton, Somerset, 1975 *Somerset Archaeol. Natur. Hist.* **121**, 38-39.

Aston, M. 1982. The medieval pattern 1000-1500 AD *in* M. Aston and I. Burrow (eds) *op cit*, 122-133.

Aston, M. and Burrow, I. 1982. The early Christian centres 600-1000 AD *in* M. Aston and I. Burrow (eds) *op cit*, 118-121.

Aston, M. and Burrow, I. (eds) 1982. *The Archaeology of Somerset* Bridgewater: Somerset County Council.

Aston, M. and Viner, L. 1984. The study of deserted villages in Gloucestershire *in* A. Saville (ed) *op cit*, 270-293.

Bell, M. 1984. Environmental archaeology in south west England *in* H. Keeley (ed) *Environmental archaeology: a regional review* London: H.M.S.O. (Directorate of Ancient Monuments and Historic Buildings Occasional Paper no. **6**), 43-133.

Bovey, R.M.L. 1984. Mammalian and bird remains *in* D.M. Griffiths and F.M. Griffith, An excavation at 39 Fore St., Totnes *Proc. Devon Archaeol. Soc.* **42**, 95.

Colley, S.M. 1984. Fish remains *in* D.M. Griffiths and F.M. Griffith, An excavation at 39 Fore St., Totnes *Proc. Devon Archaeol. Soc.* **42**, 95-98.

Coy, J. 1977. Bird bones from Wirral Park Farm, (The Mound), Glastonbury: Ancient Monuments Laboratory Report no. 2326.

Darvill, T. and Coy, J. 1977. Report on the faunal remains from Wirral Park Farm, (The Mound), Glastonbury: Ancient Monuments Laboratory Report no. 2431.

Douglass, J. in prep. The vertebrate remains from Dundas Wharf, Bristol.

Ellis, P. 1982. Excavations at Silver Street, Glastonbury, 1978 *Somerset Archaeol. Natur. Hist.* **126**, 17-31.

Grant, A. 1979. The animal bones (Citizen's House) *in* B. Cunliffe (ed) *Excavations in Bath, 1950-1975* Bristol: Committee for Rescue Archaeology in Avon Gloucestershire and Somerset (Excavation Monograph **1**), 60-70.

Harcourt, R. 1970. Animal bones *in* P. Rahtz, Excavations on Glastonbury Tor, Somerset, 1964-6 *Archaeol. J.* **127**, 56-60.

Harcourt, R. 1980. Animal bones *in* A.D. Saunders, Lydford Castle, Devon *Medieval Archaeol.* **24**, 182-184.

Higgs, E., Greenwood, W. and Garrard, A. 1979. Faunal report *in* P. Rahtz *The Saxon and Medieval palaces at Cheddar* Oxford: British Archaeological Reports (British Series **65**), 354-362.

Hunter, A.G. 1963. Excavations at the Bon Marchè site, Gloucester 1958-9 *Trans. Bristol Gloucestershire Archaeol. Soc.* **82**, 25.

Jones, G. 1980. Animal bone *in* P.J. Weddell, Excavations at 3-5 Fore Street, Exmouth *Proc. Devon Archaeol. Soc.* **38**, 112.

Jones, G. 1981. Animal bone *in* P.J. Weddell, Excavations at Southgate Cottages, Lydford *Proc. Devon Archaeol. Soc.* **39**, 138.

Jones, R.T. 1975. Animal bone: Launceston Castle: Ancient Monuments Laboratory Report no. 2967.

Langley, P. 1978. Animal bone report; Taunton Priory, Somerset: Ancient Monuments Laboratory Report no. 2460.

Leach, P. 1982. *Ilchester volume I. Excavations 1974-75* Bristol: Western Archaeological Trust (Excavation Monograph **3**).

Leach, P. 1984. *The archaeology of Taunton* Bristol: Western Archaeological Trust (Excavation Monograph **8**).

Leach, P. in prep. *Ilchester volume 2.*

Leech, R. 1984. Medieval urban archaeology in Gloucestershire *in* A. Saville (ed) *op cit*, 294-303.

Levitan, B. 1982a. Mammal bones and bird bones *in* P. Leach 1982, *op cit*, 69-284.

Levitan, B. 1982b. The animal bones *in* P. Ellis, *op cit*, fiche 42-71.

Levitan, B. 1983. Steep Holm Priory - animal bone interim report *in* S.D. and J.N. Rendell, Archaeological report - 1982 *Steep Holm Island* Weston Super Mare: Kenneth Allsop Trust, 29-36.

Levitan, B. 1984a. Faunal remains from Priory Barn and Benham's Garage *in* P. Leach 1984, *op cit*, 167-193.

Levitan, B. 1984b. The vertebrate remains *in* S. Rahtz and T. Rowley, *op cit*, 108-148.

Levitan, B. 1985. The animal bones *in* A. Saville, Salvage recording of Romano-British, Saxon, Medieval and Post-Medieval remains at North Street, Winchcombe, Gloucestershire *Trans. Bristol Gloucestershire Archaeol Soc.* **103**, 130-5.

Levitan, B. forthcoming a. Spicer's Almshouse animal bone report *in* B. Williams, Excavations at Spicer's Almshouse.

Levitan, B. forthcoming b. The vertebrate remains from Exe Bridge and St. Katherine's Priory, Exeter *Proc. Devon Archaeol. Soc.*.

Levitan, B. forthcoming c. The vertebrate remains from Manor House sites A and B, Limington Road South and Church Street, Ilchester *in* P. Leach in prep., *op cit*

Levitan, B. forthcoming d. A method for calculating body-part representation.

Levitan, B. forthcoming e. Bone analysis and urban economy: examples of selectivity and a case for comparison *in* D. Serjeantson and T. Waldron (eds) *Diet and crafts from towns: the evidence from animal remains* Oxford: British Archaeological Reports.

Maltby, M. 1979a. *Faunal studies on urban sites: the animal bones from Exeter 1971-75* (Exeter Archaeological Report Vol. 2), Sheffield: University Press.

Maltby, M. 1979b. The animal bones *in* C.M. Heighway, A.P. Garrod and A.G. Vince, Excavations at 1 Westgate Street, Gloucester 1975 *Medieval Archaeol.* **23**, 182-185.

Maltby, M. 1982. Animal and bird bones *in* R.A. Higham, J.P. Allan and S.R. Blaylock, Excavations at Okehampton Castle, Devon: Part 2 the Bailey *Proc. Devon Archaeol. Soc.* **40**, 114-135.

Maltby, M. 1983. The animal bones *in* C. Heighway, *The East and North Gates of Gloucester* Bristol: Western Archaeological Trust (Excavation Monograph 4), 228-245.

Noddle, B. 1969. The animal bones *in* P. Rahtz, Upton, Gloucestershire 1964-68 second report, *Trans. Bristol Gloucestershire Archaeol. Soc.* **88**, 124-126.

Noddle, B. 1985. The animal bones *in* L. Watts and P. Rahtz *Mary-le-Port, Bristol. Excavations 1962-3* Bristol: City Museum and Art Gallery (Monograph 7).

Noddle, B. forthcoming a. The animal bones from Bristol Castle *in* M. Ponsford (ed) *Bristol Castle*, Bristol: City Museum and Art Gallery.

Noddle, B. forthcoming b. The animal bones from Greyfriars, Bristol.

Noddle, B. forthcoming c. The animal bones from Bush Marsh Village, Crandon Bridge, Somerset.

O'Connor, T.P. 1982. *Animal bones from Flaxengate, Lincoln c. 870-1500* (The archaeology of Lincoln vol 13-1) London: Council for British Archaeology.

O'Connor, T.P. 1984. *Selected groups of bones from Skeldergate and Walmgate* (The archaeology of York 15/1), London: Council for British Archaeology.

O'Connor, T.P. 1985. On quantifying vertebrates - some sceptical observations *Circaea (Bull. Ass. Envir. Archaeol.)* **3(1)**, 27-30.

Rahtz, P. and Hirst, S. 1974. *Beckery Chapel, Glastonbury, 1967-8* Glastonbury: Antiquarian Society.

Rahtz, S. and Rowley, T. 1984. *Middleton Stoney excavation and survey in a North Oxfordshire parish 1970-1982* Oxford: Oxford University Department for External Studies.

Saville, A. (ed) 1984. *Archaeology in Gloucestershire* Cheltenham: Museum and Art Gallery and Bristol and Gloucestershire Archaeological Society.

Smith, P.D.E., Allan, J.P., Hamlin, A., Orme, B. and Wooton, R. 1983. The investigation of a medieval shell midden in Braunton Burrows *Proc. Devon Archaeol. Soc.* **41**, 75-80.

Turk, A. 1965. Notes on Cornish mammals an annual report - no 7 *Roy. Cornwall Polytech. Soc.* (Annual Report 132), 49-55.

Turk, A. 1969. Notes on Cornish mammals in prehistoric and historic times: 2 *Cornish Archaeol.* **8**, 100-104.

Turk, A. 1970. Notes on Cornish mammals in prehistoric and historic times: 3 *Cornish Archaeol.* **9**, 121-125.

Watson, J.P.N. 1979. The estimation of the relative frequencies of mammalian species: Khirokitia 1972 *J. Archaeol. Sci.* **6**, 127-137.

Yealland, S. and Higgs, E. 1966. The economy *in* R. Hilton and P. Rahtz, Upton, Gloucestershire, 1959-64 *Trans. Bristol Gloucestershire Archaeol. Soc.* **85**, 139-143.

A REGIONAL APPROACH TO FAUNAL DATA

C.M. Gerrard

In his review of the available faunal evidence for medieval sites in the south-west of England, Levitan (this volume) has stressed the value of a "regional approach". This is intended to sidestep the "parochialism" of single-site reports by developing a coherent framework for understanding faunal data from local communities at different levels of social organisation. Until now the comparative value of faunal data has been lost because sites have been chosen from too large an area and from a wide variety of contexts. The "multi-locality reports" approach is intended to remedy this by focusing on relations within a hinterland and between settlements of a specified date.

This brief paper will consider the possible value of such an approach within the historical framework of late Anglo-Saxon Somerset (pre-1974 boundaries), since this is the period for which the most faunal evidence is presently available. In particular, some preliminary statements are made about the population-agrarian characteristics of different settlement types in order to build up a typology of livestock-economy types within a "regional framework". The term region is taken to refer to a small area (< 30km radius), perhaps the size of a hundredal unit or early Anglo-Saxon estate.

HISTORICAL FRAMEWORK

In a recent article Aston introduced the concept of the "central place" to the study of post-Roman settlement study in Somerset (Aston, 1986). This is a useful bracket term which can be used to describe centres which combine a number of administrative, judical, exchange, or ecclesiastical "focal place" functions (they might be hundredal meeting places and markets, for example). Many of these "central places" act as centres for early estates whose arrangements (at least in 8th or 9th century Somerset) may have corresponded in many ways to the blocks of land later formalised as hundreds (eg the Brent estate (Aston, 1984), and may have contained several other isolated "focal places" (eg a minster). Such territories may conform to the "multiple estate" model proposed by Glanville Jones (Jones, 1971; 1976). Working from this historical model it is difficult to be explicit about the economy of such an estate in Somerset but, from examples taken elsewhere in the country, some degree of economic specialisation might be expected in outlying vills over which a "central place" had control (for discussion of terms see Gregson, 1985). Types of service rendered by these estates may not have been restricted to agricultural produce of course, but that is what concerns us here.

In the 11th century the picture of estates given to us in the Domesday book is quite different. The compact estates have gone, to be replaced by a number of scatters of manors across the county in the same ownership (Gerrard, 1987). In the 11th century there are rarely two adjacent manors in the same lay ownership, a pattern which appears to have been generated through a lack of primogeniture in late Anglo-Saxon inheritance. Instead, blocks of land were subdivided equally amongst heirs and surviving relatives (Costen, forthcoming) and the only evidence for consolidated estates is around major ecclesiastical centres (eg around Glastonbury or Wells), who rarely had to divide their estates and were constantly accumulating new land. It is possible to see economic specialisation in the 11th century too, not only because of the Domesday entries for fisheries or vineyards, but more generally, in the distribution of agricultural resources. For example, 1000 acres of pasturage are recorded at Litton, on the Mendips, 550 acres at Withycombe on Exmoor, while at Wemberton, near Yatton, pasture was important enough to be separately named. At Carhampton, 16821 acres of pasture and meadowland were available, recorded and apparently in use, where not more than 2000 acres could possibly have been needed to support the needs of the manor. In contrast, on the Somerset Levels recorded acreages of pasture and meadow are low and populations must easily have outstripped immediate resources (Gerrard, 1987, using O'Connor, 1982; 1983).

In short, the Domesday evidence illustrates that agricultural surpluses were being generated in some parts of the county in the late Anglo-Saxon period. These surpluses cannot be balanced out by proposing similar estate management practises to those which took place in the earlier Anglo-Saxon compact estates. Exchange must have been taking place between estates after the 10th century, mainly through the burgeoning market network which was located at central places, to the benefit of the Crown and wealthy ecclesiastical foundations (85% of all markets with charters before 1200 in Somerset were granted to Crown or Church) (Britnell, 1978).

The volume of exchange also implies that the market system was far more extensive than the meagre number of seven markets recorded in Somerset at Domesday. Furthermore it raises the question as to whether inter-regional economic interdependence was novel in the 11th century. Does it appear to us at this date only because there is documentation to show it? There are certainly echoes here of the later medieval/early modern pattern of regional interdependence, in which markets and fairs at major centres appear to draw upon specific areas within their hinterlands for specialised resources (Gerrard, 1985). In that case, it might be suggested that individual vills continued a relatively stable pattern of exploitation at least until 1400 (although there must have been significant changes in the numbers of markets and their functions), when earlier demographic and social changes began to have a profound effect (Bolton, 1980).

Archaeology can help to show just how wide the radius of exploitation must have been for a late Anglo-Saxon central place. Using O'Connor's "technique" it is possible to convert ratios of bones found on 11th century sites (32% cattle, 38% sheep, 10% pig) into estimates for the proportion of meat in the diet from each species (through educated but tentative speculations). The results, based on figures from Harcourt (1964) (using carcass weights of 181kg for cattle, 11kg for sheep, 36kg for pigs), reveal that 80% of the meat yield is provided by cattle, 7% by sheep, and 5% by pig (O'Connor, 1982; 1983).

The landscape impact, in terms of the required acreages of certain landuses by stock consumed, can also be defined approximately. This can be done by estimating the populations of major centres, multiplying this by the average meat consumption per year (taken to be 25kg), and thereby computing the total annual meat consumption for each settlement (O'Connor, pers comm). Further assumptions can allow us to calculate the total number of sheep eaten, postulate the number of sheep in the core flock, and the number of acres of grass needed to support them!

At first glance, the acreages required for sheep by major centres seem to be low. Even a recognised late Anglo-Saxon "town" such as Bath or Taunton would require an area no larger than a typical parish (6 square km) to be given over to pasture for its sheep. In real terms the impact is huge because each manor generally contains less than 30 acres of pasture. A settlement like Bath in the 11th century would effectively have to absorb at least 50 parishes to pasture its sheep, affecting an area in excess of 300 square km within a 10km radius around the town. The impact is really even greater because this calculation does not allow for the villages within that 10km radius to use any acreage for sheep.

Of course, it may not be reasonable to assume that all sheep were grazed on grass/pasture rather than on fallow-course arable and the post-harvest shack. In a Somerset example (Podimore), Fox (1986) has shown how, in a medieval open field system, sheep were overwhelmingly supported on the fallow, with the permanent pasture and mown meadow being reserved for draught animals and dairy cattle. It is not certain how typical this pattern might be of the late Anglo-Saxon period, and the general implication must remain - that much of the county was absorbed into the "catchment areas" for sheep for central places, and that these animals were providing dung, milk and wool, as well as meat. A varied use of animal products is perhaps the best way to explain why vast acreages of agricultural resources might be used to provide such a small percentage of the diet.

A REGIONAL APPROACH TO FAUNAL DATA WITHIN THE HISTORICAL FRAMEWORK

Beyond these very general statements there are more specific questions concerning late Anglo-Saxon economy which only archaeology can help with, but which have not yet been seriously approached. For example, how often do we find concentrations of non-local agricultural resources at early Anglo-Saxon estate centres, which might be thought of as being at the heart of "multiple estates"? In other words, are "multiple estates" really self-sufficient, "spatially widespread, yet bounded" units in which essential human and material resources are related strongly, through a mesh of service relations to specific settlement and tenurial characteristics" (Gregson, 1985)?

On another point, what changes might be expected, and at which focal places, as the switch occurs from an estate management system to a market system operating through fragmented estates? Should we expect differences in the faunal data recovered from Crandon Bridge (as a rural vill) and Cheddar palace (as a focal place)? How should these two differ from the assemblages from Bath and Taunton (as central places)? The answer is that they should differ in many subtle ways. At rural vills there might be specialisation of local resources, while at focal and central places there might be a concentration of a variety of exotic (non-local) specialised resources. This contrast is more likely to be apparent in focal places, since these tend to be isolated single sites (eg monasteries, minsters or palaces), whereas central places will inevitably have larger concentrations of population whose consumptions will vary according to rank, wealth, fashion, etc. In these cases, concentrations of exotic (non-local) specialised resources will be apparent only at certain sites, and even then may not be any reflection on the status of the site. Pottery distribution studies for the early medieval period in this area have shown how finds of imports and exotic glazed wares were related to the proximity of the site to markets and fairs, rather than being related to status (Gerrard, 1987).

In Table 1 various settlements with differing topographical and environmental characteristics from the Anglo-Saxon estate of Somerton are taken in order to illustrate the range of economy types at different stages in the early medieval period (Aston, 1985b; Dunning, 1984). Roughly speaking, the 7th-9th century bracket represents an estate economy, whereas from the 9th century onwards a marketing system was operative. For each site type, an attempt has been made to specify the type of faunal remains which could be expected.

The table shows quite well that the most interesting site types (from the faunal angle) are likely to be isolated focal places *ie*, those which were once the centres of estates (like Somerton) and failed to develop as commercial or population centres (it was Langport that later developed as a town and port in this region). This is because these sites will retain their rural character and still have dramatic changes in their status which may be represented in the faunal remains. In contrast, the most difficult sites to understand economic changes in are the "urban" sites (central places), where every assemblage will not be representative of that settlement's classification in the table (see below).

The table stresses how sites relate to each other and demonstrates how a "regional approach" might help both to broaden the range of site types excavated and to explore possible links between them at different periods (as Pearson (1982) has done with pottery in the same area). The table illustrates that most information can be extracted from the faunal evidence by comparing assemblages from known site types at specific dates. In effect, constructing such a table helps to give priority to certain assemblages. For example, it might not be as interesting to compare between Aller and Beere after the 9th century as it would before, nor might any useful conclusions be

Table 1. ***Expected faunal assemblages at a number of type localities within one early Anglo-Saxon estate in Somerset (Somerton) between the 7th and 14th centuries.***

Key:
a. Settlement Type
b. Goods/resources rendered to the estate centre or local market
c. Goods/resources received
d. Faunal evidence

	7th-9th C	9-10th C	14th C onwards
BEERE			
a.	small rural woodpasture vill on edge of Levels	stable, but possible encroachment into wood	some colonisation of marginal Levels land
b.	fish, fowl, pigs, wood, timber	stable	stable
c.		variety of local produce from market	stable
d.	fish, fowl, pigs,	fish, fowl, pigs, more cattle, sheep, more variety.	
LONG SUTTON			
a.	large vill near Somerton, the estate centre	nucleated village with common fields	stable
b.	mixed	cereals	more pasture
c.		variety of local produce	stable
d.	cattle, sheep	greater variety, more meats with older sheep later (for wool)	
SOMERTON			
a.	estate centre early focal place	hamlet	town founded but shortlived
b.		mixed	mixed
c.	whole variety of local produce from Beere, Long Sutton etc	similar to Long Sutton	
d.	non-local resources	similar to Long Sutton	

Table 1 (cont). ***Expected faunal assemblages at a number of type localities within one early Anglo-Saxon estate in Somerset (Somerton) between the 7th and 14th centuries.***

Key:
a. Settlement Type
b. Goods/resources rendered to the estate centre or local market
c. Goods/resources received
d. Faunal evidence

	7th-9th C	9-10th C	14th C onwards
ALLER			
a.	late focal place putative monastic site on Levels	village	village
b.	christian services	similar to Beere	
c.	specialised produce	similar to Beere	
d.	fish, fowl, good quality non-local meats	similar to Beere	
LANGPORT			
a.	in existence?	town/port and centre of local commercial activity	declines
b.	--	mixed	mixed
c.	--	whole variety of local and non-local produce	
d.	--	Range of different assemblages at different locations, reflecting social structure and nature of local markets. Unpredictable.	

expected to come from the comparison of pre-9th century assemblages from Somerton and Aller. However, any opportunity to examine a long sequence of faunal data from Somerton (7th to 14th centuries), particularly in comparison with an assemblage from Beere, should be snapped up.

Naturally, the present system whereby faunal specialists are presented with assemblages from archaeological sites which have already been dug, is hardly conducive to exploring local patterns of production and consumption. Comparisons of faunal assemblages between such sites are likely to be unrepresentative and confusing. However, the table does show how the existing faunal record might be complemented. For example, a similar table could be drawn up for the area around Cheddar, where faunal data from a focal place has already been recovered at the Palaces, and where the next priority might be to examine faunal data from a rural vill (the equivalent of a Beere in Table 1).

In short, a "regional approach" to faunal data would enable faunal specialists to build upon the evidence they have already assidously collected, and to contribute to research designs for future sites in any one region. Finally, they would be adding their voice to others who have argued for some while that medieval Britain was a complex society, stratified economically, socially, and intellectually, and consisting of a "series of inter-related regional systems" (Jope, 1972). More than this, through a regional approach, they would be offering a unique opportunity to test historical hypothesis with empirical archaeological data.

ACKNOWLEDGEMENTS

I should like to thank Mick Aston, Mike Costen, and Dr Paul Glennie for their constructive criticism and help in preparing this paper.

REFERENCES

Aston, M.A. 1984. The towns of Somerset *in* J. Haslam (ed) *Anglo-Saxon towns in southern Britain*, Chichester: Phillimore, 167-202.

Aston, M.A. 1985. *Interpreting the landscape: landscape archaeology in local studies*, London: Batsford.

Aston, M.A. 1986. Post-Roman central places in Somerset *in* E. Grant (ed) *Central places - archaeology and history*. Sheffield: University of Sheffield, Department of Archaeology and Prehistory, 49-78.

Bolton, J.L. 1980. *The English medieval economy 1150-1500*, London: Dent.

Britnell, R.H. 1978. English markets and royal administration before 1200, *Econ. Hist. Rev*. **34**, 183-198.

Costen, M.D. forthcoming. *Hides and ploughlands in Somerset Domesday Book*.

Dunning, R.W. (ed) 1974. *Victoria County History of Somerset Vol 3*, London: University of London, Institute of Historical Research.

Fleming, R. 1985. Monastic Lands and England's defence in the Viking Age. *English Hist. Rev*. **100**, 247-265.

Fox, H.S.A 1986. The alleged transformation from the two-field to three-field system in medieval England, *Econ. Hist. Rev*. **39**, 2nd series, 526-548.

Gerrard, C.M. 1985. Taunton fair in the 17th century: an archaeological approach to historical data,*Somerset Archaeol. Nat Hist. Soc*. **128**, 65-75.

Gerrard, C.M. 1987. Trade and settlement in medieval Somerset. Unpublished PhD. thesis, University of Bristol, Department of Classics and Archaeology.

Gregson, N. 1985. The multiple estates model: some critical questions, *J. Hist. Geog*. **11** (4), 339-351.

Jones, G.R.J. 1971. The multiple estate as a model framework *in* F. Dussart (ed) *L'habitat et les passages ruraux d'Europe*, Liege: Les congres et colloques de l'Universite de Liege **58**, 257-267.

Jones, G.R.J. 1976. Multiple estates and early settlement *in* P. Sawyer (ed) *Medieval settlement: continuity and change*, London: Arnold, 9-35.

Jope, E.M. 1972. Models in medieval studies *in* D. Clarke (ed) *Models in archaeology*. London: Methuen, 963-990.

O'Connor, T.P. 1982. *Animal bones from Flaxengate, Lincoln 870-1500* (The archaeology of Lincoln **18/1**), London: Council for British Archaeology.

O'Connor, T.P. 1983. Feeding Lincoln in the 11th century - a speculation *in* M. Jones (ed) *Integrating the subsistence economy* (Symposia of the Association for Environmental Archaeology No. 4), Oxford: British Archaeological Reports (International Series **181**), 327-330.

Pearson, T. 1982. Medieval ceramics *in* P. Leach (ed) *Ilchester Volume 1, Excavations 1974-75*. Bristol: Western Archaeological Trust (Excavation Monograph **3**), 182-215.

PAST AND PRESENT LOCAL VARIABILITY IN ENVIRONMENTAL PHENOMENA

David J. Maguire

INTRODUCTION

As techniques of palaeoenvironmental analysis become increasingly sophisticated it is possible to make more detailed reconstructions of past environments. The objective of much recent palaeoenvironmental research has been to use the increase in detail to improve the spatial resolution of reconstructions. In order that such research can proceed further it is necessary to examine both past and present local variability in the environment (defined here as the degree to which phenomena change over distances less than about 100m).

Consideration of the level of local variability is fundamental to any scientific testing (Edwards, 1983) since knowledge of the variability of palaeoenvironmental data is necessary to assesss the reliabilitty of reconstructions. (Although only spatial variability is considered here to gauge the reliability of reconstructions, temporal variability should also be assessed, but this is beyond the scope of this paper.)

The normal practice in palaeoenvironmental reconstruction (some examples are given in Edwards, 1983) is to attempt reconstructions on the basis of a single site or at best a few sites. In the case of pollen analysis Edwards (1983: 588) suggests that this is "a combination of prohibitive pollen counting times, and, possibly, a belief in the general applicability of the patterns revealed in a single core". But, as he goes on to point out, few researchers have considered how appropriate single cores are for reconstructing the patterns of wide areas. Palaeoenvironmentalists in other areas of the discipline also use this strategy, for similar reasons, but this paper is concerned with the variability of vegetation and soils.

The reconstruction of spatial variation in palaeoenvironments has been the focus of several recent research projects. The so-called "three dimensional approach" to vegetation reconstruction exemplified by Turner's (1975) paper and the innovative book by Huntley and Birks (1983) are good examples. These investigations, however, have all been concerned with variability at relatively large scales (Turner's sites for example were at least 500m apart and Huntley and Birks' were separated by several kilometres). The level of variability also has important implications for the selection of sites for palaeoenvironmental reconstruction (Jacobsen and Bradshaw, 1981). Generally speaking those features of the environment (included here are animals, plants, soils and archaeological phenomena) which exhibit a high degree of variability will require more detailed sampling than those features which exhibit relatively little variability.

Clearly the amount of detail required in the final reconstruction and the resource constraints of the project are the major considerations used to plan palaeoenvironmental research. If reconstruction of only the general features of palaeoenvironments is the project goal or if resources such as time, money, and expertise are limited then single site, or few sample, studies are undertaken. However, in order to determine the amount of confidence we can place upon the results of such studies the level of local variability in palaeoenvironments must be estimated.

This paper examines the local variability of blanket peat deposits at a site, Broad Amicombe Hole, which is characteristic of the upland areas of the South West. The first section introduces the site and the methods employed in the study. The second section considers present day local variability in the vegetation, soils and peat deposits. The third section is concerned with the past local variability in vegetation and peat. The fourth section discusses many of the points raised in the first sections and draws some general conclusions from the work.

METHODS

Only the salient features of the methods are considered here. A fuller discussion may be sought in Maguire (1983).

The three dimensional position and stratigraphy of over six hundred sites was recorded using a quickset level, staff, tape and soil auger. The percentage cover of plants was recorded in sixty three one metre square quadrats arranged in three longitudinal transects. The terminology for flowering plants and ferns follows Clapham, Tutin and Warburg (1962) and for mosses and lichens, Phillips (1980). Soil samples were collected in monolith boxes for further laboratory analysis.

Pollen extraction and concentration was achieved using the standard procedures (Moore and Webb, 1978) of potassium hydroxide digestion, acetolysis and hydrofluoric acid treatment, with silicone fluid as the mountant. Terminology follows Moore and Webb (1978). A pollen sum of 500 dry land pollen and spores was selected; all pollen values are expressed as a percentage of this pollen sum. The four radiocarbon dates presented were subjected to standard preparation procedures and were dated at the NERC Scottish Universities Research Reactor Centre, East Kilbride (Harkness and Wilson, 1972). The dates are reported, in accordance with *Radiocarbon* practice, as conventional radiocarbon years BP with one standard deviation level for analytical confidence.

SITE DETAILS

Broad Amicombe Hole lies almost at the centre of Dartmoor's northern moor (Fig. 1) at an elevation of 481m OD; national grid reference SX 578 857. It is drained to the north by an unnamed tributary of the West Okement River and to the south by the Amicombe Brook, a tributary of the River Tavy. It is bounded by the steep slopes of Amicombe Hill (573m OD) to the west and Great Kneeset (568m OD) to the east.

The site comprises an area of valley peat surrounded by hill peat. The valley peat is about 320 x 60m in maximum extent and roughly boat shaped. At the north eastern end of the valley peat an area of hill peat, the hill peat survey area (Fig. 2), was selected for detailed study. The site was chosen because of the close juxtaposition of the hill and valley peat deposits which was essential for other work conducted at the site. Unfortunately at some time in the past peat cutting took place in part of the hill peat survey area (see below for details). Although this does not invalidate the results presented in this paper it is a factor which must be considered when the results are interpreted. The hill peat survey area is the subject of this paper; further details of the valley peat will be published elsewhere.

The hill peat survey area is 200 x 60m in extent and has a concave-convex slope form with the lowest section close to the valley peat. It may be subdivided into four zones on the basis of the slope characteristics (Table 1). The first and highest zone extends from 200 to 130m. It is relatively flat with an average angle of slope less than 1°. The second zone extends from 130 to 55m. This zone includes the shoulder of the slope and has an average angle of slope of 6°. The third and steepest zone extends from 55m to 5m and is characterised by outcropping rock. It has an average angle of slope of 14°. The fourth and lowest zone extends from 5m to 0m. It has an average angle of slope of 2° and covers the transition between the hill and basin peat deposits.

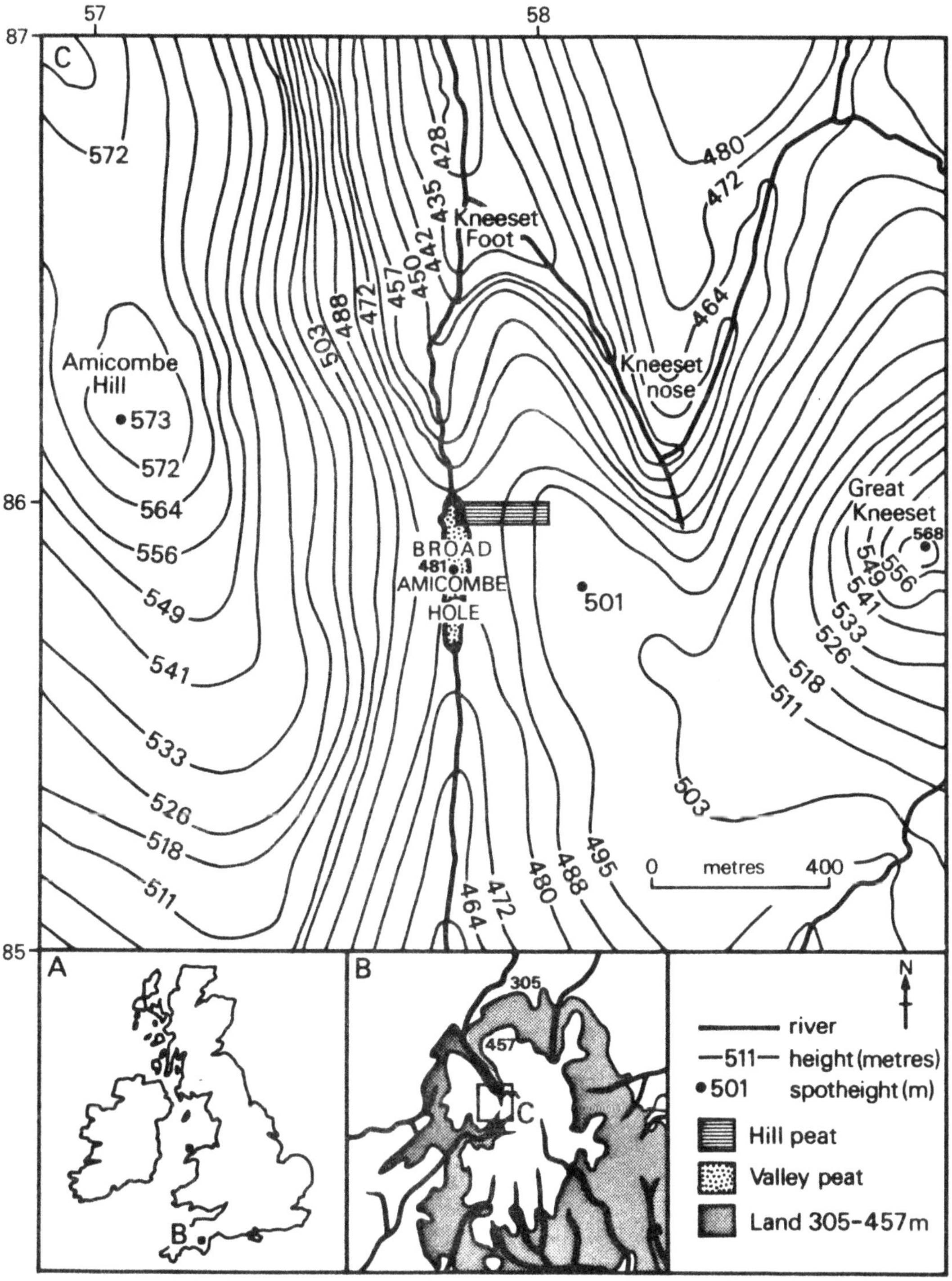

Figure 1. Location of Broad Amicombe Hole.

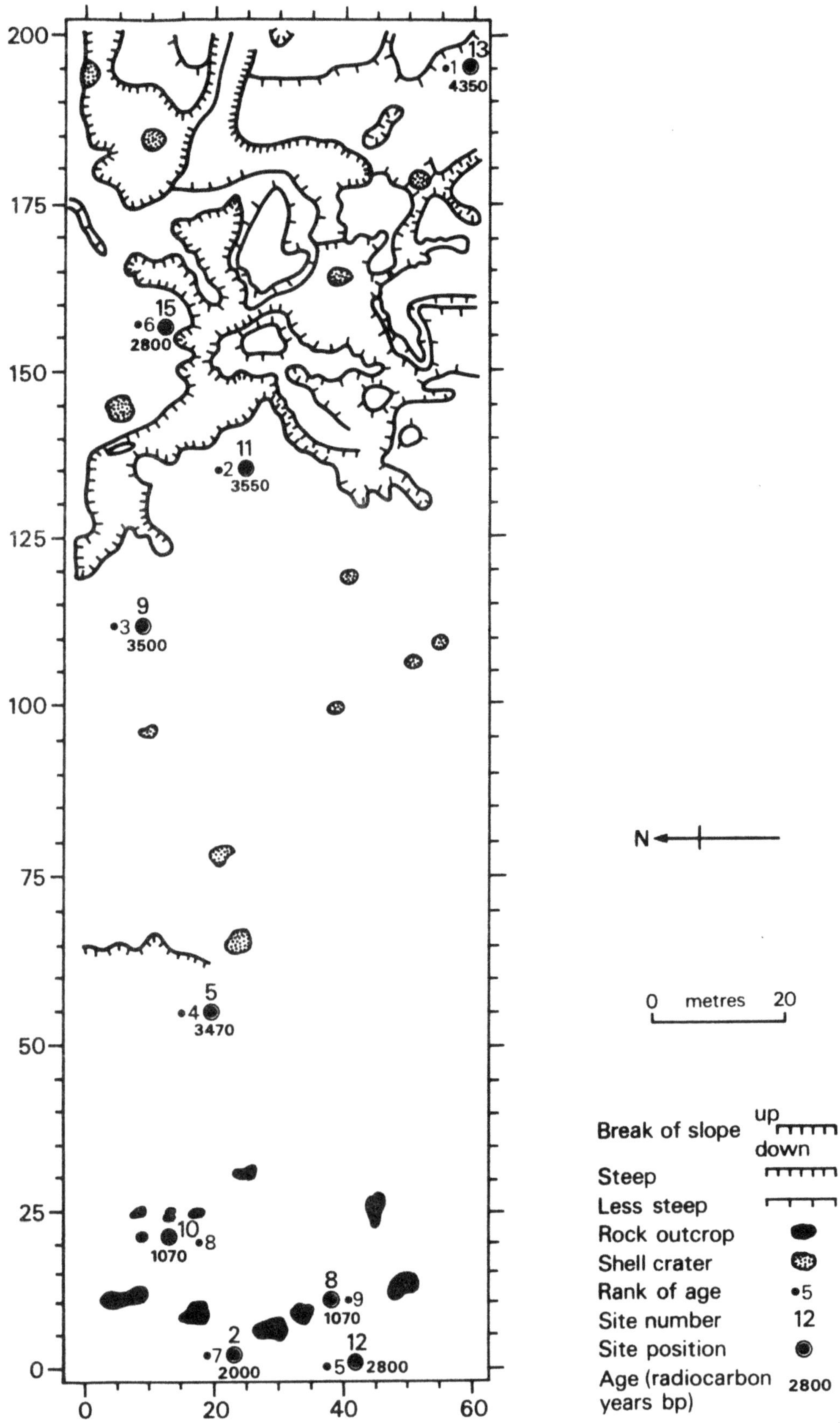

Figure 2. General map of hill peat survey area, showing dates for blanket peat formation.

Table 1. Distribution of certain features in the hill peat survey area. (See text for derivation of groups)

FEATURE	DISTANCE UP SLOPE (METRES)			
Slope form	200-130	130-55	55-5	5-0
Vegetation	200-50		50-5	5-0
Soils	200-50		50-10	10-0
Peat depth	200-65		65-5	5-0

PRESENT DAY ENVIRONMENT

Vegetation

It is convenient to discuss the present day vegetation of the study area in the context of the vegetation scheme outlined for Dartmoor by Ward *et al* (1972). In the hill peat survey area three of their units may be identified and mapped.

In the upper parts of the hill peat survey area, from about 200 to 50m (Fig. 2; Table 1) the vegetation closely resembles the blanket bog unit of Ward *et al* (1972). *Trichophorum caespitosum* is present in the majority of the quadrats, to a maximum of 50%. Also occurring in the majority of quadrats are *Calluna vulgaris*, *Erica tetralix*, *Eriophorum angustifolium*, *Molinia caerulea* (the dominant species in many quadrats), *Vaccinium myrtillus*, *Sphagnum* spp. and other mosses (*Hypnum cupressiforme* and *Dicranum scoparium* were noted in several quadrats). In wetter areas mossses (especially *Sphagnum* spp. and *Polytrichum* sp.) and sedges increase at the expense of the other plants.

The steeper areas of the slope, between 50m and 5m, carry a variable unit which most closely resembles the grass heath category of Ward *et al* (1972). As they point out, it is an extremely variable group of stands, the only common feature being that it is transitional between heath and grassland. The unit at Broad Amicombe Hole is characterised by consistent values for the grasses *Nardus stricta* (maximum cover 30%), *Festuca* sp. and *Agrostis* sp. with *Potentilla erecta*. *M. caerulea* is again the dominant species occurring with *C. vulgaris*, *E. angustifolium*, *V. myrtillus*, *Galium saxatile* and other mosses (*Pleurozium schreberi*, *H. cupressiforme* and *D. scoparium* were all noted). This unit extends higher upslope in the southern part of the hill peat survey area.

Where the survey area encroaches onto the valley peat at the base of the slope, between 5m and 0m, the vegetation is similar to the Valley Bog unit. It is characterised by large quantities of *M. caerulea*, *C. vulgaris* and *E. vaginatum* with a noticeable absence of *E. angustifolium*.

Two other habitats are worthy of mention. On the hill slope there are at least twelve shell craters, which it is thought result from recent army shells (the area is used as an army firing range). Once formed they are normally colonised by *Sphagnum* spp., followed by *E. angustifolium*, *Juncus effusus* and by other sedges, mosses (particularly *Polytrichum* sp.) and *M. caerulea*. Where the slopes are steepest, rock outcrops at the surface. These granitic blocks are commonly encrusted with lichens (especially *Cladonia impexa*), mosses and *V. myrtillus* with *C. vulgaris* occasionally present. In crevices beneath the blocks, in the shade, the fern *Blechnum spicant* may be found growing in association with *V. myrtillus*.

More specifically the variability of the vegetation may be emphasised by the fact that in the hill peat survey area *Trichophorum caespitosum* has a range of cover from 0 - 50%, *Festuca ovina* ranges from 0 - 39% and *Sphagnum* ranges from 0 - 49%. Such values are partly an artefact of the size of the quadrat used. If a larger quadrat had been used (eg two or four square metres) the variability would be reduced to some extent. Nevertheless, the data demonstrate that substantial variability in vegetation is observable even within an area as small as the hill peat survey area.

Soils

Much of the early work on the soils of the Dartmoor granite is summarised in Hogan (1978). This work indicates that the main soil types (using the terminology of Avery (1980)) are raw oligo-amorphous peat soils (Crowdy series), loamy skeletal iron pan and ferric stagnopodzols (Hexworthy and Rough Tor series) and loamy skeletal typical brown podzolic soils (Moretonhampstead series). Several other types are also locally present at high elevations. On steep slopes humic rankers (un-named) can be common and where organic matter levels are high loamy skeletal ferro-humic cryptoopodzols (un-named) may occur.

The soils in the hill peat survey area have developed in a skeletal parent material which is dominantly a granitic head deposit. They are described as loamy skeletal because they have between 35% and 70% stone content by volume and because the fine earth fraction (less than 2mm) has a loamy constituency. The exceptions to this general rule are the humic ranker soils in which upper peaty horizons rest on an R horizon of hard granitic rock with few large cracks. Large and very large stones are frequently encountered when digging into these soils. The stones normally have a subangular shape and a medium platy structure.

The fine earth fraction is dominantly sandy silt loam or sandy loam. The large silt component (commonly 40% to 60%) in some horizons may be of loessial origin. The soils also contain large amounts of organic matter both above (peat), and incorporated within, the mineral matter.

The soils of the hill peat survey area can be split into two groups, for the purposes of discussion, on the basis of the amount of superficial organic matter present. Those with greater than 40cm are classed as peat soils and have a distribution restricted to areas with an angle of slope less than two degrees. These areas are above 50m and below 10m (Table 1, Fig. 3). Between these areas the soils are dominantly stagnopodzols with small areas of humic cryptopodzols and humic rankers.

The soil profiles examined from the upper parts of the survey area (above 50m) are virtually all examples of Crowdy series soils. They are characterised by a layer of raw oligo-amorphous peat which, except where the profile has been cut over (termed cut phase: see following section on peat for explanation) is greater than 40cm and commonly exceeds 80cm in depth.

Crowdy series soils are also located in the lower part of the study area, where the survey extends on to the valley peat. The peat tends to be thicker than analagous profiles at the upper elevations of the study area and there is often more mineral matter present within the peat.

The more varied nature of soils on the central part of the survey area may be a product of the classificatory system employed. In this area, where the organic matter content is low, the relatively simple criterion of humification of peat is replaced by the presence of diagnostic mineral horizons. In this relatively small area of 12,000m^2 seven different types of soil have been identified (excluding bare rock where there are *ipso facto* no soils).

A small proportion of the profiles examined (Table 2) are of the Crowdy series type and are similar in many respects to those discussed above. However, Crowdy series soils from the central area often have a podzolic B horizon.

The most commonly occurring soil types in the central area are the skeletal ferric stagnopodzols of the Rough Tor series and the loamy skeletal iron pan stagnopodzols of the Hexworthy series. The main distinction between the two types is the lack of an iron pan in the Rough Tor series. The Hexworthy soils are further differentiated depending on whether the iron pan is indurated or non-indurated.

Closely associated, both genetically and spatially, with the stagnopodzol soils are the humic cryptopodzols. Their characteristic feature is the large accumulation of incorporated organic matter which obscures the morphological expression of eluvial and illuvial horizons. About 3% of the profiles examined during the course of the survey are of this type.

Table 2. Percentage cover of soil types in hill peat survey area. The data were collected using a soil auger sampling every five metres on a regular basis.

SOIL TYPE	NUMBER OF OBSERVATIONS (x)	%
Crowdy uncut phase	342	64.17
Crowdy cut phase	50	9.38
Hexworthy	33	6.19
Rough Tor	46	8.63
Un-named humic crypto podzol	17	3.19
Un-named humic ranker	35	6.66
Bare rock	5	0.94
Shell crater	5	0.94
	Σ x=533	Σ % =100.1

The humic ranker soils, which constitute about 7% of profiles examined, are typical immature upland soils. They have a peaty upper horizon, less than 40cm deep, above granitic rock.

A final soil type worthy of mention is the so-called "leached" stagnopodzols. These occur on the steepest areas of the slope and are closely associated with the stagnopodzolic soils. The lower mineral horizons do not meet the requirements for classification of the soil as a stagnopodzol although the other features are very similar. These soils are classed as leached stagnopodzols since they most closely resemble the stagnopodzol group. In Table 2 this group has been included in the Hexworthy or Rough Tor series depending on whether or not an iron pan is present.

Peat

Peat may be defined simply as the accumulated remains of plants and animals in various stages of decomposition. Except where granite outcrops at the surface, peat covers the entire hill peat survey area. The surface of the peat, especially in the upper regions of the study area, is dissected into sharp faces and hollows (Fig. 3). These features probably result from several processes including deposition (which is related in part to the sub-peat surface form), erosion and peat cutting; the latter probably being the most important. Statistics summarising the data for 533 survey points located on a 5m grid are presented in Table 3. The mean depth of 58.3cm is exaggerated slightly by the inclusion of a small amount of basin peat at the base of the slope. Of greater significance, however, is the reduction in the size of the mean value as a result of peat cutting in the upper part of the survey area.

Table 3. Descriptive statistics of peat depth data (see text for explanation). Depth is measured in centimetres.

Number of observations	533
Mean depth	58.3
Standard deviation of depth	31.6
Minimum depth	0
Maximum depth	185
Range of depths	185
Coefficient of variability of depth	54.2

The variability of the peat depth values can be assessed by considering the range, standard deviation and coefficient of variability. The depth ranges from 0cm, at several locations on the steeper central parts of the survey area, to 185cm at local grid reference 05 000 (Fig. 3) at the base of the slope. The relatively large standard deviation of 31.6 indicates that there is a high degree of variability of peat depth. For example, there is only a 68% probablility of peat depth at any location in the hill peat survey area falling in the range 26.7 to 89.9cm (mean $\pm$ 1 standard deviation). This high degree of variability is confirmed by the relatively large value of 54.2% for the coefficient of variability.

Discounting the effects of peat cutting, the survey area can be divided into three regions on the basis of depth. At the upper end of the survey area, from 200 to 65m, the peat is always deeper than 40cm and usually deeper than 80cm (Table 1). At the base of the slope, from 5 to 0m, the peat increases quickly to 185cm where the survey area extends into the valley peat. Between these two extremes, on the steeper central portion of the slope, the peat is less than 40cm in depth except for very occasional samples.

PAST ENVIRONMENT

Nine sites which represent the range of vegetation, soils and peat types (Table 1) were selected for examination of the past environment of Broad Amicombe Hole. The location of the sites is shown in Figure 2 and their salient features are outlined in Table 4.

Date of Peat Formation

The age estimates for the date of blanket peat formation are derived from radiocarbon dates, where available, and cross correlations between profiles using the local assemblage zones (laz). Three major problems were encountered in deriving these estimates. First, because of variations in spatial location, the sites received different pollen spectra. Other things being equal, those sites closest (in space) to an event should have exhibited the greater response. Secondly, it was sometimes difficult to correlate between diagrams because many of the changes in the pollen record were

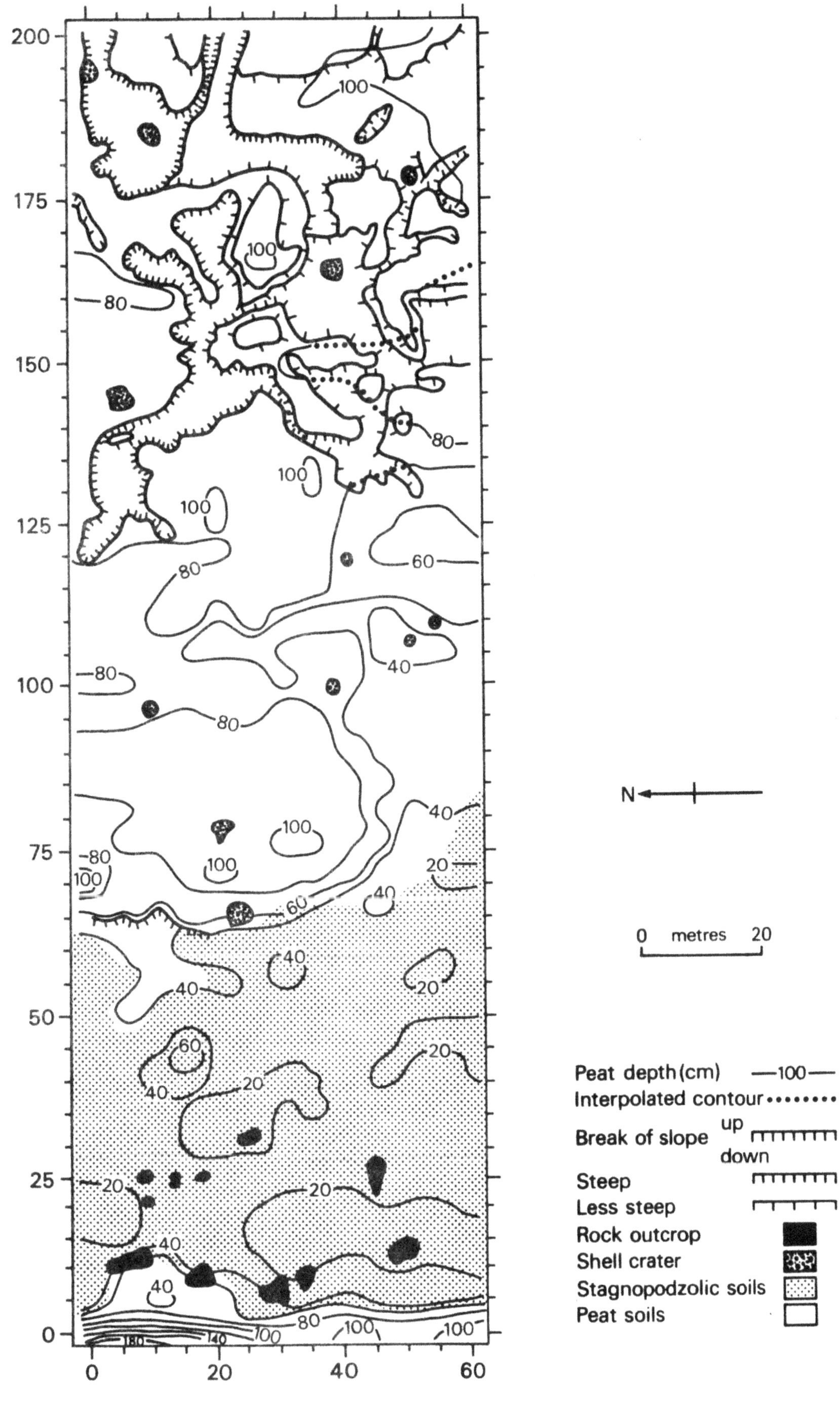

Figure 3. The distribution of soils and peat depth in the hill peat survey area

Table 4. Estimated age and characteristics of peat at sites in the hill peat survey area. The average percentage and range (in brackets) of each taxon are for the four basal samples.

Site No.	13	15	11	8	12	2	10	5	9
Peat depth (cm)	115	75	90	15	64	35	17	35	80
Basal C-14 date (years BP)	4350±50		3570±60		2810±60			3440±50	
Estimated age of basal peat (years BP)	4350	2800	3550	1070	2800	2000	1070	3450	3500
Ranked age	1	6	2	9	5	7	8	4	3
Basal laz	E-t--C/M--Q	E-t--C--G --C/M--Q--A	G--C/M--E-t G--C--Q	G--C--C/M --E-t	G--C/M-- E-t--Q--A--C	G--Q--C	G--C--C/M --E-t--Q--A	C/M--Q--C --P-u	G--C/M--C --E-t--Q--A
%A	3(3-4)	2(1-3)	3(1-3)	3(2-5)	4(2-6)	5(4-6)	4(4-4)	4(2-8)	4(1-5)
%Q	7(6-7)	7(6-8)	7(6-8)	5(4-6)	8(5-11)	11(9-13)	5(4-6)	9(3-13)	7(6-8)
%C/M	22(18-24)	15(10-18)	25(22-31)	19(14-23)	24(21-26)	11(8-12)	15(13-18)	24(22-29)	23(22-25)
%E-t	45(42-48)	25(5-48)	15(12-16)	8(5-13)	15(10-22)	8(7-11)	8(6-10)	12(7-16)	20(12-24)
%C	6(2-8)	27(14-41)	12(11-13)	21(20-23)	8(6-10)	10(9-12)	25(21-30)	15(11-19)	14(12-17)
%G	11(9-12)	20(10-30)	24(12-32)	34(32-37)	23(21-31)	30(28-35)	29(28-32)	12(6-18)	21(14-29)
%P-u	1(+-1)	+(+-+)	4(2-5)	3(1-6)	4(4-5)	1(+-1)	3(1-4)	13(4-16)	2(+-2)
Slope (degrees)	2	3	2	20	2	2	20	5	3
Distance up slope (m)	195	157	132	13	1	3	21	55	112

A = Alnus C = Cyperaceae C/M = Corylus/Myrica P-u = Polypodiaceae undif.
Q = Quercus G = Gramineae E-t = Ericaceae type + = less than 1 %

relatively subtle. The third problem encountered related to the fact that a large proportion of the pollen at the sites was derived from local plants whose distribution and abundance varied even over the relatively small distances between sites.

In spite of these problems it is possible to show that the age of peat inception at the sites does have a pattern (Fig. 2). The site (number 13) with the earliest date for inception (rank 1 = 4350 years BP) is situated at the eastern end of the survey area. This site and the sites ranked (site numbers in brackets) 2 (11), 3 (9), 4 (5) and 6 (15) are all situated above 50m on flatter ground. Peat inception at these sites, with the exception of site 15, began before 3400 years BP. The slightly later date of 2800 years BP for site 15 may be explained by the fact that the peat rests directly on rock which is slightly higher than the surrounding mineral matter. Sites ranked 5 and 7, with dates of 3800 and 2000 years BP, are situated below 5m. It appears, therefore, that peat inception on the lower part of the survey area started much later than on the upper part. It is possible that peat inception in the lower part was coincident with upward and outward spread of the valley peat. The data indicate that the shallowest peat, on the steeper, central parts of the survey area, is also the youngest peat. The sites with ranks 8 and 9 are both dated less than 1070 years BP (the date of the last major clearance episode at Broad Amicombe Hole). A Spearman's Rank Correlation Coefficient between peat age and slope angle was calculated as -0.701, which is significant at the 90% level using a Student's *t* test. This suggests that the date of blanket peat inception in the study area is strongly influenced by the angle of the slope.

Early Vegetation Succession

Three taxa (Cyperaceae, Ericaceae-type and Gramineae) dominate the pollen spectra of the basal peat samples in the hill peat survey area (Table 4); they always constitute greater than 38% of the dry land pollen and spores (ΣDLP). The lowest value is from site 5. It is possible that this value is low because the Polypodiaceae-undif. values are fairly high in the basal samples, at 13% ΣDLP; perhaps indicating poor pollen preservation. Also, the high values for mineral matter could indicate contamination of the basal samples with old pollen.

It is possible to argue that three families (sedges, heathers and grasses) either singly or in combination, constitute the major peat forming plants. That these families comprised the local plants growing at the time of peat inception, is shown by the fact that pollen samples of the same age, which are separated by less than 25m, have pollen spectra dominated by different taxa of these families (*cf* profiles 11 and 5). The fact that the peat forming species occur in the sub-peat deposits indicates that peat inception was not a rapid event in which heathers sedges and grasses colonised the soil surface. instead it was probably a gradual process and this is further substantiated by the slow average accumulation rate of 0.024cm/yr for the hill peat. Taylor and Smith (1972) have identified similar, but in their case distinct, basal layers which they term "pedogenic peat". They suggest that "pedogenic peat" layers are characteristically compact and well humified and contain large quantities of Ericaceae-type pollen. They assert that pedogenic peat accumulates as mor humus over long time periods and that accumulation is controlled by soil processes.

DISCUSSION

The previous sections of this paper have emphasised the variability in environmental phenomena within the hill peat survey area at Broad Amicombe Hole. The data about present day vegetation, soils and peat clearly indicate that there is a high degree of local variability even within such a comparatively small area (12,000 square metres). Three distinctly different vegetation units (Blanket Bog, Grass Heath and Valley Bog) are represented. There is also variability within individual units. Eight different soil types were identified (from a total sample of 533 observations). The major soil type, raw oligo-amorphous peat of the Crowdy series, covers about two thirds of the area. On the basis of depth three areas of blanket peat were differentiated.

The limited data available for reconstruction of the palaeoenvironment of the hill peat survey area similarly exhibit substantial variability (Table 4). There is considerable temporal and spatial variation in the date of blanket peat formation. There are also variations in the local peat forming plant communities, although the three families Cyperaceae, Ericaceae and Gramineae are the major components.

The high degree of variability exhibited by each of the past and present day environmental phenomena considered in this paper may in part be due to the influences of the slope (Table 1). The present day environmental phenomena may be clearly differentiated into three major zones. At the upper elevations, above about 50m, the hill peat survey area is characterised by generally flat terrain (Zone 1). The peat is heavily dissected and generally deeper than 40cm. The soils are virtually all raw oligo-amorphous peat soils (Crowdy series) and the vegetation is usually of the Blanket Bog type, being characterised by *Calluna vulgaris*, *Trichophorum caespitosum* and *Molinia caerulea.* From about 50 to 5m the hill peat survey area is characterised by steeply sloping terrain with frequent rock outcrops (Zone 2). The peat is generally shallow (less than 40cm). There is a wide range of soil types represented, with loamy skeletal ferric stagnopodzols (Rough Tor series) and loamy skeletal iron pan stagnopodzols (Hexworthy series) the most common types. The vegetation of the central area is classed as Grass Heath, a variable unit featuring *Nardus stricta* and *Calluna vulgaris.* The lowest part of the hill peat survey area, below 5m, includes the transition between hill peat and valley peat (Zone 3). The peat is generally deep (always greater than 40cm, increasing to 185cm) and is covered by valley bog vegetation dominated by *Molinia caerulea*, *Calluna vulgaris* and *Eriophorum vaginatum.* The soils are classified almost exclusively as raw oligo-amorphous peat soils (Crowdy series).

The palaeoenvironmental data also exhibit a broadly similar pattern. Figure 2 demonstrates that the oldest sites for peat formation are at the upper end (above 50m) of the hill peat survey area. These have an age range of 2800 to 4350 years BP. Dates of medium age are found at the lowest elevations (below 5m). These have an age range of 2000 to 2800 years BP. The youngest dates are situated on the steeper central portion of the hill peat survey area. The two dates for this area are both younger than 1070 years BP.

The realisation that the distribution of ecological phenomena could be related to slope form was prevalent in Tansley's studies of vegetation (Tansley, 1939). For soils this association was first formally noted by Milne (1935) who termed it a catena. He defined a catena as a group of soils which "are linked in their occurrence by conditions of topography and are repeated in the same relationship to each other wherever the same conditions are met with". The concept is reviewed fully in Gerrard (1981) and Huggett (1982).

Gerrard (1981) suggests that the main processes of catenary differentiation are surface wash, soil creep, solution and mass movement. It is the mode and efficacy of these processes which are influenced both directly and indirectly by slope properties. The processes of catenary differentiation are influenced directly by translocation, soil creep and mass movement and directly through the influence upon vegetation (particularly influenced by soil moisture), the depth and type of soil and the microclimate.

Although there is greater variation downslope than across slope at Broad Amicombe Hole, as Figures 2 and 3 demonstrate, there is still a good degree of variation across the slope. The relatively narrow width of the hill peat survey area, however, makes detailed estimations of across slope variability difficult. Because of the downslope and across slope variability in soils, Huggett (1973) and Vreeken (1973) have proposed that soils should, if possible, be viewed three-dimensionally rather than as a two-dimensional catena. This idea indicates once again that even over relatively small areas there may be substantial variability in environmental phenomena.

The results presented in this paper clearly indicate that high levels of local spatial variability occur in both present day and past environments. At Broad Amicombe Hole this is explained in part by the fact that the hill peat survey area is moderately steeply sloping. In future palaeoenvironmental research this high degree of variability should be taken into consideration when the general applicability of single site studies is being assessed.

REFERENCES

Avery, B.W. 1980. *Soil classification for England and Wales (Higher Categories)*, Harpenden: Soil Survey of England and Wales (Technical Monograph 14).

Clapham, A.R., Tutin, T.G. and Warburg, E.F. 1962. *Flora of the British Isles* Second edition. Cambridge: University Press.

Edwards, K.J. 1983. Quaternary Palynology: multiple profile studies and pollen variability *Progress in Physical Geography* 7, 587-609.

Gerrard, J. 1981. *Soils and landforms: an integration of geomorphology and pedology* London: Allen and Unwin.

Harkness, D.D. and Wilson, H.W. 1972. Some applications in radiocarbon measurement at the Scottish Research Reactor Centre, *Eighth International Conference on Radiocarbon Dating Proceedings* New Zealand: Lower Hutt.

Hogan, D.V. 1978. *Programme and guide to excursions, British Society of Soil Science* Exeter: University Press.

Huggett, R.J. 1973. Soil landscape systems: theory and field evidence, Unpublished PhD thesis, University of London.

Huggett, R.J. 1982. Models and spatial patterns of soils, *in* E.M. Bridges and D.A. Davidson (eds) *Principles and Applications of Soil Geography* London: Longman.

Huntley, B. and Birks, H.J.B. 1983. *An atlas of past and present pollen maps for Europe: 0-13000 years ago* Cambridge: University Press.

Jacobson, G.L. and Bradshaw, R.H.W. 1981. The selection of sites for palaeovegetational studies *Quaternary Research* **16**, 80-96.

Maguire, D.J. 1983. The inception and growth of blanket peat: A study of northern Dartmoor, Unpublished PhD thesis, University of Bristol, Department of Geography.

Milne, G. 1935. Some suggested units of classification and mapping particularly for East African soils, *Soil Research Berlin* **4**, 183-198.

Moore, P.D. and Webb, J.A. 1978. *An illustrated guide to pollen analysis* London: Hodder and Stoughton.

Phillips, R. 1980. *Grasses, ferns, mosses and lichens of Great Britain and Ireland* London: Pan.

Tansley, A.G. 1939. *The British Isles and their vegetation* London: Cambridge University Press.

Taylor, J.A. and Smith, R.T. 1972. Climatic peat - a misnomer? *Proceedings of the Fourth International Peat Congress* **1**, 471-484.

Turner, J. 1975. The evidence for land use by prehistoric communities: the use of 3-dimensional pollen diagrams *in* J.G. Evans, S. Limbrey and H. Cleere (eds) *The effect of man on the landscape: the highland zone*, London: Council for British Archaeology (Research Report No 11), 86-95.

Vreeken, W.J. 1973. Soil variability in small loess watersheds: clay and organic matter content *Catena* **1**, 181-196.

Ward, S.D., Jones, A.D. and Manton, M. 1972. The vegetation of Dartmoor *Field Studies* **4**, 505-533.

A COMPARISON OF QUALITATIVE AND QUANTITATIVE PEAT CLASSIFICATIONS FOR WEST SEDGEMOOR

S.M. Ross and A.L. Heathwaite

INTRODUCTION

The palaeobotany of the Somerset Levels has been documented since the 1940's (eg Godwin, 1941; 1955) but, until recently, such records have been concentrated in the Brue valley, on the area around Shapwick Heath and Meare Heath. The most southerly stratigraphic survey to date was carried out on King's Sedgemoor in 1982 (Alderton, 1983). A large number of pollen diagrams have been constructed for the Central and Northern Levels, particularly for the sites of several prehistoric wooden trackways found within the peat, notably the Abbot's Way track, the Sweet track and the Meare Heath track. Together with a very large number of radiocarbon dates for peat and timbers in these areas (almost 200 dates are listed by Orme, 1982), the pollen evidence has been used to propose regional pollen assemblage zones for the Somerset Levels (Beckett and Hibbert, 1979). This has allowed the development and spread of vegetational changes to be traced for the northern parts of the Levels since their transition from saltmarsh around 5500BP (Beckett and Hibbert, 1979).

Problems of interpreting pollen data in locally variable substrates (Beckett, 1978) and in those with a high water content, indicate the advantages of using *in situ* peat stratigraphy and macro-floral remains for palaeoenvironmental reconstruction. Despite the possible potential for environmental interpretation, virtually no chemical analyses have been carried out on the peats of the Somerset Levels. This paper examines the relationship between peat stratigraphy and peat chemistry using the previously unstudied example of West Sedgemoor, the most southerly of the Somerset moors.

West Sedgemoor lies within an enclosed basin, underlain by Triassic Keuper Marl which forms the North Curry - Stathe ridge to the north-west and, together with overlying Rhaetic and Lower Lias shales and limestones, forms the Curry Rivel ridge to the south-east. The Somerset Levels area was subject to marine inundation around 6000 BC and a thick layer of blue-grey silty clay was deposited. Since then, about 5.5 metres of fen peat have formed in West Sedgemoor, with alluvial material overlying the peat around the basin peripheries. Findlay (1965) mapped West Sedgemoor simply into two soil groups: (i) the Sedgemoor Series in the centre of the moor, of pre-Roman fen peat with high base status due to flooding with base rich water and (ii) the Midelney Series on the moor peripheries, where, on average, 40cm of alluvium, derived mainly from the surrounding Keuper Marl uplands, overlies the peat. In West Sedgemoor particularly, but possibly in other parts of the Somerset peatlands, the mineralogy and geochemistry of the Keuper Marl are thought to have an important influence on the current chemistry of the peat.

There were two main aims of the present study: (i) to characterise the surface peat stratigraphy of an unmanaged section of West Sedgemoor and (ii) to relate peat types, classified in the field according to Troels-Smith (1955) and von Post and Granlund (1926), to chemical characteristics determined in the laboratory. A subsidiary aim was to attempt to correlate West Sedgemoor stratigraphy with the stratigraphy and assemblages given for the Northern Levels.

MATERIALS AND METHODS

Field Site

A location on the south western part of West Sedgemoor (ST 360 250), on land owned by the Royal Society for the Protection of Birds, was chosen for the siting of a peat sampling transect. Sites were located at 100m intervals along the 1200m transect, running from the edge to the centre of the moor. Peat stratigraphy was examined to a depth of 2m using an Ejkelkamp 3cm gouge auger and described using the Troels-Smith (1955) classification, which was adopted by the Soil Survey of England and Wales in their most recent national peat survey. The degree of peat humification was evaluated using the von Post ten point scale (von Post and Granlund, 1926). The high degree of humification of the plant remains necessitated microscopic identification of the floristic composition. Peat samples for chemical analysis were taken at depths of 25, 50, 100 and 150cm using a 7cm dutch auger. At each depth, five pH readings were taken directly on fresh material using a WPA portable pH meter.

Chemical Analyses

Subsamples of peat were air dried and then ground to pass through a 0.5mm mesh. 0.1 g portions of this fraction were digested using a modified micro-kjeldhal technique with concentrated sulphuric acid and hydrogen peroxide (Allen *et al*, 1974) and were subsequently used for the determination of total N, P, K, Na, Ca, Mg and Fe in the peat. N and P in solution were determined using automated versions of the colorimetric techniques of Crooke and Simpson (1971) and Murphy and Riley (1962) respectively. K and Na in solution were determined by flame emission spectrophotometry and Ca, Mg and Fe in solution by atomic absorption spectrophotometry using a Pye Unicam SP9. Total S in the peat was determined directly by X-ray spectrophotometry. Loss on ignition (LOI), determined at 450° C for 4 hours, was used as an index of soil organic matter content.

RESULTS AND DISCUSSION

Field Stratigraphy

The detailed stratigraphy for the transect is given in Figure 1. Each peat sampling location was described in the field by three standard organic deposit descriptions: (i) von Post degree of humification, (ii) Troels-Smith sediment description system and (iii) floristic composition. Several problems arise in the use of these descriptors in the field. The von Post degree of humification classification was designed for the description of fresh *Sphagnum* peats and thus may be limited in its application to sedge and reed peats (Puustjarvi and Robertson, 1975). In the case of the Troels-Smith classification, both the environment of formation and the identifiable plant remains are considered. The major problem in this study was the predominance of amorphous peat material, which precluded the easy identification of plant remains in the field. In view of these problems, four field categories were assigned to the peat of West Sedgemoor, using a combination of the above data (Table 1).

The upper stratigraphic sequence surveyed in this study can be tied into the Somerset Levels peat survey published by the Soil Survey of England and Wales (1980). This survey indicates that in West Sedgemoor 560 - 590cm of peat have developed over the blue-grey marine clays which line the basin, indicating that it is one of the deepest peat deposits in the Somerset Levels with about 1.5m more peat than the Shapwick and Meare Pool sites surveyed by Godwin (1955). Figure 2 illustrates the relation of surveyed peat stratigraphy to the West Sedgemoor basin and the underlying clays.

Although neither absolute dates nor the full stratigraphic sequence are yet available for West Sedgemoor, it seems likely that, as in the central parts of the Somerset Levels, the surface layers of the blue-grey estuarine clay correspond to the boundary between Godwin's (1948) pollen Zones VI and VII; the Boreal/Atlantic boundary. This boundary corresponds to the base of Beckett and Hibbert's (1979) regional pollen zone A and has been dated at around 5600±70 BP (Beckett, 1979a). A gradual change in conditions can be identified in the top 2m of peat stratigraphy. The sequence of deposits depicted in Figure 1 and summarised in Figure 2, indicates three possible stages in the later

Figure 1. Stratigraphic cross-sectional transect for West Sedgemoor using the Troels-Smith notation (Troels-Smith, 1955).

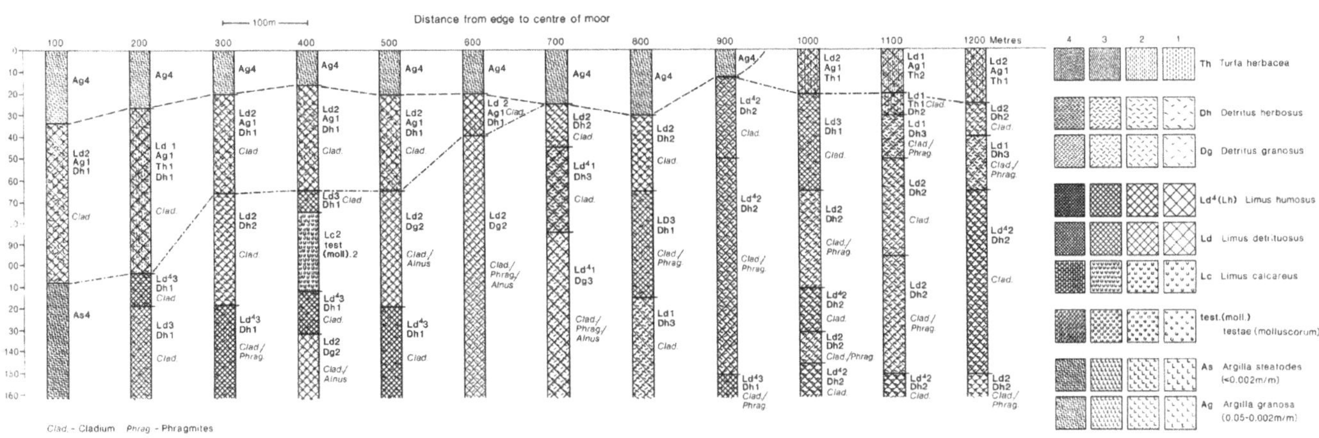

Please note that a full-size version of this figure is available to download from www.barpublishing.com/additional-downloads.html. The original foldout has been reduced in size to match the A4 format of this book, the image is therefore not as clear as the original foldout. The image might also be cut off or not complete in the printed book. Please refer to the original foldout via the download for the original content.

development of the moor. These three stages represent: (i) wet, increasingly freshwater reedswamp conditions, with *Phragmites communis* and, in shallower waters, *Cladium mariscus*, bathed in base-rich ground waters, (ii) slightly drier conditions around the moor peripheries, with *Alnus* becoming established up to 800m into the moor and representing fen-carr colonisation and (iii) a return to slightly wetter conditions with *Cladium* as the dominant species, leading up to the current vegetation, a base-rich meadow of grasses and sedges. These sequences are clearly very similar to those reported by Alderton (1983) for sites on King's Sedgemoor where *Cladium* peat is also common and is associated with *Phragmites* and *Alnus*. As in King's Sedgemoor, the West Sedgemoor peats are interdigitated with and overlain by clay lenses. In both King's Sedgemoor and West Sedgemoor, the peat is overlain by clay topsoils. Both sites have remained under the influence of high water tables throughout their history, indicating that rises in water level have kept pace with sediment deposition (Alderton, 1983).

Table 1. Characteristics of Four Categories of Peat Deposits Identified by Field Survey.

Group	Von Post Degree of Humification	Troels-Smith Category	Troels-Smith Description	Floristic Composition
1	HP(1-3)	Ld/Ag/Th	humose/loamy peat	not identifiable
2	M(4-6)	Ld/Th/Dh	herbaceous peat	*Cladium*
3	A(7-10)	Ld/Dg Ld4/Dh	herbaceous/woody peat	*Cladium* *Phragmites* *Alnus*
4	ZC/C	Ag/As	silty clay, silty loam	-

Key for peat groups is given in Figure 3.
Key for von Post categories is given in Figure 3.
Key for Troels-Smith categories is given in Figure 1.
There were only two cases of fibrous peat (F) and these were dominantly herbaceous detritus (Ld/Dh), containing *Cladium* and *Phragmites* remains.

The King's Sedgemoor and West Sedgemoor stratigraphy indicate that in the southern parts of the Somerset Levels, the transition from saltmarsh to freshwater fen and then to fen-carr probably occurred much later than in northern parts of the Levels. Beckett and Hibbert (1979) point out that at the Abbot's Way site fen-carr replaced open fen below the base of their pollen diagram and certainly before their lowest date of 4800 BP. They suggest that the earliest development of fen-carr in the Somerset Levels occurred around the Abbot's Way and migrated outwards. Some localised areas of the northern fens, as well as south of the Polden Hills, remained open and free of carr development, as indicated by Hibbert (1977) at Walton Heath. The further development in and around the northern Levels of mixed deciduous woodland, dominated by oak, elm and some alder, with

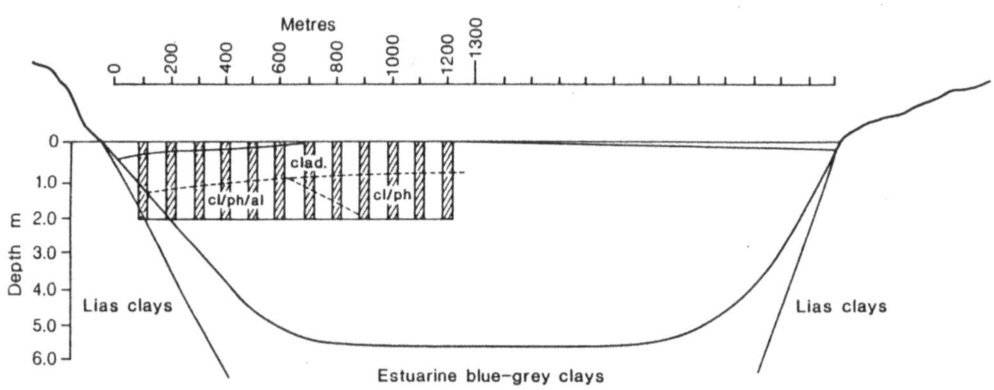

Figure 2. Schematic cross-section of West Sedgemoor, illustrating the relationship of the surveyed section to the valley clays.

phases of cultivation, cannot be corroborated south of the Polden Hills since pollen data are not yet available. The raised bog development characteristic of the central Levels around Shapwick and Meare Heaths, first identified by Godwin (1955) and subsequently studied in greater detail (eg Beckett, 1979a), is not present in West Sedgemoor.

At a distance of 400m on the West Sedgemoor transect (Fig 1), the *testae molluscorum* deposit occurs at a depth of 100cm, within stratigraphies dominated by *Cladium*. It is a freshwater assemblage, dominated by *Lymnaea peregra*, indicating well aerated, probably flowing water, conditions all the year round at this time in the fen to fen-carr transition.

Peat Chemistry

The field determined pH values ranged from 6.0 to 6.5, with little variation between peat type categories.

Figures 3 and 4 illustrate the distribution of total elements in the peat section surveyed. As expected, trends in both LOI and C:N ratio closely resemble the categorisation of peat outlined earlier, with highest values in category 3, the highly humified, herbaceous, woody peat and lowest values in category 1, the humose peat. Category 4, the mineral group, is lowest overall in LOI and C:N ratio.

Distributions of total P, K, Mg and Fe show the same trend (Fig 4), of decreasing values with depth and distance into the moor. The lowest values were recorded for category 3 and the highest values for category 1 of the peat types. Mineral deposits in category 4 are highest overall.

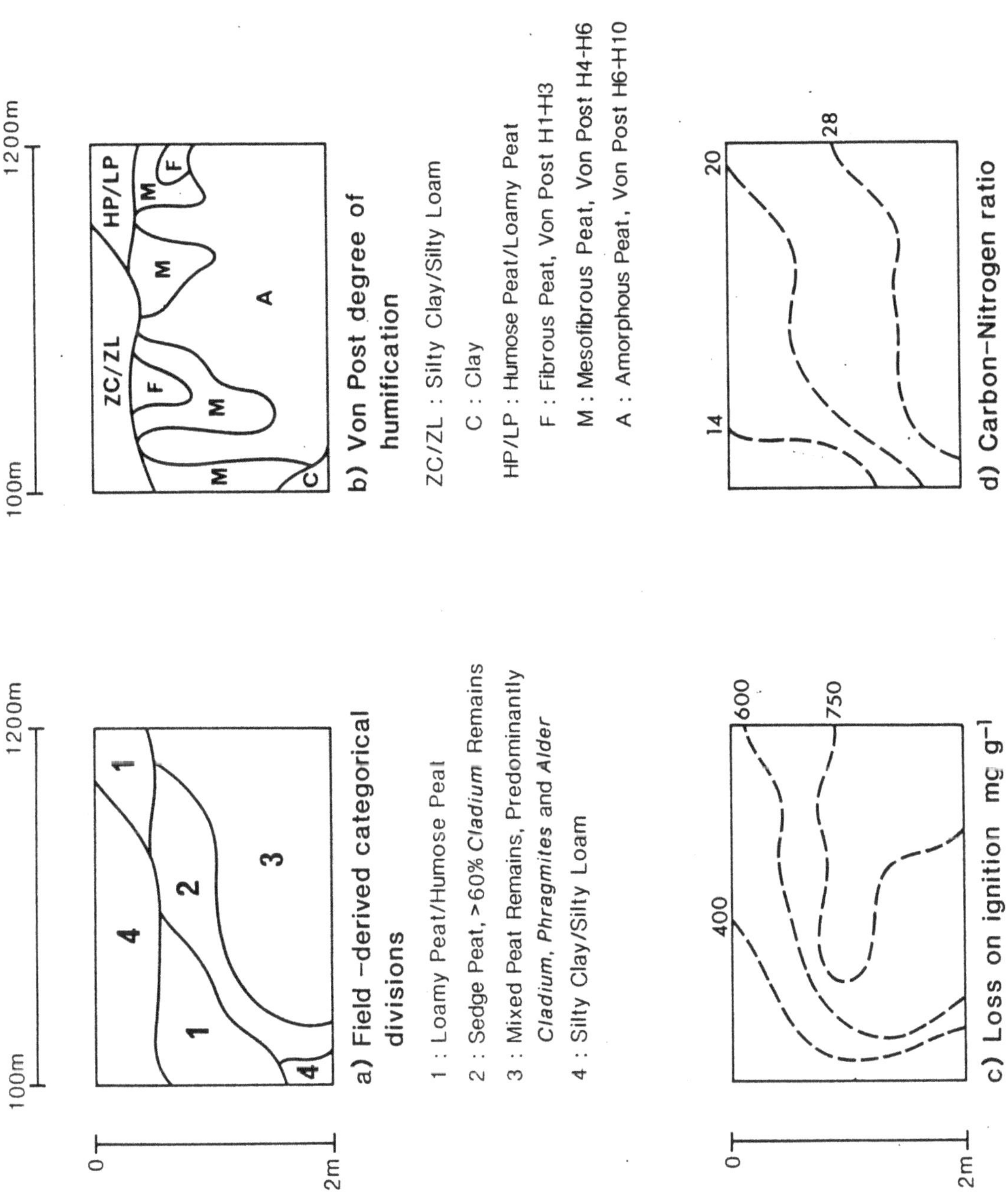

Figure 3. *Schematic cross-section of West Sedgemoor, illustrating peat humification and organic characteristics.*

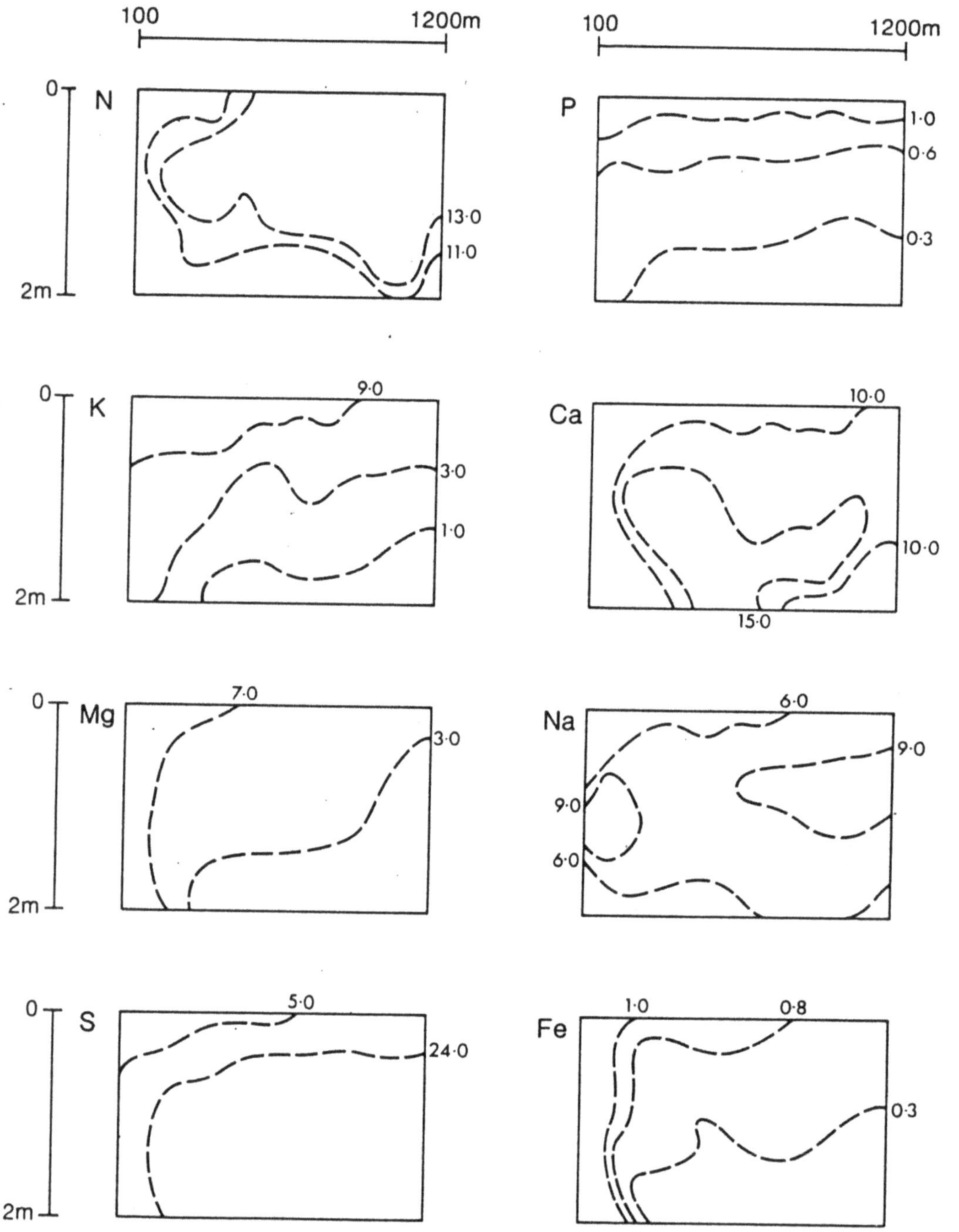

Figure 4. Schematic cross-section of West Sedgemoor, illustrating total peat content of N, P, K, Ca, Mg, Na, S and Fe. (all isopleths calibrated in mg/g)

Less clear patterns are seen for the other chemical elements. Further discussion of these and other chemical analyses are being prepared for publication elsewhere.

Multivariate Classification

In view of the doubts expressed about the use of qualitative peat descriptions in field survey, the total chemical data, together with the LOI values, were used to classify the peat types quantitatively. An hierarchical, agglomerative clustering method was used (Gower, 1967) to produce the dendrogram depicted in Figure 5. The similarity levels at which clusters were fused were calculated using a centroid algorithm in GENSTAT (Alvey *et al*, 1977).

Three clear groups are identified by the clustering technique based on total chemical composition of the peat (Fig 5). There are three outlier samples, all belonging to the core taken at 100m into the moor which has mineral sediment influences throughout. The clearest division in the dendrogram is between purely organic deposits and those deposits incorporating substantial mineral material; these two broad groups being fused only at the low similarity level of 25 per cent. Overall, there is less relationship between dendrogram clusters and Troels-Smith classes than there is with simple floristic composition.

Two main groups of organic deposits are distinguished in the dendrogram: (i) an amorphous and mesofibrous peat group in which *Cladium* is the only clearly identifiable species and (ii) an amorphous and fibrous peat group in which *Cladium* is associated with *Phragmites* and *Alnus*. One might anticipate nutritional differences with ecological change and expect higher levels of chemical elements in sediments with greater diversity. The opposite is true in this case. *Cladium* dominated sediments contain higher levels of P, K, Mg and Fe, with lower C:N ratios, indicating a more advanced degree of mineralisation than in sediments containing *Phragmites* and *Alnus*. There may be several explanations for this finding. The simplest explanation is that the *Cladium* deposit was originally more species diverse, probably with at least some *Carex* species present, but that other species remains are no longer identifiable. A second and equally probable explanation is that chemistry is a function of depth in West Sedgemoor: the *Cladium* deposits, nearer the surface of the moor, being influenced by chemical inputs to the moor in surface runoff and sediments during flooding. The seasonal and artificial fluctuations of the groundwater table due to local drainage schemes may also change the peat chemistry, particularly in the surface 1.5-2.0m.

CONCLUSIONS

Peat stratigraphy in West Sedgemoor indicates that the vegetation transition from saltmarsh to freshwater reedswamp and then to fen-carr occurred much later than in central and northern parts of the Somerset Levels. The smaller range of peat types identified in West Sedgemoor suggest a simple three-stage vegetation transition from *Phragmites* and *Cladium* dominated reedswamp to a drier fen-carr where *Alnus* was important around the moor peripheries and then back to a *Cladium* dominated sedge fen, terminating in the current base-rich meadow of grasses and sedges. The raised bog development characterised by *Calluna*, *Eriophorum* and *Sphagnum*, which is present on Shapwick Heath and Meare Heath, does not occur in West Sedgemoor.

In this study, the Troels-Smith description of sediments was initially thought to have advantages over other methods of description for several reasons: (i) its ease of use in the field, (ii) sediments are frequently mixtures of materials and (iii) its facility in representing sediments diagrammatically (Birks and Birks, 1980). At first sight the strength of the technique for organic deposits would appear to be in combining

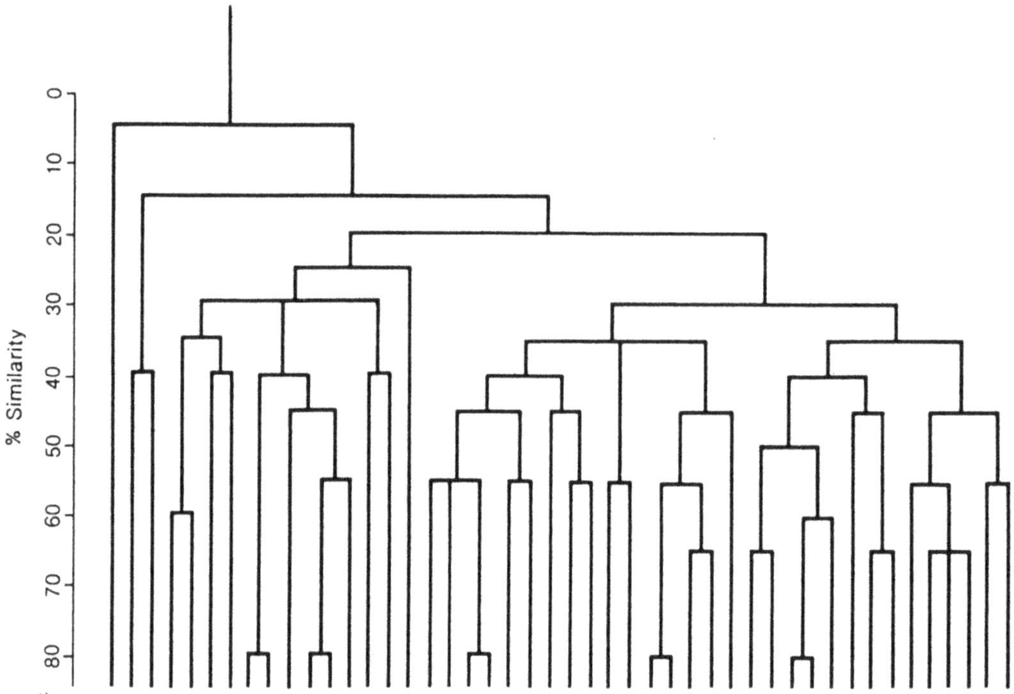

FIELD STRATIGRAPHY

Category	Dist. into moor (m)	Depth of core (cm)	Troels-Smith Classification	Floristic Composition	von Post Class
4	100	25	Ag 4		ZC
1	100	50	Ld 2 Ag 1 Dh 1	Cl	5
1	100	100	Ld 2 Ag 1 Dh 1	Cl	5
1	200	25	Ag 4		ZC
2	300	25	Ag 4		ZC
3	600	25	Ag 4		ZC
1	1100	25	Ld 1 Ag 1 Th 2		HP
3	200	50	Ld 1 Ag 1 Th 1 Dh 1	Cl	9
1	300	50	Ld 2 Ag 1 Dh 1	Cl	5
3	500	25	Ag 4		ZC
3	900	25	Ag 4		ZC
3	1000	25	Ld 2 Ag 1 Th 1		HP
2	1200	25	Ld 2 Ag 1 Th 1		HP
3	400	25	Ag 4		ZC
2	700	25	Ag 4		ZC
3	800	25	Ag 4		ZC
2	200	100	Ld 1 Ag 1 Th 1 Dh 1	Cl	9
3	800	150	Ld 1 Dh 3	Cl	8
2	900	150	Ld^{4}3 Dh 1	Cl /Ph	9
2	1000	150	Ld^{4}2 Dh 2	Cl	9
3	500	100	Ld 2 Dg 2	Cl /Al	8
2	1100	50	Ld 1 Dh 3	Cl /Ph	3
2	200	150	Ld 3 Dh 1	Cl	7
3	1100	100	Ld 2 Dh 2	Cl	8
1	1200	50	Ld 2 Dh 2	Cl	3
1	600	150	Ld 2 Dg 2	Cl /Ph/Al	7
3	700	150	Ld^{4}1 Dg 3	Cl /Ph/Al	10
1	300	150	Ld^{4}3 Dh 1	Cl /Ph	9
2	500	50	Ld 2 Ag 1 Dh 1	Cl	6
3	400	150	Ld 2 Dg 2	Cl /Al	7
3	500	150	Ld^{4}3 Dh 1	Cl	9
2	1200	150	Ld 2 Dh 2	Cl /Ph	7
3	300	100	Ld 2 Dh 2	Cl	4
2	1100	150	Ld^{4}2 Dh 2	Cl	7
3	400	50	Ld 2 Ag 1 Dh 1	Cl	3
2	1000	100	Ld 2 Dh 2	Cl /Ph	7
4	1200	100	Ld^{4}2 Dh 2	Cl	9
4	600	100	Ld 2 Dg 2	Cl/Ph/Al	7
4	700	50	Ld 2 Dh 2	Cl	9
4	700	100	Ld^{4}1 Dh 3	Cl	10
4	600	50	Ld 2 Ag 1 Ph 1	Cl	7
1	900	50	Ld^{4}2 Dh 2	Cl	9
1	900	100	Ld^{4}2 Dh 2	Cl /Ph	9
4	1000	50	Ld 3 Dh 1	Cl	5
4	800	50	Ld 2 Dh 2	Cl	6
4	800	100	Ld 3 Dh 1	Cl /Ph	8

Figure 5. Hierarchically clustered dendrogram, using a centroid similarity measure, expressed as a percentage. Key for Troels-Smith categories is given in Figure 1. Key for von Post categories is given in Figure 3. Key for floristic composition: Cl = Cladium, Ph = Phragmites, Al = Alnus.

information about peat humification and plant remains. Problems arise in predominantly amorphous peat deposits. In this study, differentiation into four qualitative categories involved microscopic identification of floristic remains indicating the limitations in the field use of the Troels-Smith system in these types of organic deposits.

By classifying the deposits according to quantitative chemical variables three distinct categories were identified. These categories were not clearly related to Troels-Smith types directly but bore closer resemblance to von Post values and to floristic composition independently. This would suggest that the value of the Troels-Smith classification in combining humification and floristic composition information is limited in amorphous fen deposits such as those in West Sedgemoor. Its efficiency and relation to peat chemical composition in more diverse materials such as those of the Central Somerset Levels is yet to be evaluated.

The use of chemical data in differentiating peat types must be interpreted not simply in relation to degree of humification and floristic composition but also in the context of contemporary environmental factors. In West Sedgemoor these influences are (i) catchment run-off into the moor, thought to be influenced by the geochemistry of the surrounding Keuper Marl, (ii) seasonal flooding and (iii) control of peat hydrology by local pump drainage schemes, leading to alteration of the ground water table and its dynamics. In West Sedgemoor, these factors combine to increase levels of most elements at the surface and peripheries of the moor, perhaps masking the original chemical characteristics of the peat.

ACKNOWLEDGEMENTS

Part of this work was funded by a NERC Studentship (ALH). Thanks are also due to Simon Godden, Tony Philpott and Steve Kelly of the University of Bristol, Phil Turton for field assistance and John Humphreys of the RSPB. Vanessa Straker and Caroline Ellis kindly helped in macrofossil identification.

REFERENCES

Alderton, A.M. 1983. The Sedgemoor Survey 1982: Environmental Results. *Somerset Levels Papers* **9**, 9-15.

Allen, S.E., Grimshaw, H.M., Parkinson, J.A. and Quarmby, C. 1974. *Chemical Analysis of Ecological Materials* Oxford: Blackwell Scientific Publications.

Alvey, N.G. *et al.*, 1980. *GENSTAT A General Statistical Program.* (Lawes Agricultural Trust, Rothamsted Experimental Station). Oxford: Numerical Algorithms Group Ltd.

Beckett, S.C. 1978. Palaeobotanical investigations at the Difford's 1 site *Somerset Levels Papers* **4**, 101-106.

Beckett, S.C. 1979a. The environmental setting of the Sweet Track, drove site *Somerset Levels Papers* **5**, 75-83.

Beckett, S.C. 1979b. Pollen influx in peat deposits: values from raised bogs in the Somerset Levels, South-Western England *New Phytologist* **83**, 839-847.

Beckett, S.C. and Hibbert, F.A. 1979. Vegetational change and the influence of prehistoric man in the Somerset Levels *New Phytologist* **83**, 577-600.

Birks, H.J.B. and Birks, H.H. 1980. *Quaternary Palaeoecology* London: Edward Arnold.

Crooke, W.M. and Simpson, W.E. 1971. Determination of ammonium in kjeldahl digests of crops by an automated procedure *J. Sci. Fd. Agric.* **22**, 9-10.

Findlay, D.C. 1965. *The Soils of the Mendip District of Somerset* Harpenden: Soil Survey of England and Wales (Memoir).

Godwin, H. 1941. Studies of the post-glacial history of British vegetation. VI correlations in the Somerset Levels. *New Phytologist* **40**, 108-132.

Godwin, H. 1948. Studies of post-glacial history of British vegetation. X correlation between climate, forest composition, prehistoric agriculture and peat stratigraphy in Sub-Boreal and Atlantic peats in the Somerset Levels *Phil. Trans. Roy. Soc. London B* **233**, 275-286.

Godwin, H. 1955. Studies of the post-glacial history of British vegetation. XII The Meare Pool region of the Somerset Levels *Phil. Trans. Roy. Soc. London B* **239**, 161-190.

Gower, J.C. 1967. A comarison of some methods of cluster analysis. *Biometrics* **23**, 626-637.

Hibbert, F.A. 1977. Peat stratigraphy and pollen analysis from Walton Heath *Somerset Levels Papers* **3**, 30-31.

Murphy, J. and Riley, J.P. 1962. A modified single solution method for the determination of phosphate in natural waters *Anal. Chim. Acta* **27**, 31-36.

Orme, B.J. 1982. The use of radiocarbon dates from the Somerset Levels. *Somerset Levels Papers* **8**, 9-25.

Puustjarvi, V. and Robertson, R.A. 1975. Physical and chemical properties of peat *in* D.W. Robinson and J.G.D. Lamb (eds) *Peat in Horticulture* London: Academic Press.

Soil Survey of England and Wales 1980. *Thickness of Peat in the Somerset Moors* (Part of Ordnance Survey Map Sheets No. 182 and 193, 1:50,000). Harpenden: Lawes Agricultural Trust.

Troels-Smith, J. 1955. Karakterisering af lose jordarter. [Characterization of Unconsolidated Sediments.] *Danm. geol. Unders.* **6 Bd.3, Nr.10.**

von Post, L. and Granlund, E. 1926. Sordra Sveriges torvtillgangar I *Sveriges geol. Unders.* **C335**.

THE HINTERLAND OF ILCHESTER: ARCHAEOLOGY, ALLUVIATION AND THE ENVIRONMENT

Peter Leach

INTRODUCTION

The Somerset wetlands as a whole form an immense resource, with an environmental and archaeological potential which is by no means exhausted, as continuing research demonstrates (*Somerset Levels Papers* 1975-1986; McDonnell, 1980). Lying immediately above the Levels proper, and mainly to the south, a system of river valleys drain into them and play a major role in the creation and maintenance of the wetland environment (Fig. 1).

Above Langport, these valleys, the headwaters and tributaries of the River Parrett, replicate much of the character and regime of the Levels. The region is essentially an inland basin comprising broad floodplains separated by low islands or promontories of higher ground, subject to frequent, periodic inundation, and mainly lying below 50 feet (*c* 15m) OD. Despite its similarity to and relationship with the Levels, this floodplain basin has its own distinctive character, and one which is to a great extent shared more widely with some river bottom environments elsewhere in lowland Britain (Fig. 1).

Major river valleys have long been recognised as foci for prehistoric and later settlement, notably through utilisation of the often extensive and well drained environments provided by terrace gravel deposits (eg Benson and Miles, 1974). Until quite recently the potential of overbank alluvial deposits comprising silts, loams and clays in the valley bottom, at least as areas with any significant potential for settlement, has been largely overlooked, if not positively discounted by archaeologists. Recent and increasingly widespread demonstrations that this is not necessarily the case, allied to a potential for good, well stratified environmental preservation, is focussing fresh attention upon such areas and the sites which they contain (eg Robinson and Lambrick, 1984; Limbrey, 1983). The value of archaeological sites preserved in such environments is further enhanced by their role in elucidating the palaeohydrology of floodplains; and thus for the archaeologist the provision and refinement of yet another tool to guage the impact of man upon the post-glacial environment (Bell, 1981: 75-9).

THE YEO VALLEY

At and below Ilchester, the Yeo valley is characterised by a broad, level floodplain, mantled by extensive deposits of alluvium. Today these support an agricultural regime based upon exceptionally rich and productive valley pastures, much of it permanent. Ploughing is rarely undertaken, and opportunities for the examination of underlying depositional sequences or buried horizons are generally few. Archaeological excavation has provided one such opportunity, and data arising from major projects undertaken in 1974 and 1975 at Ilchester first focussed attention upon some of the problems as well as the potential of floodplain alluviation, and its relationship with the history of agricultural exploitation in this region (Leach, 1982: 59-60). Occupying a slight "island" towards the head of the broad lower Yeo valley floodplain (Fig. 2), Ilchester's archaeology is bound up with the palaeohydrology and sedimentology of this valley (Thew, forthcoming).

The recent geomorphological history of the Yeo valley as discerned so far in occasional deeply excavated exposures through the valley bottom depositional sequence is outlined as follows.

1) The valley floor, cut into Lower Lias shales and limestones, is mantled by uneven deposits of sands and predominantly chert/flint gravels at *c* 10m OD. They are generally between 0.5m and 1.5m thick, and are characterised by internal cross-bedding, upward

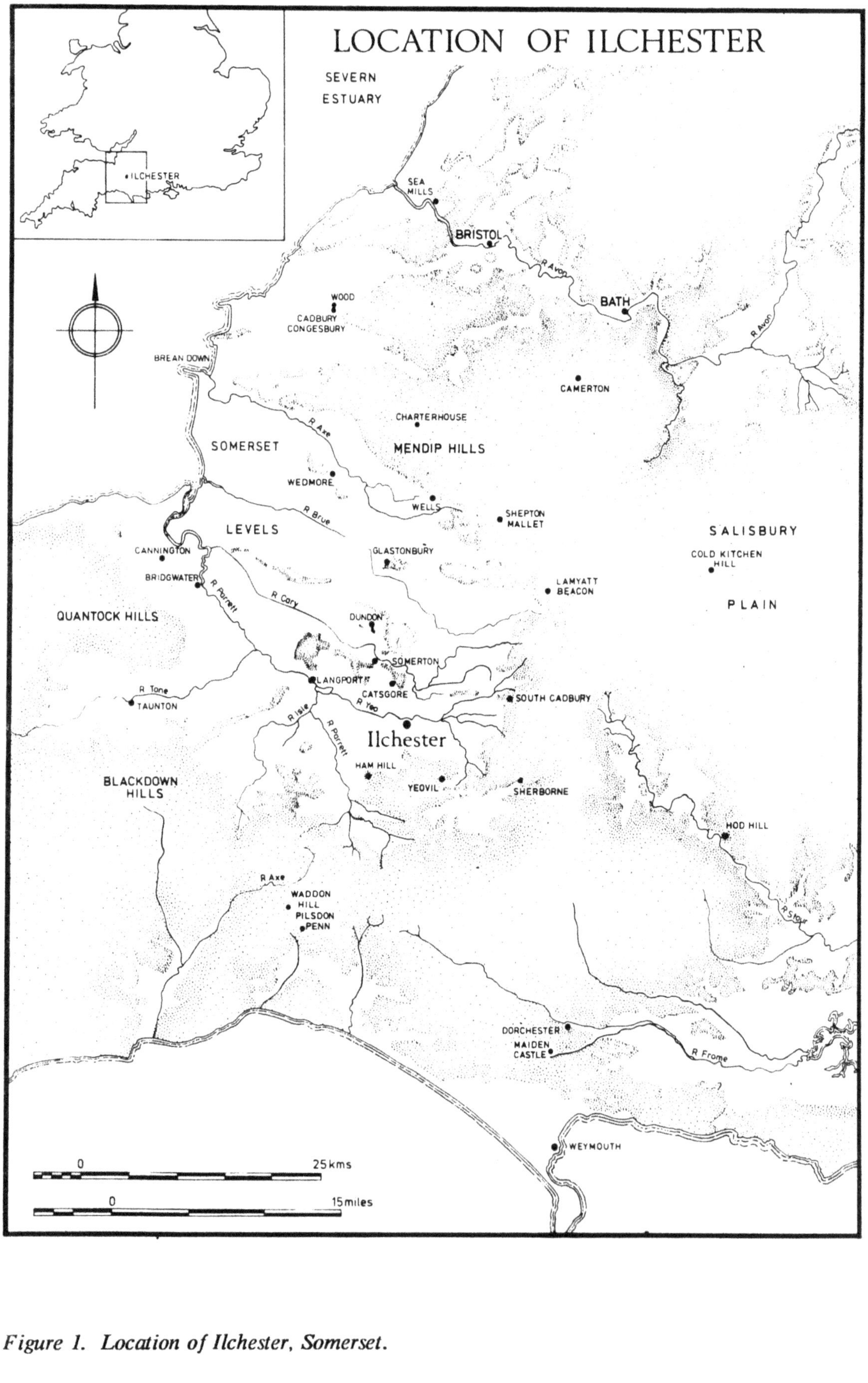

Figure 1. Location of Ilchester, Somerset.

grading, and occasional organic lenses. Such deposits probably originate under the periglacial conditions of a freeze-thaw regime, high and rapid seasonal meltwater discharges, and thus the movement and deposition of large quantities of coarse, unconsolidated sediment. Valley-wide reworking in a braided river channel would explain the uneven basal and upper contact planes of the sediments. In some other lowland river systems (eg Thames), similar terrace gravels originated during the Devensian glacial maximum (Gibbard *et al*, 1982).

2) An incised palaeochannel of the River Yeo has been detected beneath later alluvium, following a course mainly to the east and south west of Ilchester. This channel, *c* 200-300m wide and containing *c* 1.5m thickness of gravel, was eventually filled with waterlogged clays and silts, rich in molluscan and plant remains. Beneath the present course of the River Yeo lies another deeply incised channel cutting into both the valley bottom gravels and the underlying Lias. During this stage the river evidently breached a spur of the Pleistocene gravels between Northover and Ilchester to attain this course. These channels may represent the establishment of a meandering river regime with the ameliorating climate at the end of the Devensian.

3) Masking the coarse deposits and earlier geomorphology of the valley floor are extensive deposits of finer alluvial clays and silts, generally in excess of 1.0m thick. These appear to be mainly fairly homogeneous, exhibiting subtle variations in structure, texture and colour. Horizontal differentiation is most readily perceived relative to archaeological horizons or material occasionally incorporated therein. It is suggested that the bulk of this finer alluvium originated as Holocene deposition, largely reflecting human activities and impact upon the region.

During the early Flandrian sea levels rose virtually to their present day position. With the climatic amelioration came the development of a maximum forest cover by *c* 7000 BC, which would have resulted in minimal surface runoff and even greater groundwater flows. In the Yeo valley a relatively small, incised, gravel-bottomed, meandering stream can be postulated at this time. Sediment loads and flooding would be at a minimum, and thus relatively little deposition is likely (*cf* Blackwater River, Clarke and Dixon, 1981).

Progressive human interference with the forest/vegetation cover, leading eventually to the creation of pastoral and arable agricultural regimes, has a dramatic effect upon erosion, surface runoff and groundwater levels (Limbrey, 1978). The resulting increase in quantity and variability of surface discharge is commensurate with increased suspended silt loads, and ensuing sedimentation with greater likelihood of flooding. This process includes river channel aggradation which tends to promote further flooding and creation of both levee and overbank floodplain deposits. These effects and their results are reflected by the extensive valley bottom deposits, an overall rise in the valley floor level and consequent burial of the palaeochannel and the early Flandrian Yeo river channel. Today the river bed lies up to 2.0m higher than the original, wholly within the subsequent sediments.

A number of exposures have been documented, with archaeological material ranging from the early second millenium BC to the late medieval period, stratified within the fine sediment floodplain deposits. All the evidence so far points to most of the alluviation occurring during the Flandrian, subsequent to forest clearance in both the Yeo catchment and on the floodplain itself. The evidence from three sites in the Ilchester area is briefly reviewed, illustrating the above, and by analogy with observations and hypotheses derived from study areas outside this region, some more general implications for a history of man's impact upon this area become apparent.

King's Moor

The earliest site so far recorded lies downstream from Ilchester amid the extensive floodplain meadows of King's Moor (Fig. 2: A). The excavation of a new field drainage

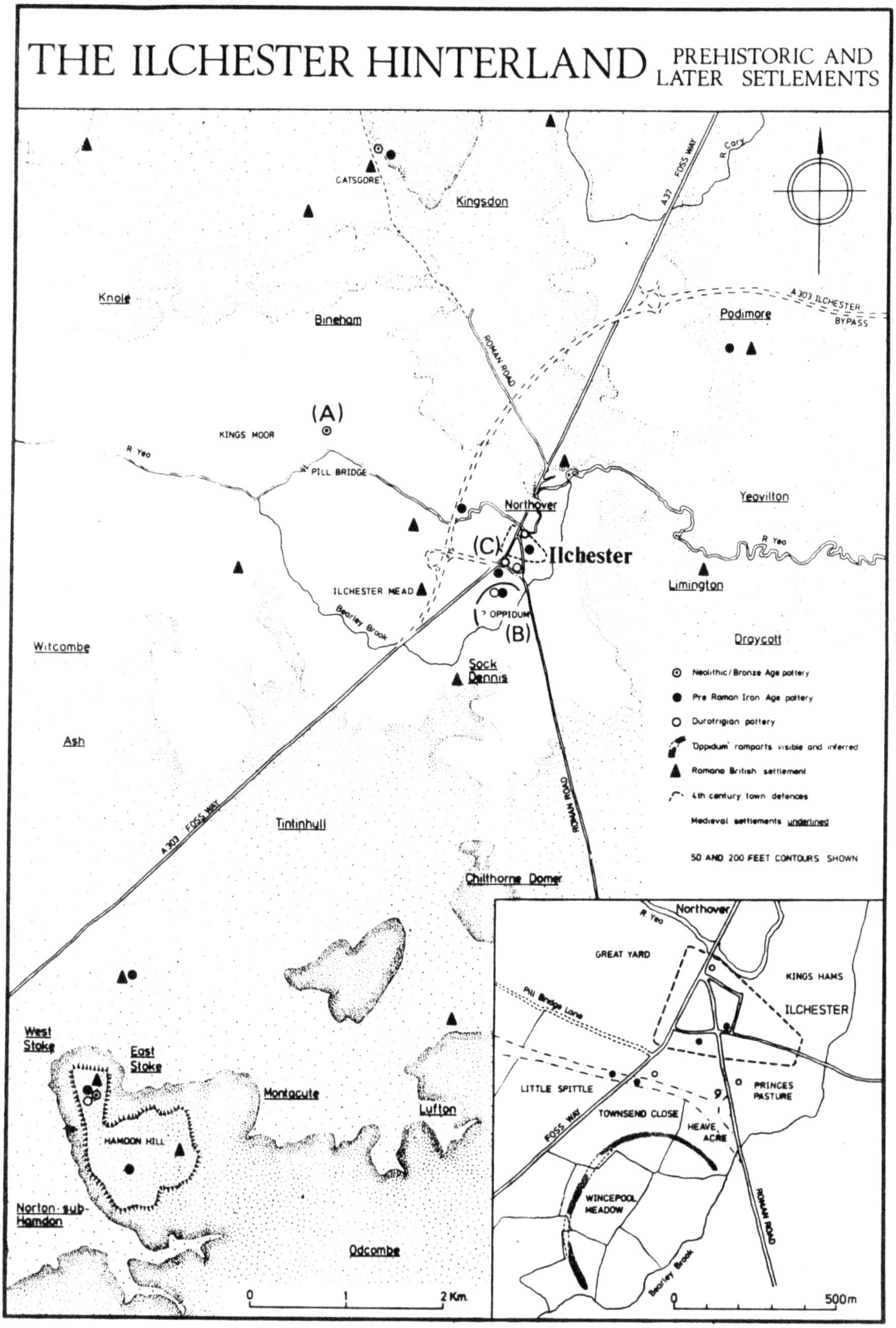

Figure 2. The Ilchester hinterland: prehistoric and later settlements; locations for Figure 3.

ditch (rhyne) in 1981 exposed an horizon of human activity sealed beneath approximately 0.80m of alluvium. The cleaning and re-cutting of these rhynes provide virtually the only opportunity for discoveries in such localities. A fortuitous observation of the cut by Nigel Thew in the course of a Bristol University undergraduate research project, enabled both the recovery of material and a record of the stratigraphy to be made (Leach, forthcoming).

Within the alluvium an horizon could be traced for at least 100m on either side of a more clearly defined zone of sandy clay and gravel spreads, ditch cuts and fills, and charcoal. This was associated with at least one stone-packed post-hole, some animal bone, and early Bronze Age pottery types (Fig. 3A). It appeared that a pre-second millenium BC alluvium was identifiable which had masked the underlying gravels before the site was utilised (higher undulations of gravel were observed nearby, cut by the rhyne). Interpretation was difficult because formal excavation was not possible and the full depth of alluvium was not exposed. A domestic character is indicated by an extensive spread of charcoal and some bone within the buried soil/occupation horizon, although at the point illustrated in Figure 3A the stratigraphy and the prehistoric funerary pottery suggest the remains of a round barrow. Examination of the alluvium above and below the early Bronze Age horizon showed the lower to be non-calcareous, extensively iron-stained, and with some pea grit, whereas the upper was non-calcareous, heavily gleyed, and little oxidised.

South Meadows

Aerial photographs taken during an episode of severe flooding at the end of 1979 drew attention to the segments of an intermittent semicircular bank standing just above the floodwater among the meadows south of Ilchester. Careful field survey and the coincident opportunities provided by excavations for public utility pipelines across the area revealed a very large man-made enclosure of at least 16 hectares (Fig. 2: B). Arising from these preliminary observations, a research programme, formulated and directed with Nigel Thew, was initiated in 1984 to attempt a more thorough evaluation of the site's archaeological and environmental potential.

Results so far suggest a monument of late Pre-Roman Iron Age date comprising an enclosure defined by a massive, though degraded clay bank, now largely obscured by subsequent alluviation. Its exterior is revetted by low drystone walls of lias limestone and Ham Hill stone, beyond which lies a broad, relatively shallow, concentric outer ditch. Within, activity and occupation has so far only been located in zones close to the rampart tail. Here, quantities of late Pre-Roman Iron Age pottery, daub, industrial slag, bone and building remains have been recovered from (as yet) very limited sampling excavations. To the south, the enclosure appears to have been defined by other means, possibly by a former channel of the River Yeo, now a modern water course - the Bearley Brook - also marking the approximate course of the now buried and obliterated palaeochannel (above) (Fig. 2).

Sections through the defences and interior reveal horizons of both construction and use, sealed by up to 0.50m of subsequent floodplain alluvium. At least a further 1.0m of deposits lie beneath, sealing, at variable depths, an uneven surface of the chert gravels (Fig. 3B). When archaeological material was not present detection of the old land surface was almost impossible. This may imply a virtual absence of soil formation under conditions of continuing alluviation. Once again the alluvium beneath the late Pre-Roman Iron Age horizon was more heavily iron-stained than the gleyed alluvium above it.

Physically this site bears closest comparison with the late Iron Age "Oppida" of the early first century AD, more familiar in the south east of England, a suggestion somewhat enhanced by the subsequent status and development of adjacent Roman Ilchester. No earlier dated horizons are yet recognised interstratified within the pre-Oppidum

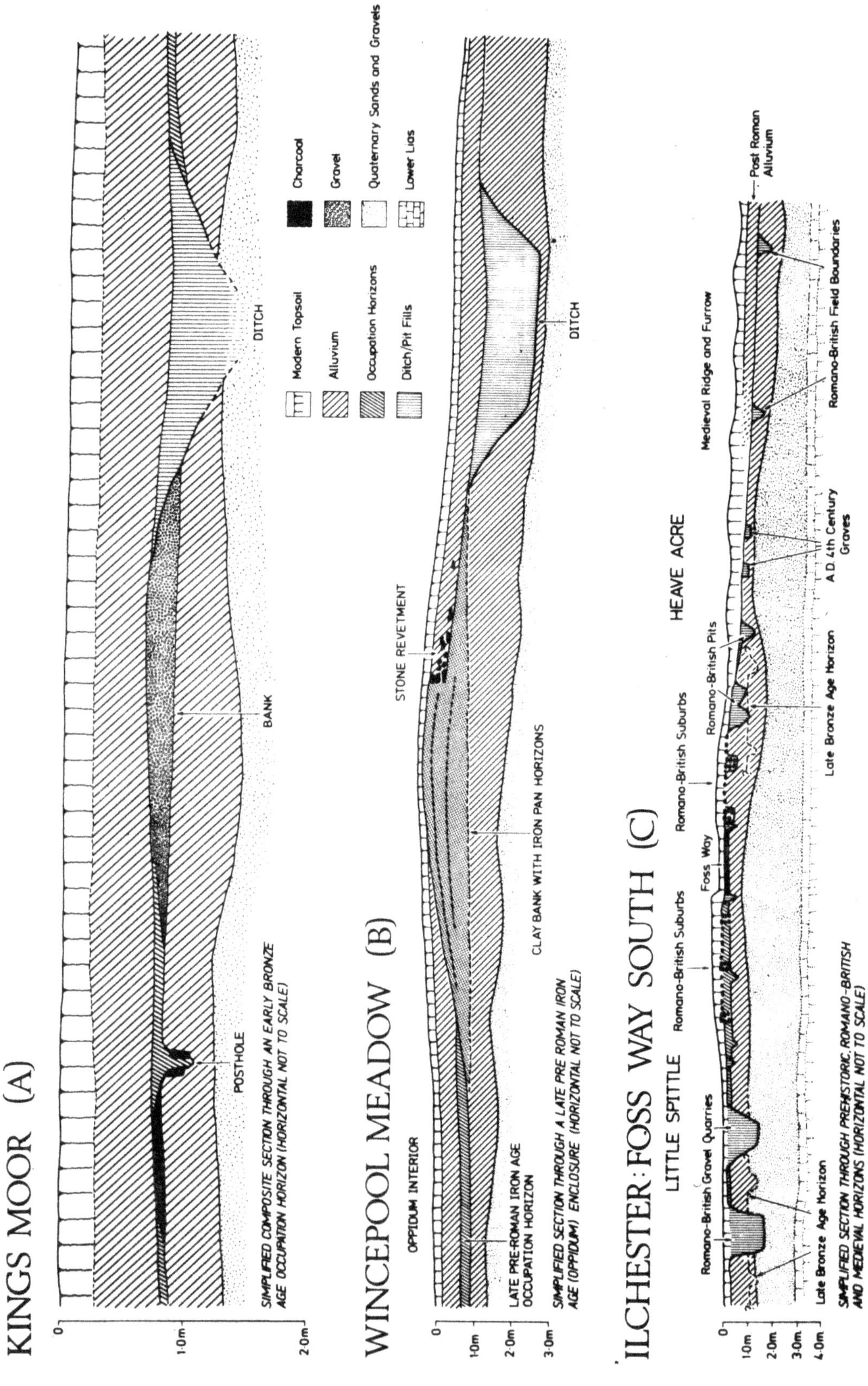

Figure 3. Ilchester, archaeology and alluviation, three examples:–
(A) Kings Moor section
(B) Whincepool meadow section
(C) Fosse Way South section

alluvium, but residual Romano-British and early medieval material is incorporated within the overlying alluvial strata (Leach, 1985 and forthcoming).

Foss Way South

Closer to Ilchester lie other archaeologically excavated sites where observations suggest further refinements to the depositional sequence. The most extensive observations were made in connection with the Ilchester Bypass where the potential of inter-relationships between archaeology, alluviual sedimentology and past environments were first recognised (Fig. 2: C). Excavations beside the Foss Way revealed traces of late Bronze Age occupation beneath the Romano-British suburbs, but separated from the later horizon by 0.5m or more of sterile alluvium (Leach, 1982: 59-60). Only the deeper excavation of Romano-British pits and field boundary ditches gave a clue to this earlier settlement, and no extensive investigation of its character or relationship with preceding deposits was possible. Once again, however, it is evident that occupation horizons, this time of the earlier first millenium BC, lie interstratified within the Holocene alluvium (Fig. 3C).

The relationship of later archaeology to subsequent regimes of alluviation and land use is also detectable in this area (Fig. 2: C). The location at Ilchester of a Roman military base, to be succeeded by a small but thriving Romano-British town, during the first four centuries AD, provides a wealth of archaeological evidence and many opportunities for relevant data collection.

Romano-British suburban developments alongside the Foss Way from the first century AD were matched by agricultural developments, probably involving an intensification in the utilisation of the floodplain. The effects of suspected widespread drainage and exploitation of the Somerset Levels at this time upon the Yeo basin are difficult to assess, but the layout of fields, bounded by drainage ditches, can certainly be demonstrated. Analysis of ditch fills and crop cleaning residues indicate not only the existence of hedges, but of arable farming perhaps more widespread than today, and, above all, the cultivation of spelt as the main cereal crop (Murphy, 1982).

Overall, the analysis of plant remains from field ditches indicated a wet and ill-drained environment, a suggestion also supported by the alluvial silting with which these boundaries were eventually choked. Post-Roman cultivation had truncated the Romano-British horizon in some areas, along with any overlying depositional sequence. Elsewhere, built-up areas, notably within the town or flanking the main roads, largely remained clear of all but the most severe flooding. A late first to early second century phase of desertion in some areas was emphasised through a brief phase of continued alluviation (Leach, 1982: 65).

The clearest demonstrations of Post-Roman alluviation were recorded in the vicinity of the Ilchester Mead villa, and east of the Foss Way (Fig. 2). In the latter area up to 0.30m of silt had obliterated Romano-British field boundaries and fourth century cemetery inhumations. Towards the top of the sequence medieval ceramics make their appearance, and the final seal upon the archaeological/alluviation cycle is set by the evidence of medieval and later cultivation, preserved in the form of ridge-and-furrow earthworks (Fig. 3C).

DISCUSSION

By the second millenium BC a considerable thickness of alluvium had accumulated at King's Moor. Human influence upon hydrological and alluvial regimes is now well established, but to what extent Mesolithic, Neolithic or Early Bronze Age activities, as opposed to natural agencies, were responsible for this deposit cannot yet be assessed. A local Neolithic presence at sites within the hinterland, eg South Cadbury (Alcock, 1972:

108-113), Ham Hill (Somerset County Museum Collections) or Catsgore (Ellis, 1983: 23) is certainly attested. Early Bronze Age activity is represented by sparse flint and pottery distributions and a scatter of round barrows and ring ditches, mainly on the limestone hills immediately to the north west of Ilchester. The less gleyed conditions of the early alluvium suggests at least seasonally low ground water in the valley bottom.

The clearance of woodland results in increasing, less evenly distributed surface run-off causing a rise in ground-water levels and an increase in the liklihood of periodic flooding of valley bottoms. The exposure of soil by arable farming greatly increases the potential of soil mobilisation by water, and thus of alluviation.

From the later Bronze Age and throughout the first millenium BC there are indications that settlement exploitation and the organisation of the landscape in Britain entered a new and more intensive phase (Bradley, 1978). Crop mark sites and pottery, the latter most often associated with the evolution of major hill-forts, and subsequent Romano-British settlements, all hint at a considerable expansion of population and the consequent intensification of agricultural exploitation in the Ilchester hinterland. An accumulation of 0.50m or more of alluvium separated first century AD Romano-British suburbs from underlying Late Bronze Age occupation horizons, while the more gleyed nature of the alluvium suggests a rise of the water table. It is likely that this was a consequence of the agricultural expansion. This effect may have been intensified further by a phase of climatic deterioration postulated during this period, whose effects included lower average temperatures and higher rainfall (Turner, 1981: 256-264).

The scale of this impact seems to have been maintained and probably intensified during the succeeding four centuries of Roman control. The settlement density and evident prosperity of this region during that time are well demonstrated by research, and was undoubtedly based upon agriculture (Leech, 1977). A reflection of this is once again found in the continuing alluviation cycle, resulting in partial obliteration through burial of some horizons during this period.

Post-Roman economic collapse is more difficult to chart with certainty through cycles of alluvial sedimentation. A drastic decrease in population and thus agricultural exploitation between the fifth and seventh centuries should provide some reflection in lower levels of alluviation. Any such effect may be masked by subsequent developments, while the virtual absence of readily dateable artefacts in this period further obscures the picture.

The last great upturn in arable cultivation prior to the Industrial Revolution, approximately between the nineth and thirteenth centuries AD, lies fossilised in the ridge-and-furrow earthworks of medieval field systems on the floodplain around Ilchester. The scale of this upturn and its intensification is also apparent in documentary references (eg Domesday ploughlands), or in the density of local, formerly medieval settlements. The corresponding alluviation of this phase obscured in turn portions of the Romano-British landscape, before a late medieval and subsequent reversion to pasture in the region brought the cycle of alluviation, if not the flooding, virtually to an end. This last phase is evidenced by the survival of unburied medieval earthworks and in the development of a well defined modern "A" horizon on the floodplain soils. Shrunken or deserted medieval settlements (eg Ashington, Bineham, Draycott, Mudford, Speckington, Sock Dennis, or Yeovilton), and, sometimes, documentary evidence may also reflect this phase.

Comparable depostional sequences are still rare in the literature. Herriot's Bridge in the Chew Valley, however, may be cited as a good and relatively local example illustrating later depositional sequences which can be linked with archaeologically detectable episodes in a small river valley (Rahtz and Greenfield, 1977: Fig. 36).

Further afield more extensive work in the Thames Valley provides a useful source for comparisons with the whole range of deposits encountered in the Yeo Valley (Robinson and Lambrick, 1984). Parallels between the two areas can be drawn in major phases of alluviation during the later pre-Roman Iron Age and Romano-British periods, and in late Saxon-early medieval times. Where a divergence appears most marked, however, is in the delayed commencement of Thames Valley alluviation relative to the Yeo Valley. Several factors may be relevant here, notably the predominance of much freer-draining interfluves, mainly limestone or gravels in the Thames Valley; a much larger catchment area; and the broadness of the sample. In the Yeo Valley a much smaller catchment area, a lower ground water/higher run-off potential, and more localised sampling, serve to emphasise better the effects of past landscape utilisation upon observed patterns of alluvial accumulation.

For an effective test of palaeohydrological response to land utilisation by man, the requirements seem to be a relatively small, well defined catchment with largely impermeable interfluves, in association with an adequate range and abundance of interstratified archaeological data. The Yeo Valley and, in particular, the immediate Ilchester hinterland, appears to provide an ideal testing ground.

Within the floodplain deposits the preservation of vertebrate, molluscan, pollen and waterlogged plant remains is locally very good. Their association with successive archaeological sites and defined horizons provides an invaluable opportunity for establishing a dated environmental sequence for the floodplain.

The principal aim of this brief resume has been to demonstrate the study potential and implications of work in the River Yeo floodplain. In presenting this review, I hope also to provide a stimulus for further research.

ACKNOWLEDGEMENTS

I am grateful to Dr Ann Ellison, Bruce Levitan, Mark Robinson and Nigel Thew for reading and commenting on this paper, and to Trevor Pearson for the illustrations.

BIBLIOGRAPHY

Alcock, L. 1972. *"By South Cadbury is that Camelot ..." Excavations at Cadbury Castle 1966-70* London: Thames and Hudson.

Bell, M. 1981. Valley sediments and environmental change *in* M. Jones and G. Dimbleby (eds) *The environment of man: the Iron Age to the Anglo-Saxon period* Oxford: British Archaeological Reports (British Series **87**), 75-79.

Benson, D. and Miles, D. 1974. *The Upper Thames Valley: an archaeological survey of the Thames gravels* Oxford: Oxford Archaeological Unit (Survey no. 2).

Bradley, R. 1978. *The prehistoric settlement of Britain* London: Routledge and Kegan Paul.

Clarke, M.R. and Dixon, A.J. 1981. The Pleistocene braided river deposits in the Blackwater Valley area of Berkshire and Hampshire, England *Proc. Geol. Ass.* **92**, 139-157.

Ellis, P. 1983. *Catsgore 1979* Bristol: Western Archaeological Trust (Excavation Monograph **6**).

Gibbard, P.L., Coope, G.R., Hall, A.R., Prece, R.L. and Robinson, J.E. 1982. Middle Devensian deposits beneath the "Upper Flood Plain" terrace of the River Thames at Kempton Park, Sunbury, England *Proc. Geol. Ass.* **93**, 275-289.

Leach, P.J. 1982. *Ilchester volume 1. Excavations 1974-5* Bristol: Western Archaeological Trust (Excavation Monograph **3**).

Leach, P.J. 1985. *A late Iron Age "Oppidum" at Ilchester, Somerset* Bristol: Western Archaeological Trust.

Leach, P.J. forthcoming. *Ilchester volume 2: excavations and field work to 1984.*

Leech, R.H. 1977. Romano-British settlement in South Somerset and North Dorset. Unpublished PhD. thesis, University of Bristol, Department of Classics and Archaeology.

Limbrey, S. 1978. Changes in the quality and distribution of the soils in lowland Britain. *in* S. Limbrey and J.G. Evans (eds) *The effect of man on the landscape: the lowland zone* London: Council for British Archaeology (Research Report **21**), 115-123.

Limbrey, S. 1983. Archaeology and palaeohydrology *in* K.J. Gregory (ed) *Background to palaeohydrology* London: Wiley, 189-212.

McDonnell, R. 1980. The Upper Axe Valley, an interim statement *Somerset Archaeol. Natur. Hist.* **123**, 75-82.

Murphy, P. 1982. Plant remains from Roman deposits at Ilchester *in* P.J. Leach *op cit*, 286-290.

Rahtz, P.A. and Greenfield, E. 1977. *Excavations at Chew Valley Lake, Somerset* London: H.M.S.O. (Department of the Environment Archaeological Report **8**).

Robinson, M.A. and Lambrick, G.H. 1984. Holocene alluviation and hydrology in the Upper Thames basin *Nature* **308(26)**, 809-814.

Thew, N. forthcoming. The geological setting of the Yeo valley and Ilchester *in* P.J. Leach *op cit.*

Turner, J. 1981. The Iron Age *in* I. Simmons and M. Tooley (eds), *The environment in British prehistory* London: Duckworth, 250-281.

DOZMARY POOL, BODMIN MOOR, CORNWALL: A NEW RADIOCARBON DATED POLLEN PROFILE

I.G. Simmons, J.I. Rand and K. Crabtree

This paper reports on a part of the work carried out by one of the authors (JIR) while employed on an SERC research contract to the other two authors to study "man and the environment in the terminal Mesolithic of the Highland zone". Already published in this context is a reappraisal of the Blacklane site on Dartmoor (Simmons *et al*, 1983). The record of the Mesolithic microliths and the earlier pollen work of Conolly *et al* (1950) and Brown (1977) suggested to the authors that Dozmary Pool would be a good location to try to relate pre-elm decline vegetation changes to possible Mesolithic influences.

THE SITE: PREVIOUS WORK

Dozmary Pool is a small relatively shallow lake at 265m above OD on the southern side of Bodmin Moor (Grid Ref SX 192744). The pool lies in a shallow col between Fowey and St Neots Rivers; it is approximately 18ha in extent and has a catchment of about 40 ha. The maximum elevation of the catchment basin is only about 285m above OD and hence the slopes are relatively shallow. There is a small outlet stream at its south-eastern edge but no clear inflow stream. On its south-western shore there is a raised mire with a vegetation community of *Eriophorum vaginatum*, *Rhynchospora alba*, *Erica tetralix*, *Calluna vulgaris*, *Molinia caerulea*, *Juncus effusus* and *Narthecium ossifragum*. Cores taken by Brown were from this mire area as were the cores used in the present study.

Wainwright (1960) describes the microlith industries found at Dozmary, mostly in the nineteenth century, and now in museum collections. He notes some 2500 flints in Plymouth museum and various others scattered around other museums which are ascribed to Dozmary Pool or Dozmary Pool area but none are from a stratigraphic context. Although Wainwright gives no indication of absolute age to the Mesolithic flints, Brown (1977) suggests that they may be from a period when he records burnt peat within the stratigraphy just prior to the elm decline, ie relatively late Mesolithic. However, Jacobi (1979) in a reassessment of the Mesolithic flints of south west England, links the Dozmary microlith assemblage with Thatcham in the Kennett valley which would suggest an early eighth millennium date, ie very early Mesolithic. This, says Jacobi, places the Dozmary flints in a period with *Empetrum* heath and juniper scrub on the hillslopes prior to the spread of birch woodlands. His view locates the Dozmary microlith assemblage as very much earlier in the profile than suggested by Brown (1977).

The first pollen work from Dozmary Pool was published by Conolly *et al* (1950) and recorded pollen from 90cm to 230cm of a 270cm profile. The lowest deposits, with a *Corylus* maximum, high *Betula* and low *Quercus* and *Ulmus* and no *Alnus* and *Tilia*, were placed in pollen zone V of the Godwin scheme. The authors noted that layers of ash and charcoal in the deposit correspond to sharp temporary disturbances in the run of tree pollen. It is unclear from their limited diagram how these layers relate to the elm decline.

Brown (1977) in his profile description records the presence of carbonised fragments and "black layers" down to depths of 207cm at a level radiocarbon dated to 7925±100 BP (Q-1023). More distinctive signs of burnt material occur at a level when bracken spores show a big increase from horizons dated to 6793±70 BP (Q-1024) and a major fire horizon with burned bases of *Eriophorum vaginatum* occurs just prior to the elm decline.

The pollen diagram presented by Brown (1977) covers the levels 180cm to 238cm in detail, radiocarbon dated to 6451±75 BP (Q-1025) and 9053±120 BP (Q-1021). Hence his detail covers the early and middle Mesolithic periods only. It does show the maintenance of

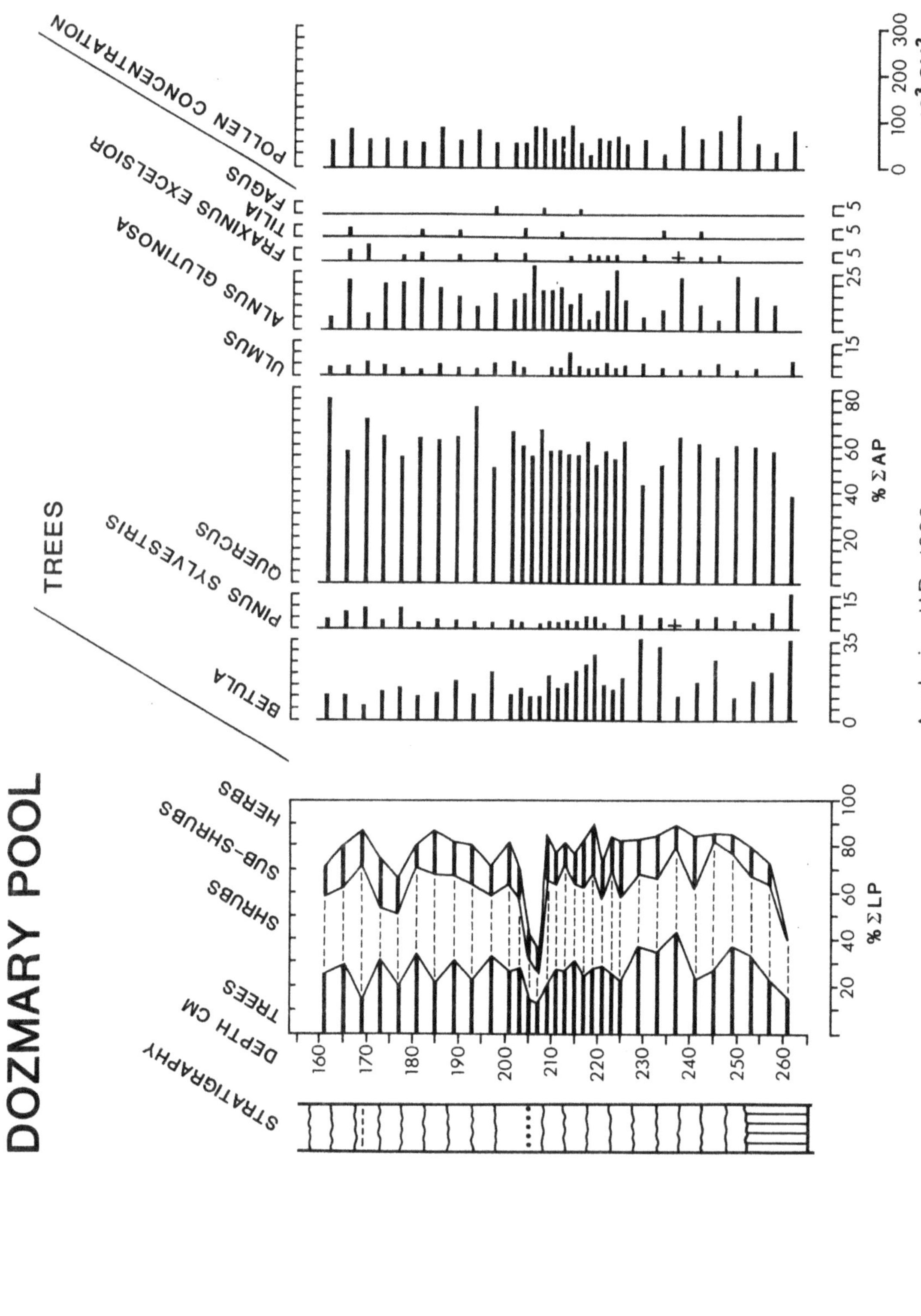
DOZMARY POOL
STRATIGRAPHY
DEPTH CM
TREES
SHRUBS
SUB-SHRUBS
HERBS
% ΣLP
TREES
BETULA
PINUS SYLVESTRIS
QUERCUS
ULMUS
ALNUS GLUTINOSA
FRAXINUS EXCELSIOR
TILIA
FAGUS
POLLEN CONCENTRATION
% ΣAP
x 10³ CM³
Analysis J.I.R. 1982

Figure 1.

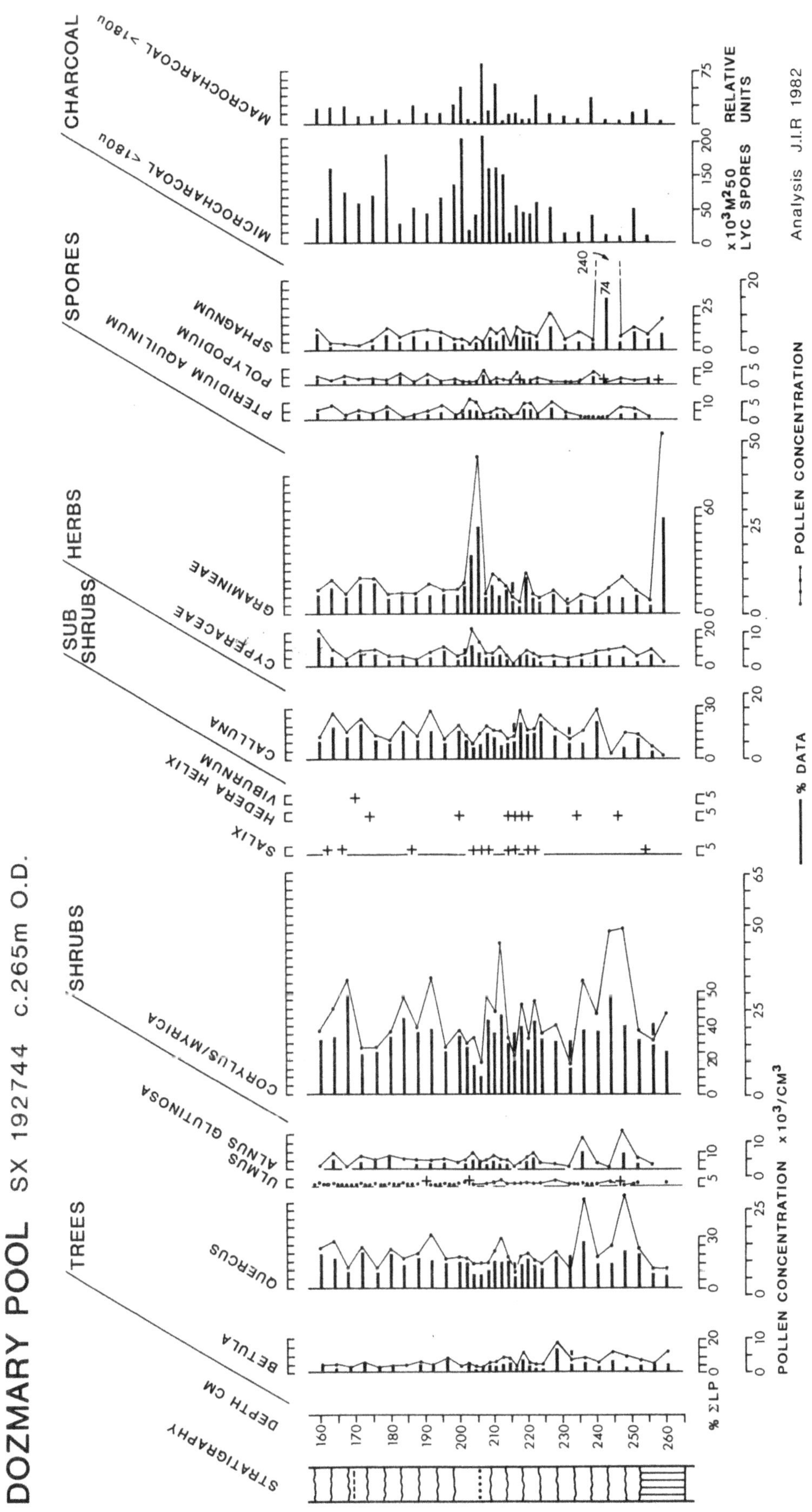
DOZMARY POOL SX 192744 c.265m O.D.
STRATIGRAPHY
DEPTH CM
TREES
BETULA
QUERCUS
ULMUS
ALNUS GLUTINOSA
SHRUBS
CORYLUS/MYRICA
SALIX
HEDERA HELIX
VIBURNUM
SUB SHRUBS
CALLUNA
HERBS
CYPERACEAE
GRAMINEAE
SPORES
PTERIDIUM AQUILINUM
POLYPODIUM
SPHAGNUM
CHARCOAL
MICROCHARCOAL <180u
MACROCHARCOAL >180u
% ΣLP
POLLEN CONCENTRATION x10³/CM³
% DATA
POLLEN CONCENTRATION
x10³M²50 LYC SPORES
RELATIVE UNITS
Analysis J.I.R 1982

Figure 2.

some openness of the vegetation at this elevation throughout the period, which Brown attributes to the exposure of the site. He also notes a treelessness in the later part of the period with increases in bracken spores and the pollen of *Potentilla* and Gramineae. He suggests that the pattern of pollen curve fluctuations may record the restriction by fire of *Quercus* and *Ulmus* in the woodland (found on better drained soils) with an associated rapid spread of *Betula*, Gramineae and *Pteridium*. This feature is at a level close to the radiocarbon date of 6451±75 BP. Later in his text Brown (1977) notes Simmons' (1964) view that similar rises in *Pteridium* spores and Gramineae pollen at Blacklane on Dartmoor not only record a recession of the forest vegetation, but that it may be due to human (Mesolithic) influence. Brown, however, is of the opinion that deliberate clearance of woodland by later Mesolithic people would have been unnecessary in an open landscape, and therefore could not have taken place to any measurable extent.

THE SITE: PRESENT INVESTIGATIONS

From this earlier work it seemed that Dozmary would repay detailed analysis of the interesting level containing charcoal and also showing fluctuations in the tree pollen curves. Accordingly cores were taken from the mire surface at a location fairly close to that of Brown (1977). A summary of the stratigraphy of the core is given below:

0-80cm	Unhumified coarse fibrous light reddish brown, *Sphagnum/Eriophorum vaginatum* peat.
80-90cm	Darker, slightly humified fibrous *Sphagnum/Eriophorum vaginatum* peat.
90-112cm	Unhumified coarse fibrous light reddish brown *Sphagnum/Eriophorum vaginatum* peat.
112-252cm	Dark brown/black well humified *Sphagnum/Eriophorum vaginatum/Calluna* peat. Frequent horizontal dark bands with distinct ones at 124-128cm, 151-156cm and 169-171cm. Quartz grains up to 2mm in diameter at 204cm. Frequent carbonised material throughout. *Calluna* stems, leaves and flowers common.
252-265cm	Highly humified monocotyledonous peat containing carbonised material. *Juncus* seeds present towards the base.
265-281cm	Detritus mud containing occasional *Sphagnum* leaves, frequent *Juncus* seeds and *Isoetes* megaspores.
281- ?	Kaolin clay

This was a deeper profile than those recorded by Brown (235cm) or Conolly *et al* (*c* 270 cm).

The lower detritus mud (below 265cm) and the upper peat (above 110cm) were not analysed. From the intervening peats samples were prepared for absolute pollen analysis by addition of an exotic (*Lycopodium* spores in standard Stockmarr tablets (Stockmarr, 1971)). Each pollen sample consisted of 0.5cm^3 of sample. Treatment with hydrofluoric acid was not required nor was oxidation. The final residue was mounted in silicone oil. Counts of at least 500 land pollen grains were made at each horizon, and at the same time an assessment of the charcoal content was made. The <180 micron fraction passed through the sieve and was present on the count slide. The area of blackened fragments encountered during counting was estimated until 50 *Lycopodium* spores had been counted. This method will overestimate the amount of fine charcoal fragments present. The >180 microns charcoal was retained on the sieve and a relative assessment made of its area by placing a petri dish over graph paper. The squares covered by charcoal were recorded as viewed under a low power microscope. It is a relative measure of charcoal only as area and not volume is measured. However, all the fragments can be identified as charcoal, unlike the <180 micron fraction.

The pollen diagrams presented with this paper are abbreviated pending full publication, they cover only the critical layers between 261cm and 160cm. Figure 1 simply records tree pollen percentages plus the total pollen concentration. Figure 2 shows the tree shrub and selected herb pollen curves and pollen concentration levels. Table 1 lists the occurrence of pollen and spore types not included in the diagrams. Pollen assemblage zones (PAZ) have not been defined for the diagram, mainly because the diagram presented was thought to largely cover only one conventionally definable PAZ within which fluctuations of a partly rhythmic character occured.

Figure 1 shows that AP (total tree pollen) percentage never exceeds 42% and was generally between 18% and 32% of Σ LP (total land pollen) during the period. Shrub pollen (mostly *Corylus/Myrica*) rose to a peak of *c* 50% and fell to a low of *c* 10%. Combined tree and shrub pollen had a maximum value of 82% at 245cm but generally lay between 60% and 70% except at 207cm where it fell to just under 30%. Together with the sporadic but regular occurrence of a range of herbs and open land indicators these figures all point to the persistence of some open landscape in the vicinity of the Dozmary site; this accords with Brown's (1977) interpretation of his diagrams. Some quite large fluctuations in the relative contributions of trees, shrubs, small (Ericaceous) shrubs and herbs occur, with a very major fluctuation between 209cm and 203cm around the level of much charcoal and mineral fragments. These fluctuations will be considered in more detail later in this paper.

Figure 1 also provides an indication of the dominant tree pollen types in the deposits. *Quercus* is the major tree pollen type with percentage values varying between 40% and 65% of ΣAP. *Alnus* is lacking in the lowest sample but rises steadily to peaks of around 25% before falling and rising again several times to similar values. This lack of a single clear and sustained rise in *Alnus* is fairly typical for other sites in south west England and contrasts markedly with the behaviour of the *Alnus* curves in many other parts of Britain. *Pinus* pollen also behaves in a pattern characteristic of south west England being present only in small quantities throughout the period: *Pinus* would not appear to have been a constituent of the woods on Bodmin Moor at any time during the period covered by this diagram. *Ulmus* is generally low, less than 5% of total tree pollen, and extremely low or absent during the major recession period associated with the charcoal layer at around 205cm. *Fraxinus* and *Tilia* pollen occur sporadically with no discernable trends. Occasional grains of *Fagus* pollen were recorded around 197-205cm.

On Figure 2 are details of the pollen and spores present in terms of their relative values in respect of total land pollen or total land pollen plus the taxon concerned. Also recorded are relative amounts of charcoal and the pollen concentration values for the major pollen and spore types. The pollen concentration in a peat will vary with the rate of accumulation of the peat, with the changes in the total inputs due to varying pollen productions over time and with the degree of humification and compression of the peat following accumulation. Chance factors such as the inclusion of an anther of a plant within the sub-sample taken for pollen preparation may cause marked fluctuations of one particular species, which on a percentage diagram would depress all other species represented in the pollen sum. If high peaks are supported by the samples on either side it is possible to be more confident in the view that they are real peaks and not due to chance inclusion of an anther.

Table 1

Depth (cm)

	161	165	169	173	177	181	185	189	193	197	201	203	205	207	209	211	213	215	217	219	221	223	225	229	233	237	241	245	249	253	257	261
Erica type			+		+	+	#	#	#				#		#				+			+		+			#	+	+			
Empetrum			+		+		+		+		+				+												+					
Ranunculus type	+				+	+			+											+		+					+				+	
CRUCIFERAE										+																						
Sinapis type					+																											
CARYOPHYLLACEAE						+		+			+	+				+	+												+		+	
CHENOPODIACEAE	+					+		+		+	+	+	+					+						+						+		
Ulex type		+					+		+																				+			
Lotus type		+																														
Vicia cracca type					+						+																					
Lathyrus type															+																	
ROSACEAE	+				+	+		+	+		+			+			+							+						+		
Filipendula	+			+	+			+	#			+	+		+			+					#		+					+	+	
Potentilla type					+	+		+				+		#	+		+		+		+			+	+	+				+	+	
Drosera intermedia									+																		+		+			
UMBELLIFERAE		+				+		#	+						+									+								
Polygonum bistorta type								+																								
Oxyria type									#	+					+																+	
Rumex acetosa/acetosella					#	+	+		+	+	+	+	+			+					+		#			+				+		+
Rumex obtusifolius			+		#				#														#									
Lysimachia vulgaris type																#														+		+
Plantago media/major			+	+					+						+		+	+														
Plantago lanceolata	+	+				#		#		#	+	#	#	#					+	#	#	#		+	+				+			
Plantago maritima																																+
Galium type						+				+			+											+	+							
Valerianella												+																				
Succisa pratensis				+		+	+													+												
Bidens type	+											+	+		#	+	+				+			+					+	+	+	
Anthemis type							+							+		+			+			+					+					
Artemisia						+				+				+						+	+							+			+	
Cirsium type										+							+							+								
Centaurea nigra type											+																					
Taraxacum type		+			+		+								+					+	+			+			+					
Narthecium					+					+											+											
Hydrocotyle																															+	3
Menyanthes																											+					
Litorella																																+
Alisma type																																+
Eupotamogeton																																+
Typha angustifolia type											+																					+
Isoetes																															+	12
Osmunda regalis																	+			+	+											

+ = less than 0.5% # = 0.5 - 2%
Figures refer to percentage LP plus taxon concerned.

The pollen concentration values (see Fig. 1) show some variation but only by a factor of three between highest and lowest recorded values. Some of the individual taxa exhibit much greater fluctuations in pollen concentrations (see Fig. 2) and these suggest that some of the variations in the pollen curves are directly relatable to changes in pollen production of the vegetation and hence to actual changes in the vegetation. The most variable pollen concentration curve is that of *Corylus/Myrica* which varies from *c* 10,000cm^{-3} to 50,000cm^{-3}. In detail the *Corylus/Myrica* concentration curve exhibits a series of peaks and troughs, with peaks at 249cm, 237cm, 221cm, 213cm, and 201cm and troughs at 257cm, 241cm, 233cm, 217cm and 207cm. Assuming that the pollen came from hazel and not bog myrtle, then these changes may represent a sequence of woodland clearances encouraging the development and flowering of hazel scrub at the margins. Inspection of the charcoal counts reveals that the peaks tend to occur during the troughs in the *Corylus/Myrica* pollen concentrations. This would suggest that burning was the cause of the fluctuations. The most distinctive of these episodes is associated with the charcoal layer which is visible in the stratigraphy at 204cm but picked out in the analyses as extending over the levels from 209cm to 201cm. Total tree pollen is seen to fall from 25% at 209cm to 18% at 207cm and to recover again to *c* 30% by 201cm, with *Betula*, *Ulmus* and *Quercus* the trees affected. *Corylus/Myrica* pollen falls to one third its original value and then recovers to about two thirds of its original value over these 8cm. Gains in both relative percentages and concentrations occur in Cyperaceae and Gramineae pollen and for *Pteridium* and *Polypodium* spores plus a number of herb pollen types including *Plantago lanceolata*, *Potentilla* and various Compositae. Several of the herbs may represent not only a more open vegetation but also the colonisation of burnt ground. In addition to the high charcoal values at 209cm and again at 203cm, mineral matter was present on the prepared slides at 205cm and 203cm. The blowingwashing in of mineral dust particles from exposed surfaces nearby is a possible explanation for this phenomenon.

RADIOCARBON DATES

The radiocarbon dates now available for the profile are:

203-204cm	1560±80	BP	(HAR-5077)
207-208cm	510±70	BP	(HAR-5078)
225-226cm	6020±90	BP	(HAR-5079)
241-242cm	2740±120	BP	(HAR-5080)
257-258cm	7590±100	BP	(HAR-5083)

The lowest sample with the oldest date is compatible with the pollen assemblage. It predates the rise in *Alnus* and *Calluna*, and the appearance of *Fraxinus*, *Tilia* and *Fagus*. Brown (1977) has the equivalent pollen assemblage zone (DP4) as dated between 7925±100 and 6793±70 BP, also occurring within a monocotyledonous peat.

The remaining radiocarbon dates are not in chronological sequence and must throw considerable doubt on the deposits being a sequential accumulation. Either the samples were contaminated or there has been wholesale disturbance of the upper 242cm with some inversion of the profile. Contamination by the penetration of modern roots is unlikely to produce the radiocarbon dates obtained. Similarly, careful removal of the material from the cores taken should have removed any chance of contamination by more recent peat or humic material. The evidence for considerable disturbance with charcoal, mineral matter and major fluctuations within the pollen assemblages has already been recorded. That disturbance primarily referred to the samples between 201cm and 209cm. Lower in the profile, although some charcoal does occur, fluctuations in the pollen assemblages are much less marked. At 241-242cm a radiocarbon date of 2740±100 BP is obtained only 16cm above the date of 7590±100 BP and 15cm below a date of 6020±90 BP. The pollen assemblage is not entirely incompatible with a much earlier date. Agricultural indicators are low or absent as is *Fagus*. High agricultural indicators and the presence of *Fagus* at 2740±120 BP would be expected. An overall low in tree pollen, reduced *Quercus* and *Alnus* concentrations and a peak in charcoal concentration all occur at 241cm (hence the decision to have a radiocarbon date at this level). Immediately prior, at 245cm, there is a peak in *Sphagnum* spores. This might support a view that infilling of a former cut-away surface was occurring, possibly naturally at first with *Sphagnum* developing but later positive infilling must have occurred to cause very much earlier material to lie above the level. As the stratigraphy was recorded from cores and not from an open face, unconformities were not visible. The cores, and indeed the pollen sequences, give an impression of a sequential accumulation of the deposits except in the 203-210cm levels.

At 225-226cm, with a radiocarbon date of 6020±90 BP, charcoal values increase again, but the number of ruderals is low and *Fagus* and *Tilia* are absent. The pollen assemblage is compatible with the radiocarbon date if the peat is *in situ* or if it includes backfill material in the form of a complete sod of peat.

The level of the major burning horizon was seen to be one of very major pollen changes and now also with an inverted radiocarbon date for the samples at 207-208cm and 203-204cm. Extremely low tree pollen and high Gramineae pollen, plus many ruderal pollen types occur. The very marked and chaotic changes in the pollen histograms is compatible with a major hiatus with burning and inversion of the peat profile. The inclusion of material of a radiocarbon date of 510±70 BP places the hiatus well within the historic period yet no suggestion of this sort of disturbance was traceable in the work of Brown (1977). This may be explained either by there being no hiatus in his profiles even though the sample site was within the same general area of the raised mire and contained a major burning horizon, or his sampling interval and lack of radiocarbon dates later than 6451 BP did not allow him to pick up such a major pollen fluctuation.

CONCLUSION

At Blacklane Brook on Dartmoor, the present authors (Simmons *et al*, 1983) suggested that, in a period which was radiocarbon dated to between 6010±90 BP (HAR-4461) and 7660±140 (HAR-4462) with charcoal layers and rises in bracken and *Polypodium* spores, forest recession occurred with grassland replacing some deciduous woodland and with shrubs such as hazel and members of the Rosaceae becoming prominent components of the local ecosystem. This phenomenon was described as being most likely a result of human activity and the dates placed it firmly in the later Mesolithic period. At Dozmary Pool, with its associated Mesolithic flint industry, it was hoped to provide more positive evidence of this type of clearance especially as charcoal layers, apparently of Mesolithic age, appeared present in the published records of the profiles sampled by Brown (1977). The palynological analyses carried out in this new study were suggestive of clearance phases but the radiocarbon dates have given evidence for the deposits being non-sequential. If these dates are accepted then disturbance of the site was not simply by fire alone but by actual physical removal of some of the peat and inversion of the profile at a date at least as recent as 2740 BP in the case of the lower disturbance and 510 BP for the higher disturbance. Both may have been caused by a single disturbance at or after the later date. The very large fluctuations in the pollen, charcoal and mineral contents in the 201-209cm levels give support to this view. The reasons for this major disturbance are unknown as there is no record, for example, of tin working within the immediate vicinity of the site, nor is there any patterning of the mire surface suggestive of former turbaries.

The results may be taken as a warning that core stratigraphical and pollen evidence alone, especially where no check from exposed peat faces is available, may be misleading and that further work with additional radiocarbon dates is needed to determine just how much disturbance of a major kind has occurred or whether any of the radiocarbon dates may be aberrant.

ACKNOWLEDGEMENTS

This work was suported by a grant from the SERC Science-based Archaeology Committee (GR/B/22857) to KC and IGS, with JIR as Research Assistant. The diagrams were drawn by S Godden of the Department of Geography, University of Bristol.

REFERENCES

Brown, A.P. 1977. Late Devensian and Flandrian vegetation history of Bodmin Moor, Cornwall *Phil. Trans. Roy. Soc. London B*, **276**, 251-320.

Conolly, A.P., Godwin, H. and Megaw, E.M. 1950. Studies in the Post-glacial history of British vegetation XI. Late glacial deposits in Cornwall *Phil. Trans. Roy. Soc. London B*, **234**, 397-469.

Jacobi, R. 1979. Early Flandrian Hunters in the South West *Proc. Devon Archaeol. Soc.* **37**, 48-93.

Simmons, I.G. 1964. Pollen diagrams from Dartmoor *New Phytologist* **63**, 65-108.

Simmons, I.G., Rand, J.I. and Crabtree, K. 1983. A further pollen analytical study of the Blacklane peat section, on Dartmoor, England, *New Phytologist* **94**, 655-667.

Stockmarr, J. 1971. Tablets with spores used in absolute pollen analysis *Pollen et Spores* **13**, 615-621.

Wainwright, G.J. 1960. Three microlithic industries from South-West England and their affinities *Proc. Prehist. Soc.* **26**, 193-201.

THE EARLY MEDIEVAL WATERFRONT AT REDCLIFFE, BRISTOL: A STUDY OF ENVIRONMENT AND ECONOMY.

J. Jones and N. Watson

HISTORICAL BACKGROUND

The settlement of Bristol, in the 10th century, was established on a peninsula of Triassic sandstone at the confluence of the river Avon and its tributary, the Frome. It commanded the lowest point at which the Avon could then be bridged and, being almost circled by water, was relatively secure. The tidal waters offered safe anchorage in a natural harbour, even though the town lay 10km from the mouth of the river. At this early date Bristol seems to have been a centre of economic importance stimulated by sea-borne trade with Ireland and France, as well as local trade along the Severn Estuary.

The 12th century saw the rapid development of Redcliffe where flat open land offered a larger site than the walled borough and a good anchorage along the east bank of the river. The area attracted many rich merchants and became a rival to Bristol. By the mid 13th century Bristol's port facilities were proving inadequate, and the citizens embarked on an ambitious programme of harbour improvement. In 1240 the old course of the Frome was diverted, and a new deeper channel cut through the marsh to join the Avon further downstream (Fig. 1). It needed an order from Henry III to force the men of Redcliffe to assist the Bristol burgesses in making the trench, "for the common good of the whole town of Bristol" (Harding, 1930). The diversion of the Frome transformed the topography of the town. A stone quay with improved port facilities was built alongside the new Frome channel and ocean going ships now docked here. The old quays along the river Avon became "the Backs" and dealt mainly with ships trading along the coasts of the South West, Wales and Ireland. With the improved port facilities the importance of Bristol as a trading centre grew and the merchants of Redcliffe shared in this prosperity. By the middle of the 14th century Bristol was the wealthiest provincial town in England and became the country's leading cloth exporter.

Over the past seven years the City of Bristol Museum and Art Gallery has carried out a series of excavations in Redcliffe as a project designed, in part, to investigate the growth of Redcliffe and its role in the development of Bristol. It is clear that the waterfront dominated Redcliffe and was the main source of its prosperity. The merchants of the area benefitted greatly from its harbour facilities, and documentary sources show that trade such as dyeing and fulling were carried on here.

The aim of the project described here has been to investigate the local environment of deposition, both before the occupation of the river banks, and the later medieval developments, and to establish the nature and origin of the organic waterfront sediments. In view of the recorded industries on the waterfront, it is also hoped to throw some light on early industrial processes and crafts, such as cloth-making, dyeing, tanning and bone working. The domestic nature of some of the deposits, with the recovery of large quantities of animal bones, fruit stones and other plant remains, will provide evidence of the diet and economy of the inhabitants of early medieval Bristol.

Three of the excavations, at Bristol Bridge (1981), Dundas Wharf (1982-3) and Canynges House (1983-1985) have shown a complicated series of quay structures gradually encroaching onto the river's course, the reclaimed land being infilled by alternating layers of urban dumps and natural fluvial deposits.

An important part of the excavation at Dundas Wharf was the implementation of an on-site environmental sampling programme to examine these waterlogged deposits. A major concern was to ensure good recovery of animal bones, and a sieving experiment undertaken during the early stages of excavation underlined this need. The experiment was based on the work from West Hill, Uley, Gloucestershire (Levitan, 1982). Samples of up to 60 litres

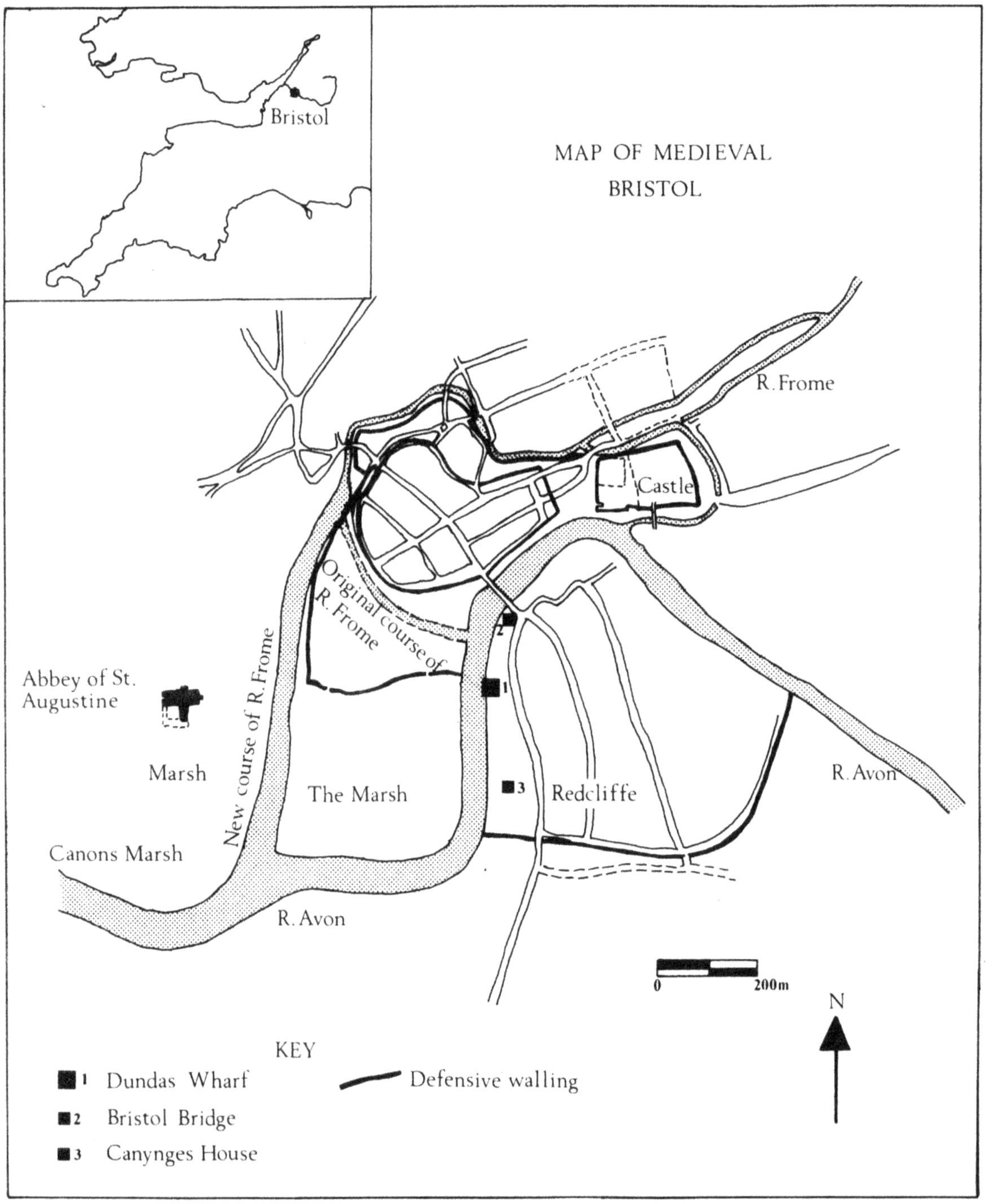

Figure 1. Map of medieval Bristol.

from all the organic deposits were sieved and sorted to 2mm, and this also ensured the recovery of the larger seeds, nuts, fruit stones and molluscs. A series of column samples were taken through all the major deposits, both organic and fluvial, for the extraction of plant macrofossils, and where possible for insect and molluscan remains. Additional samples through the original bank of the river Avon, and the fluvial deposits in the archaeological stratigraphy were taken for diatom analysis.

THE ARCHAEOLOGICAL BACKGROUND

Dundas Wharf

The excavation at Dundas Wharf, completed in 1983, uncovered three medieval tenements. One of these extended from the Redcliffe street frontage to the present quay of the Floating Harbour (constructed 1804-9). The earliest occupation on the street frontage revealed floor levels of building of early-mid 13th century. These are likely to represent shops and workshop areas, with storage space and yards behind. Between tenements 128 and 129, a narrow cobbled lane, probably of 13th century origin, provided public access to the river. To the west of the occupation levels, a series of quay structures were built onto the original east bank of the river Avon. Excavation revealed 3m of waterlogged organic and clay deposits between these quay walls (Fig. 3).

Stratigraphy

The series of plans illustrated in Figure 2 show the development of the waterfront in the 12/13th centuries. These represent only the preliminary interpretation of the deposits from tenements 127 and 128. A complete analysis must await the full publication of the site, but the following sequence can be suggested. The dating is based upon a pottery chronology which is expected to be revised in the light of dendrochronological work recently completed (R. Nicholson, pers comm).

Phase A, 12th century
Wall 5 was constructed along the crest of the original river bank to stand above MHWS (mean high water springs). Rubbish disposed of onto the bank was preserved as finely stratified lag deposits contexts CCG, BWQ and the layers ponded behind rubble CGH. There followed a period of erosion or excavation before renewed bank progradation with the deposition of the organic and clay sequence BLZ. Wall 80 was constructed by truncating the sedimentary spur BLZ and setting the battered wall against this exposed face to the height of layer AZS. This was well below MHWS. It appears that wall 80 provided a low water facility for wall 5. Rubbish was dumped behind wall 80 between periods of fluvial sedimentation (BWA). Wall 5 was refaced.

Phase B, early 13th century
Clay accretion against wall 80 formed a thick sedimentary wedge CCK and presumably hindered its use. There was renewed activity with the construction of a low, pitched retaining wall 89, several metres to the west. Associated with this two sill beams (F176) were laid along the top of CCK, held in position by timber stakes driven through the clay. The beams were morticed and one supported an upright 2.5m high. Although obviously a major feature, its exact function is unclear.

Phase C, mid 13th century
The top courses of wall 80 were robbed and its southern end demolished, probably to give easy access to the river bank. The rubble from the destruction (layer CCL) spread onto the bank. The northern half of wall 37 was built with arch F112 and a timber feature F166, contemporary with the organic layers AZN and AYZ. The southern portion was butted against wall 37 over rubble CCL, thereby setting above and on the line of wall 80. This marks the first major encroachment on the river; wall 37 succeeding wall 5 as the high water wall. Organic refuse was dumped behind wall 37 and intermittently sealed by high

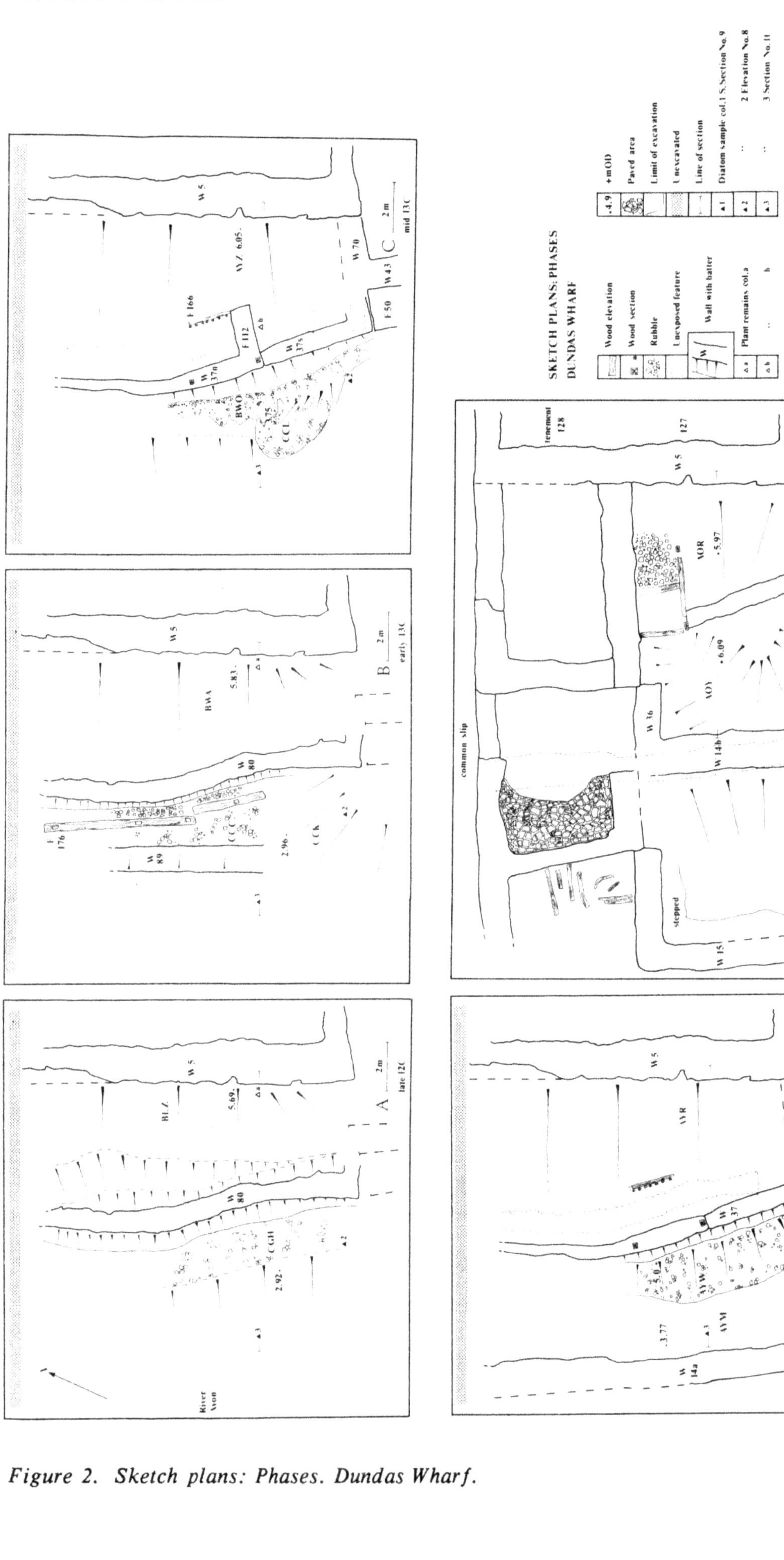

Figure 2. Sketch plans: Phases. Dundas Wharf.

water fluvial clays. Organic lag deposits BWW, BLE were sealed by clays laid down over the footings of the wall to the west. Again the process of clay accretion formed a sedimentary wedge AYW against wall 37 thereby impeding access. Wall 37 would appear to have been in an anomalous position since for three months of the year high tide did not even reach its base which stood above MHWN (see section on tidal regime). In an apparent effort to remedy this a low water wall 14a was built 4m further west to take full advantage of all the neap tides.

Phase D, late 13th century

At this point the stratigraphy is unclear because the base of wall 14a was never fully exposed. The wall could have been built from any one of a number of different levels though the available evidence suggests that, like wall 80, it was set against an exposed face where sediment had been cut away from the bank. It was built to a depth of at least +2.65m OD probably from the top of the accretionary wedge AYW which had been levelled off to the west. There is no cut further to the west to indicate trench building, nor is it likely that wall 14a was built from higher levels because of the volume of sediment which would have had to be removed. Finally, a thick sequence of organic refuse and fluvial clays accumulated behind the upstanding courses of wall 14a and against the west face of wall 37.

Phase E, 14th century

After the deposition of the organic and clay layers behind 14a the wall was built over by 14b. A new policy of individual tenement advancement developed in the 14th century. Walls were wider, mortar bonded and built above high water. Tenements 127 and 128 pushed alternately over the river bank. Wall 15 was constructed soon after 14b providing a new low water wall, and was in turn enlarged and raised above MHWS.

The tidal regime

During the excavation of tenement 127 an eleven metre deep section (section no. 11, Fig. 3) exposed a sequence of medieval sediments and stone structures overlying the early southern bank of the river Avon. From the position of this bank the original width of the river has been estimated at 100m, twice the width of the modern canalised Floating Harbour. The natural alluvium of the bank is approximately 8m thick. It is underlain by Pleistocene gravels which rest unconformably on the Triassic Mercia mudstone.

The tides experienced in Bristol are an extension of the macrotidal regime operating in the Severn Estuary where the tidal reach is 13m. This figure reduces inland to 10.3m at the Cumberland Basin, 2km downstream from Dundas Wharf. The present-day MHWS at Dundas Wharf has been estimated as +6.95m OD and is the height to which fluvial clays will accrete. Thus, the highest *in situ* medieval fluvial clays should rest approximately at the medieval MHWS. These layers are at +6.40m OD, 0.55m lower than the present-day MHWS. Post depositional compaction must account for some of this difference and the medieval MHWS was probably within 0.5m of its present-day value. A difference of this scale is confirmed by the sea level curve for South West England. 1000 years ago the sea level was about 1m lower than at present, although the data are insufficient for a conclusive deduction (Heyworth and Kidson, 1982; Donovan, pers comm). Unfortunately neither the quantity nor the quality of the medieval wooden structures are sufficient to estimate MHWN from measuring the heights of rotted timbers, as has been employed to effect at Trig Lane, London (Milne and Milne, 1982). The comparison between the present-day and medieval MHWS implies that the present-day MHWN of +3.65m OD is also within 0.5m of the medieval value. At the lowest level of water (-3m OD) at least 60% of the channel is exposed; during this time in the medieval period, river traffic ceased and most ships were forced to rest on the bottom (Sherbourne, 1965).

It is also worth considering how frequently different heights of the banks are submerged. Modern data collected by Bracher (1929) for the tidal Avon are probably a reasonable reflection of the medieval situation. The highest reaches of the bank are

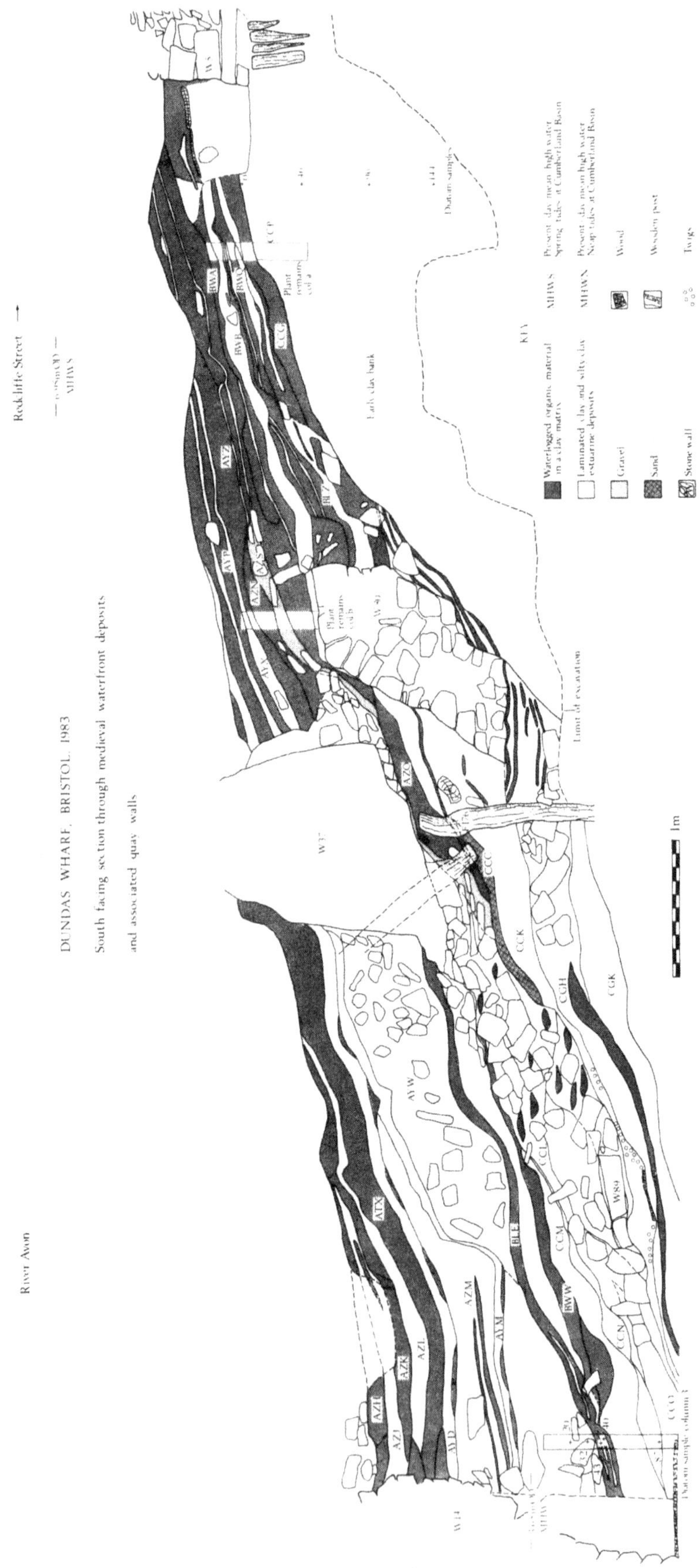

Figure 3. South facing section through medieval waterfront deposits and associated quay walls

submerged by between 59 and 123 high tides per year for about 10-20 minutes only. Although, because of the steepness of the dip of the bank, MHWS and MHWN are separated by a distance of only 12 m, the number and duration of the inundating tides at MHWN is increased to 641, and 2 to 2.5 hours respectively. The position of the quay walls is crucial to ensure an effective use of the high tides.

Sediments (Fig. 3)

The stratigraphy consists of two main types of unconsolidated sediment: clays and organic material. The mineralogenic layers can be generally described as laminated silty clays. Coarser particles increase towards the base of the section where they form thick silt layers and weakly developed flaser beds. The latter are a typical intertidal deposit, the clay laminae are incompletely developed and became trapped in the troughs of the silt ripples during slack water. The sediments represent an upward-fining lateral accretion deposit within the channel of the river (ie the river bank migrates riverwards as coarser particles are laid down at the base of the bank and finer particles at the top). The organic material is composed of fragmented plant remains set in a clay matrix which produces a matted texture of finely interdigitating fibrous layers. Preservation varies from amorphous fragments showing degrees of humification to readily identifiable macroscopic remains of leaves, straw, twigs, nuts, wood chips and charcoal. A cultural component was often present and included pottery sherds, leather, bones and metal slag. As a whole the deposit is a typical deposit of urban refuse similar to those recorded from terrigenous sediments in York (Hall *et al*, 1983) and subaqueous ones in Bergen (Krzywinsky *et al*, 1983).

There are two facies of organic material. Thick organic deposits are concentrated higher in the stratigraphy (eg AZL, AYZ, AYP, Fig. 2) and display few characteristics of a fluvial sediment; they are unsorted, contain sparse clay interlaminations, and have a generally chaotic fabric. These deposits were not subjected to rigorous tidal action, having been contained behind walls where their positions would be washed only by high spring tides. At lower levels the layers are thinned lag deposits (eg CGH, BWW, BLE) where much of the original material has been removed by scouring currents. Their preservation involves a process of substitution of the light organic particles for cohesive clays which creates a gradual increase in the clay:organic ratio through the organic layer. The deposits are frequently interlaminated with clay and have orientated fabrics.

Different modes of rubbish transport are reflected in these two types of organic deposit. The earliest is rubbish, strewn onto the bank where the bulk is reworked and eventually removed by the tides. The second involves material dumped in a cumulative backfilling process interceded by periods of fluvial sedimentation. In both cases, burial is sufficiently fast to minimise aerobic decay and maintain reducing conditions below the depositional interface.

A preliminary examination of the pottery suggests that the sequence accumulated in as little as 100 to 150 years (Fig. 3). By measuring the thickness of the clay layers, and assuming a constant rate, the rate of sedimentation can be estimated at between 1.4cm and 2.1cm *per annum*. This is probably an underestimate because it does not include the clay deposited in the organic layers. A rate of 0.18cm *per annum* has been calculated from a series of preliminary archaeomagnetic dates for the sediments below the archaeological stratigraphy in the early bank (M. Allen, pers comm). The early medieval period, therefore, was experiencing sedimentation at least 8 to 12 times greater than had been the natural rate prior to the occupation of the waterfront.

PLANT AND ANIMAL REMAINS

The bulk of the organic deposits show the accumulation of an enormous depth of well preserved material, which seems to be characteristic of the medieval period in many towns of Northern Europe. The majority of these deposits have been built up largely as the result of the deposition of plant and animal matter which has been preserved due to waterlogging and the sheer volume which has inhibited decay. A town such as Bristol, with a large population, must have relied on its hinterland for foodstuffs, timber for fuel and building and other raw materials for industrial and domestic purposes. The resultant large input into the town must have made rubbish disposal a major problem.

Diatom analysis

Diatoms are unicellular algae with a siliceous frustule or external wall which can be preserved in fresh water and marine sediments. Many species have distinct ecological tolerances and they have been used at waterfront sites to assess salinity and as indicators of cultural eutrophication. For example, Battarbee (1983) records an increase in the euhalobous (marine) component of diatom assemblages from 2% in Roman deposits to 11% in Saxon deposits at London. This is interpreted as indicating a gradual change in the position of the tidal head in the river Thames. Moss (1983) analysed a small sample of diatoms from riverine deposits in late Saxon Norwich. He reports an indigenous freshwater mud assemblage and although these deposits were overlain by organic material dumped along the river bank, there was no evidence to suggest any major pollution of the water course.

The diatom analysis at Dundas Wharf had two objectives. Firstly, to investigate any evidence for the eutrophication of the 12th and 13th century waterfront environment. This was accomplished by comparing the assemblages from the early bank with those from the archaeological stratigraphy. Secondly, to assess the relative influences of the tidal and freshwater sedimentation.

Fifteen samples were taken, eleven from archaeological layers, and four from the early bank. Diatom slides were prepared following the method described by Mannion (1975). The very low concentration of diatom valves and the quantity of inorganic debris meant that for each sample 4 coverslips were traversed 16 times at x1000 magnification to obtain counts of at least 600 frustules. Each coverslip contained a 0.1ml aliquot drawn from a solution of 1ml of sediment per 100ml. The analysis of the results is still in progress, but preliminary conclusions can be drawn.

Interim results

Figure 4 illustrates the frequencies for species of greater than 1%. The changes in salinity tolerances of the assemblages through the stratigraphy is displayed in Figure 5, and Figure 6 shows the diatom genera present. Ecological data were derived from Hendey (1964), Hustedt (1930; 1957) and van der Werff and Hulls (1958-66) and the taxonomy is based on Hendey (1964).

The major feature of Figure 4 is the occurrence throughout the stratigraphy of the euhalobian (marine) species (*Coscinodiscus* spp., *Paralia sulcata*, *Campylosira cybelliformis*, *Cymatosira belgica*, *Actinopychus senarius*, *Cocconeis scutellum* v. *parva* and *Raphoneis surirella*). They completely overwhelm the assemblages from the early bank where the sum of the mesohalobous (brackish) and oligohalobous (freshwater) species is less than 8% (Fig. 5). These same species also dominate the archaeological layers. The frustules of the epilithic (sediment associated) component, *Campylosira* sp., *Cymatosira* sp., *Cocconeis* sp. and *Raphoneis* sp., although present, are rarely found living in the modern surface samples from the banks of the tidal river Avon. This is due to the unfavourable range in the salinity of the river of from 20mg/l of Cl to 13500mg/l of Cl depending upon the state of the tide and the time of the year, (D.J. Palmer, pers comm). Along with the planktonic species *Paralia*, *Coscinodiscus* and *Actinoptycus*, valves of the epilithic species are transported upstream suspended on the flood tide from the estuary where they were living in the more consistently saline environment.

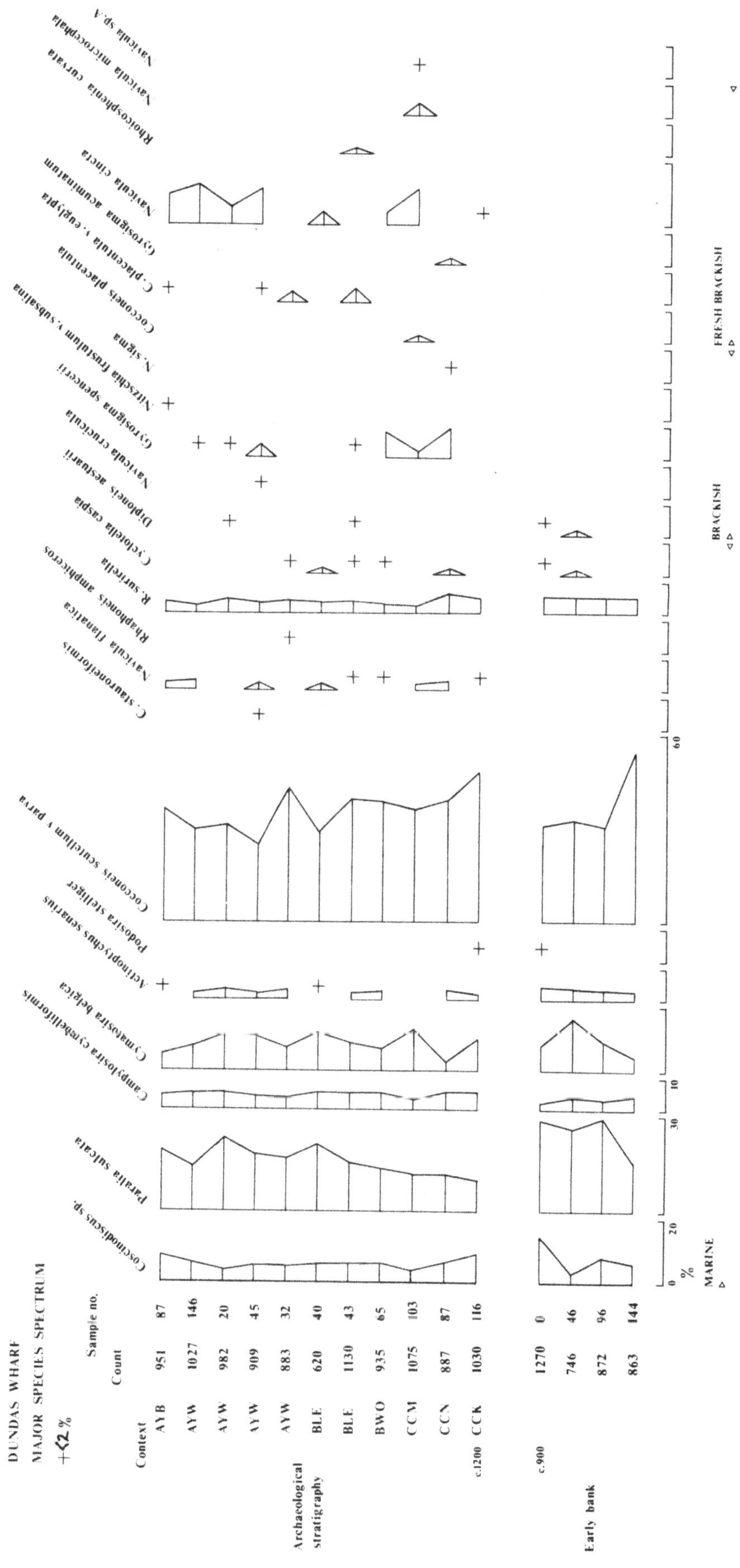

Figure 4. Diagram showing the major diatom taxa.

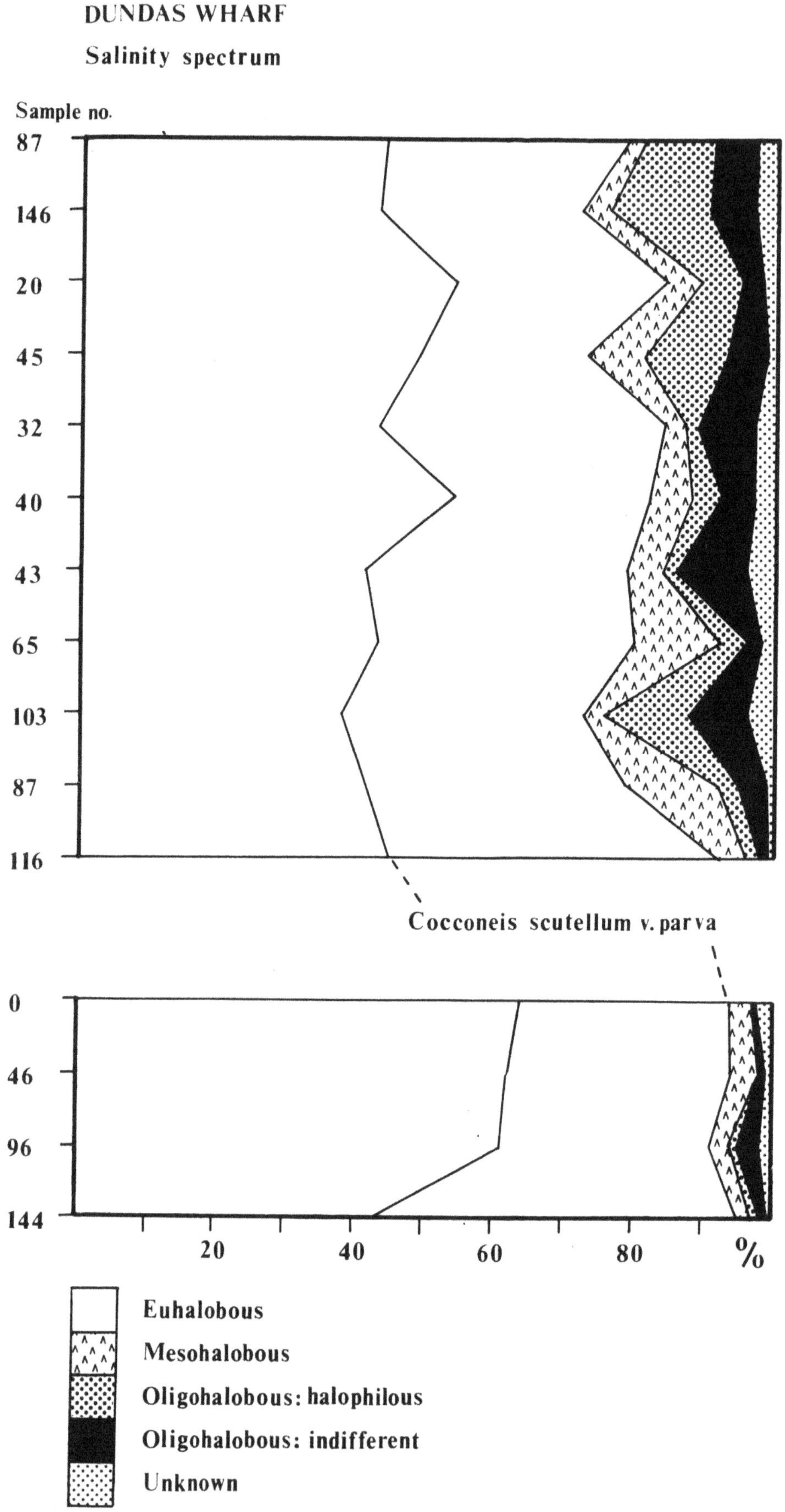

Figure 5. Diagram showing an increase in brackish (mesohalobous) and freshwater (oligohalobous) diatoms in the archaeological deposits.

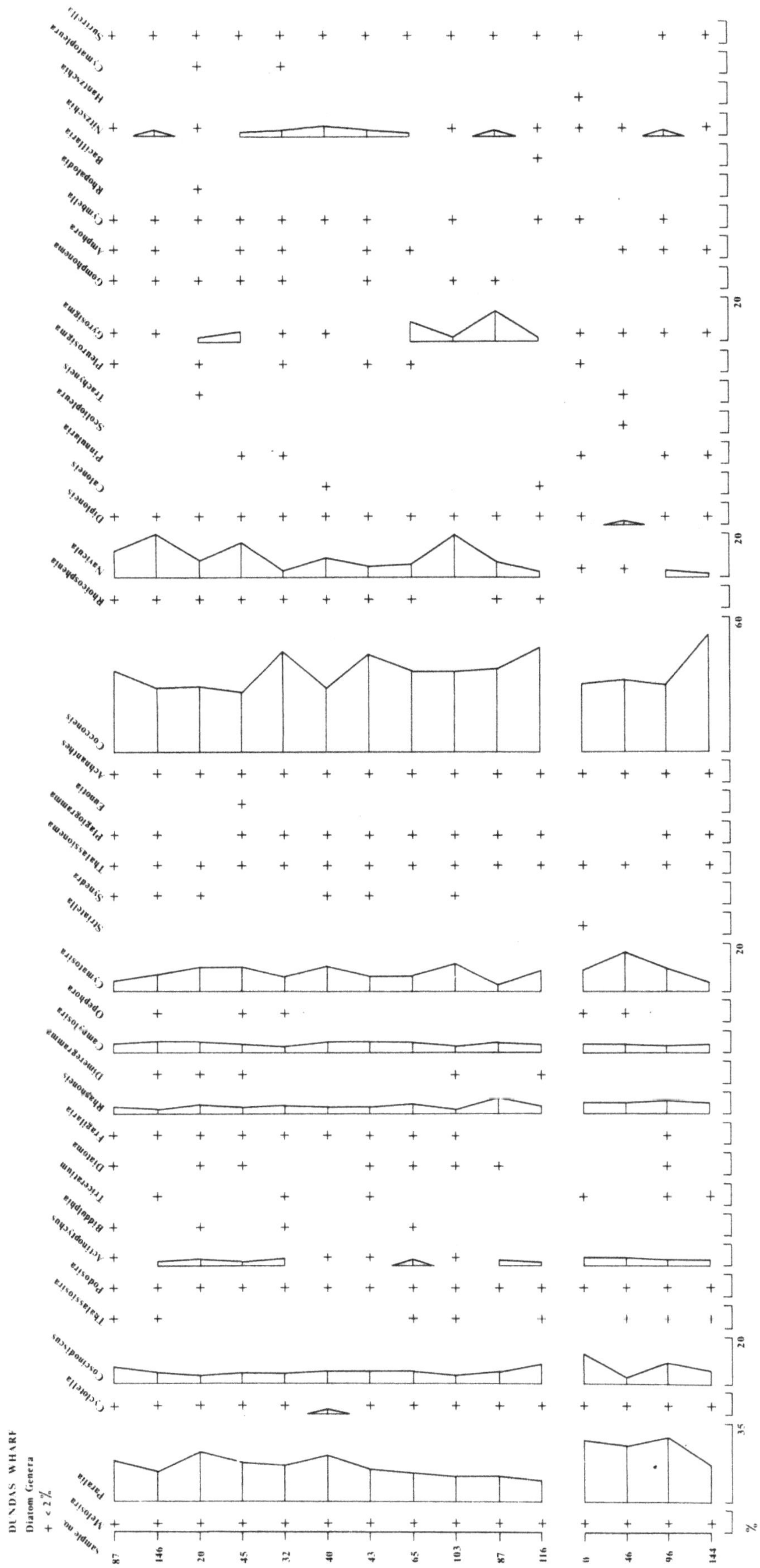

Figure 6. Diagram showing the diatom genera.

The occurrence of the euhalobous forms in such constant proportions throughout the stratigraphy, and the lack of any freshwater plankton implies that the water was estuarine in origin, and that the same process of transportation and deposition prevailed. It confirms that the freshwater flow was confined to the centre of the channel, and that its sediment burden, however great, was insignificant under the influence of the tidal regime. It is known that the sediment burden of the tributary of the Frome was easily sufficient alone to block the channel in the 19th century when the non-tidal harbour was constructed (Lord and Southam, 1983).

The second significant feature is the expansion of the mesohalobous and oligohalabous species in the archaeological stratigraphy (Fig. 5). These diatoms are believed to have grown *in situ*. The main ones are illustrated in Figure 4. They form three distinct groups. First, those only occurring in the archaeological layers, the most important of which are *Rhoicocphenia curvata, Cocconeis placentulata* v. *euglypta* and the genera *Synedra* and *Gomphonema*. Secondly, those that rarely occur in the early bank, but consistently occur in the archaeological horizons, the freshwater species of the genera *Fragillaria* and *Diatoma*. Finally, those genera which form the indigenous flora of the early bank and expand appreciably in the archaeological horizons, *Nitzschia, Navicula* and *Gyrosigma*. *Navicula cincta* is numerically the most important species (12% in sample 146). It appears to flourish in the clean accreting clays against wall 37 (layer AYW). In contrast the euryhaline epiphyte *Cocconeis placentula* v. *euglypta* is the most prolific in the organic lag deposit BLE (sample 43) and the organic concentrates at the base of AYW (sample 32) supporting the supposition of exposed organic-mud surfaces. The exact significance of the second group is unclear though they may have grown in response to local freshwater runoff from the structural surfaces on the bank.

In general many estuarine diatoms can tolerate broad changes in salinity, however, epilithic species, such as those discussed above, are distributed in part by their association with the nature of the substratum (Amspoker and McIntire, 1978). Thus the occurrence of new forms in the archaeological stratigraphy would appear to be indicative of new and varied forms of substrate together with a rising nutrient input from waste disposal rather than any significant alteration in the salinity range of the river. Nutrient enrichment from sewage pollution is believed to have caused changes in the flora and fauna of early 14th century sediments from the River Fleet in London (Boyd, 1981).

Plant macrofossils

At present, two column samples have been examined for plant macrofossils from the earliest levels of 13th century date above the original bank of the River Avon (Fig. 2). 250 gram sub-samples, sieved to 250 microns, have been looked at from the 12 samples listed in Table 1. The table lists all waterlogged seed remains. The term 'seed' is used in a wide sense to include fruits, nutlets, achenes etc. Other waterlogged plant remains are listed in Table 2. Nomenclature and habitats are based on Clapham *et al* (1962). The habitat groups, based on Robinson (1979) are as follows: aquatic; bankside; cultivated; disturbed ground; disturbed ground including arable; grassland; marsh; scrub and woodland. Preliminary identifications have also been made of the fruitstones and nuts from the bulk-sieving samples.

All the samples show an underlying uniformity with many species in common. Several distinct plant communities are represented by, for example, groups of plants associated with arable land, meadows, pasture, marsh and saltings and since not all of these would have been present locally, material was clearly being imported to the site, both by human activity and by the influence of the tide.

Arable land and cornfields

Medieval cereal crops contained a specifically cornfield weed component. *Agrostemma githago* (Corncockle), *Chrysanthemum segetum* (Corn Marigold) and *Centaurea cyanus*

(Cornflower) are all closely tied to arable agriculture, and cultivation is probably a pre-requisite for their survival in Britain. All three species are represented, *C. segetum* being particularly abundant.

In his *Booke of Husbandry*, Fitzherbert (1979) mentions a number of weeds that "doe moche harme". These include "Cockle" (*A. githago*), and "Gouldes" (*C. segetum*). *Agrostemma* seeds are poisonous and affect the physical properties of wheat flour.

Many of the other species occur in arable fields, although they also grow in other habitats, but it is likely that some *Anthemis cotula* (Stinking Mayweed); *Spergula arvensis* (Corn Spurrey); *Aphanes arvensis* agg. (Parsley Piert) and *Sonchus* spp. (Thistles) were also introduced with cereal crops. Some of these weeds are from different arable habitats, *C. segetum*, for example, being a weed of acid soils and *A. cotula* of heavy basic soils. As could be expected, therefore, cereals supplied to the town from a variety of locations are represented.

Although cereals themselves are represented by only a few charred grains, one of the tenements contained bread ovens of 14th century date. It is possible that threshing or seed cleaning refuse was being deposited.

Pasture and hay meadows

Tansley (1939) classifies grasslands according to use, into meadows which are regularly mown for hay, and pastures which are grazed. Characteristic weeds of closely grazed pasture tend to be low growing species, with rosettes of leaves which are unshaded by the short surrounding turf and so can flourish unhindered. *Bellis perennis* (Daisy) is typical of this habitat and is particularly abundant in one sample. *Hypochaeris* sp. (Cat's Ear) and *Plantago major* (Great Plantain) are also present in small numbers.

Hay meadows, in contrast, are a semi-natural crop consisting of a tall herbage of grasses and broad leaved plants, and are fairly species rich. Most of the samples contain a haymeadow element: *Chrysanthemum leucanthemum* (Ox-Eye Daisy), *Centaurea* sp. (Knapweed) and *Rhinanthus minor* (Yellow Rattle), being the most closely associated species with this habitat. Many species occur under both types of management, but meadowland conditions enable them to set seed much more profusely than in pastureland. Gramineae (Grass) seeds are exceptionally abundant in some of the samples and the most likely explanation is that they had been imported in hay. Other flowering species that are conspicuous in hay meadows include *Ranunculus repens* (Creeping Buttercup), *Leontodon* sp. (Hawkbit) and *Prunella vulgaris* (Self-heal).

The hay was perhaps imported for animals stalled in the town, while the seeds of species more characteristic of pasture could have been brought in by stock pastured elsewhere. Salisbury, (1961) mentions seeds of certain grassland plants which have been found in horse and cattle dung. These include *Achillea millefolium* (Yarrow), *Chrysanthemum leucanthemum* and *Ranunculus repens*.

Disturbed ground and waste places

Plants of waste places and disturbed ground which frequently represent the "local" flora on many urban sites occur in all the samples. The important factor for the establishment of this weed flora is a degree of soil disturbance, so colonisation can occur in a fairly competition free environment. Rubbish dumps may contain viable seeds, for example, from the animal dung mentioned previously. Others may have arrived as impurities in crops, though most weeds have a very efficient natural means of seed dispersal. *Anthemis cotula*, though generally regarded as a weed of arable land, may have been an inhabitant of urban situations in the past. Archaeological evidence suggests that it was widespread on nutrient-rich disturbed ground in medieval towns (Hall *et al*, 1983) and Kay (1971) records it in Europe from villages where it grows in farmyards, near ponds, in chicken runs and on roadsides. Other species eg *Urtica dioica* (Stinging Nettle), *Stellaria media* (Chickweed), *Solanum nigrum* (Black Nightshade) and *Chenopodium* sp.

Table 1. Results of seed analyses

	Common name	Column A						Column B						Habitat
		CCP 236 cl	CCG 237 org	BWQ 238 org	BLZ 239 org	BWB 240 cl	BWA 241 org	AZO 130 org	AZS 131 gr	AZN 132 org	AYZ 133 org	AYX 134 cl	AYP 135 org	
Achillea sp.	Yarrow	-	-	-	-	-	1	-	-	2	-	-	-	G
Aethusa cynapium L.	Fool's Parsley	-	-	-	1	-	-	1	-	1	-	1	-	Da
Argostemma githago L.	Corncockle	-	-	-	-	-	-	1f	-	5f	-	-	6f	Da
Alisma sp.	Water Plantain	2	-	-	-	-	-	-	-	-	-	-	-	A, B - shallow water
Alnus/Betula	Alder/Birch	-	-	-	-	1	-	-	-	-	-	-	-	W
Anethum graveolens L.	Dill	-	-	-	-	-	-	1	-	-	-	-	-	C
Anthemis cotula L.	Stinking Mayweed	2	88	29	101	3	194	65	23	47	31	14	20	Da, esp base rich heavy soil
Aphanes arvensis agg.	Parsley Piert	-	1	-	-	-	-	-	-	-	-	-	-	Da, G - bare places
Arctium cf *minus* L. Bernh.	Lesser Burdock	-	-	-	-	-	1	-	-	-	-	-	-	D, S, W, - margins
Atriplex sp.	Orache	-	5	4	1	-	8	3	-	11	13	78	2	Da
Bellis perennis L.	Daisy	-	-	62	2	1	3	2	-	2	5	1	5	short G
Betonica officinalis L.	Betony	-	1	-	-	-	-	-	-	-	-	-	-	G - on lighter soils, open W
Brassica sp.	Mustard, Rape, Cole, etc	-	-	-	-	-	-	-	-	-	1	-	-	Da, B, C
Caltha palustris L.	Marsh Marigold	-	-	1	-	-	-	-	-	-	-	-	-	M, wet G & W
Caryophyllaceae gen et sp indet		-	-	-	1	-	5f	9	-	5	-	1	5	
Centaurea cf *cyanus* L.	Cornflower	-	1	-	-	-	-	1	-	1	-	1	-	Da
Centaurea sp.		-	-	-	-	-	1	-	-	1	-	-	1	Da, G
Cerastium sp.	Mouse-ear, Chickweed	-	-	6	-	-	3	23	2	9	6	3	2	Da, G
Chaerophyllum temulentum L.	Rough Chervil	1	-	-	-	-	-	-	-	-	-	-	-	S, D - esp hedges
Chenopodium album L.	Fat Hen	-	5	-	-	-	1	-	2	2	2	2	-	Da
Chenopodium ficifolium Sm.	Fig-leaved Goosefoot	-	-	-	-	-	-	-	-	-	1	1	-	Da esp around manure heaps
Chenopodiaceae gen et sp indet		-	3	3	1	-	2	1	1	2	3	-	-	
cf *Chenopodium*		-	-	-	-	-	1	-	1	-	-	-	-	
Chrysanthemum leucanthemum L.	Ox-eye Daisy	-	1	1	3	1	1	2	-	2	-	-	8	G
Chrysanthemum segetum L.	Corn Marigold	-	2	9	14	-	17	15	1	4	11	19	10	Da - acid soils
cf *Cirsium* sp.	Thistle	-	-	-	-	-	11	-	-	-	-	-	-	Da, M, G, S
Cirsium/Carduus sp.		-	-	-	-	-	-	-	1	-	-	-	-	
Corylus avellana L.	Hazel	-	17f	3f	6f	-	-	-	2f	8f	-	-	-	S, W
Cyperaceae gen et sp indet	Sedge	1	15	25	14	-	80	59	16	26	47	7	9	on wetter soils
Daucus carota L.	Wild Carrot	-	-	-	-	-	-	-	-	-	2	1	-	G, dry calcareous soil, C
Eleocharis palustris/uniglumis	Spike Rush	-	4	4	15	1	26	19	10	5	54	5	-	A, shallow water, M, wet G - open vegetation
Galaeopsis cf *tetrahit* L.	Common Hemp Nettle	-	1	-	-	-	-	-	-	-	-	-	-	Da, occasionally W, M
Galaeopsis sp.		-	2	-	-	-	-	-	-	1	-	-	-	

Table 1. Results of seed analyses (continued)

	Common name	Column A CCP 236 cl	CCG 237 org	BWQ 238 org	BLZ 239 org	BWB 240 cl	BWA 241 org	Column B AZO 130 org	AZS 131 gr	AZN 132 org	AYZ 133 org	AYX 134 cl	AYP 135 org	Habitat
Gramineae gen et sp indet	grass	3	32	1128	84	20	342	776	102	192	221	80	328	
Heracleum sphondylium L.	Cow Parsnip	-	-	-	-	-	-	-	-	-	-	1	-	D, S, W
Hypericum sp.	St. John's Wort	-	1	-	-	-	-	-	-	-	-	-	-	M, G, S, W
Hypochoeris sp.	Cat's Ear	-	-	-	-	-	1	-	-	-	-	-	-	Da, G
Juncus sp.	rush	24	400	115	73	17	153	735	130	637	555	119	125	M, wet G
Labiatae gen et sp indet		-	1	-	-	-	-	-	-	-	-	-	-	
Lapsana communis L.	Nipplewort	-	-	2	3	-	4	-	-	1	-	1	1	Da
Lathyrus/Vicia sp.	Vetch	-	1c	-	1	-	-	-	-	-	-	-	-	
Leontodon sp.	Hawkbit	-	2	3	2	-	2	4	1	-	7	-	9	G
Linum catharticum L.	Purging Flax	-	1	-	4	-	17	14	6	5	1	1	19	G - esp calcareous soil
Linum usitatissimum L.	Cultivated Flax	-	-	-	-	-	5	1	-	-	-	-	2	C
Lychnis flos-cuculi L.	Ragged Robin	-	-	-	1	-	1	-	-	-	-	-	-	M, wet G & W
Lycopus europaeus L.	Gipsy-wort	-	-	1	-	-	-	-	1	-	-	2	-	M, B
Mentha cf *aquatica* L.	Water Mint	-	-	-	-	-	-	1	-	-	-	-	-	B, M, wet W
Odontites/Euphrasia sp.	Bartsia/Eyebright	-	2	-	-	-	1	13	-	2	4	-	-	Da, G
Papaver rhoeas/dubium/ lecoquii/hybridum	Poppy	-	-	-	-	-	1	-	-	-	1	-	-	Da
Papaver sp.	Poppy	-	1	-	-	-	-	-	-	-	-	-	-	
Picris echioides L.	Bristly Ox-tongue	-	-	-	-	-	4	-	-	1	2	-	1	Da
Picris hieracioides L.	Hawkweed Ox-tongue	-	1	5	-	1	3	2	-	1	-	-	-	G, D
Plantago major L.	Great Plantain	-	-	-	-	-	3	1	5	4	2	1	2	short or grazed Da, farmyards
Polygonum aviculare agg.	Knotgrass	-	1	-	-	-	2	1	-	3	2	-	-	Da
Polygonum lapathifolium/nodosum L.	Pale Persicaria	-	-	-	-	-	-	-	-	1	1	1	-	Da & B - damp
Potentilla cf *erecta* L. Rausch.	Common Tormentil	-	-	-	-	-	-	-	-	1	1	-	-	G, light acid soils
Potentilla sp.		-	2	2	-	-	-	1	-	-	2	-	-	M, G, D, S, W
Prunella vulgaris L.	Selfheal	-	2	6	8	-	3	8	-	7	1	2	6	G, D
Ranunculus acris/repens/bulbosus	Buttercup	-	5	36	5	2	50	16	5	11	4	2	12	damp G, W, D(a)
Ranunculus flammula L.	Lesser Spearwort	-	3	-	-	-	-	-	-	-	-	-	-	M
Ranunculus parviflorus L.	Small-flowered Buttercup	-	-	-	-	-	-	1	-	-	-	-	-	dry Da, G
Ranunculus sardous Crantz.	Hairy Buttercup	-	1	26	-	-	2	2	-	2	8	-	-	Da
Ranunculus sceleratus L.	Celery-leaved Crowfoot	1	-	-	-	-	-	-	-	-	-	-	-	A, B - shallow esp mineral rich water
Raphanus raphanistrum L. (seed & pod)	Wild Radish	-	-	-	-	-	-	-	-	-	-	1	-	Da, esp non-calcareous soil
Reseda luteola L.	Weld	-	-	1	-	-	-	-	-	-	-	-	-	Da, C - dyeing
Rhinanthus minor L.	Yellow Rattle	-	-	-	-	-	-	-	-	1	-	-	-	G
Rorippa microphylla (Boenn) Hyland	One-rowed Watercress	-	-	3	-	-	14	-	-	-	-	-	-	A - moving water

Table 1. Results of seed analyses (continued)

	Common name	Column A CCP 236 cl	CCG 237 org	BWQ 238 org	BLZ 239 org	BWB 240 cl	BWA 241 org	Column B AZO 130 org	AZS 131 gr	AZN 132 org	AYZ 133 org	AYX 134 cl	AYP 135 org	Habitat
Rubus cf *friticosus* agg.	Blackberry	-	1	-	1	-	1	-	-	-	-	-	-	W, S, D
Rubus cf *idaeus* L.	Raspberry	-	1	-	-	-	-	-	-	-	-	-	-	W, S
Rumex acetosella agg.	Sheep's Sorrel	-	1	2	-	-	4	1	2	2	1	-	-	G, Da
Rumex conglomeratus Murr.	Sharp Dock	2	-	1	-	-	-	-	-	1	-	-	-	G - damp
Rumex crispus L.	Curled Dock	-	-	-	-	-	-	-	-	1	1	-	1	G, Da
Rumex maritimus/palustris L.	Golden Dock	1	-	-	-	-	-	-	-	-	-	-	-	B, bare muddy ground beside lakes
Rumex spp.		2	4	3	3	-	2	3	3	10	6	6	2	Da, G, M, S, W
Sambucus nigra L.	Elder	1	-	-	-	-	-	-	-	-	-	-	-	S, W, D, esp base rich & nitrogen rich soils
Scleranthus anuus L.	Annual Knawel	-	-	-	-	-	1	-	-	-	-	-	-	Da, sandy often acid soil
Senecio cf *aquaticus* Hill.	Marsh Ragwort	-	-	4	3	1	4	1	1	2	1	-	-	M, wet meadows
Senecio sp.	Groundsel/Ragwort	-	2	2	1	-	2	-	-	-	-	2	-	Da, M, G
Silene spp.	Campion/Catchfly	-	3	3	1	-	1	1	-	1	-	-	2	
Solanum nigrum L.	Black Nightshade	-	2	-	-	-	-	-	-	-	-	-	-	Da
Sonchus asper L. Hill.	Spiny Sow-thistle	-	1	2	2	1	-	3	1	1	-	3	3	Da
Sonchus oleraceus L.	Sow-thistle	-	-	-	-	-	-	-	-	1	-	-	-	Da
Spergula arvensis L.	Corn Spurrey	-	-	1	-	-	-	-	-	-	-	-	-	Da
Stellaria graminea L.	Lesser Stitchwort	-	-	-	-	-	-	1	-	1	-	-	1	G, W
Stellaria media (L.) Vill.	Chickweed	-	-	-	1	-	4	2	-	-	-	-	-	Da
Stellaria sp.		-	-	1	-	-	1	-	-	-	-	-	-	
Suaeda maritima L. Dunn.	Herbaceous Seablite	1	-	-	-	-	-	-	-	-	-	-	-	M, salt, below HWM
Torilis sp.	Hedge Parsley	-	1	2	4	-	3	3	5	5	2	6	1	Da
Triticum sp.	Wheat	-	-	-	-	1c	-	-	-	-	-	-	-	C
Urtica dioica L.	Stinging Nettle	-	-	-	-	-	10	-	-	-	-	-	-	D, W, S, B
Urtica urens L.	Small Nettle	-	-	-	-	-	1	-	-	-	1	-	-	Da, dry light soils
Valerianella dentata (L.) Poll.	Corn Salad	-	-	1	-	-	-	-	-	-	-	-	-	Da, dry basic soils
	TOTAL	42	601	1494	350	50	992	1791	319	1015	1000	362	577	

f = fragments only; c = carbonised; org = organic layer; cl = clay layer; gr = gravel layer.

Habitats: A = aquatic; B = bankside; C = cultivated; D = disturbed ground; Da = disturbed ground including arable; G = grassland; M = marsh; S = scrub; W = woodland.

Note: *Rumex* identified to species only where perianth present, apart from *R. acetosella* which was identified from the seed.

are also characteristic of nitrogen rich conditions which exist near dung heaps and in the vicinity of farms.

Some plants seem especially tolerant of trampling and compacted soil, such as occurs along pathways and these may have been present in back yards and lanes. *Polygonum aviculare* (Knotgrass), *Rumex crispus* (Curled Dock), *Plantago major* and also *Bellis perennis* all adopt a rosette form, with a sprawling mat form of growth, which can withstand some pressure from trampling.

Many of the plants mentioned are associated with places rich in nutrients, especially phosphates and nitrogen, conditions characteristic of a large town with a huge organic input. It may then be possible to envisage a "farmyard" type environment in yards and other open spaces. Litter swept from the streets, stables and floors onto refuse heaps would have provided ideal conditions for this type of weed community to become quickly established.

Wetland plants

The wetland and maritime plants, present in all the samples, could have originated in much of the low-lying ground around the city.

The earliest sample examined from the top of the early clay bank of the Avon reflects the immediate environment of the area, possibly before any major activity occurred. It contains the only example of *Suaeda maritima* (Herbaceous Seablite), a salt marsh species, probably washed in on the tide. *Ranunculus sceleratus* (Celery-leaved Crowfoot) and *Rumex maritimus/palustris* (Golden Dock) may have lived in muddy pools on the site. Both species like wet places rich in nitrates, and *R. maritimus* can tolerate brackish soils.

Other marshland species could have been brought in on the tide, or arrived in animal dung, or on the feet of humans or stock grazed in the marsh. Hall *et al* (1983) suggest that all the aquatic and marsh plants recovered from the site at 6-8 Pavement, York, could have been imported in the mud on peoples' footwear. *R. sceleratus* is particularly abundant there, and it may be possible that the seeds were transported by the occupants on their frequent visits to the nearby Foss, perhaps to collect water or deposit waste. Amongst the species identified at Dundas Wharf, *Caltha palustris* (Marsh Marigold), *Lycopus europaeus* (Gipsywort), *Rumex conglomeratus* (Sharp Dock) and *Ranunculus acris* (Meadow Buttercup) can live in alluvial marsh not subjected to continual flooding by salt or brackish water. Farther from the maritime influence, and possibly where moving fresh water was present, the seeds suggest a community including *Mentha* cf *aquatica* (Water Mint), *Senecio* cf *aquaticus* (Marsh Ragwort), *Iris pseudacorus* (Yellow Flag Iris) and *Rorippa microphylla* (One-rowed Watercress). Many of the sedges (Cyperaceae), common in all samples, could have occurred in any of these habitats, on the wetter soils.

All samples contain a varying abundance of the calyces of *Trifolium squamosum* (Sea Clover). This lives in sub-marine pasture land and is likely to have arrived in animal dung.

Eleocharis palustris/uniglumis is also common in all samples. *E. uniglumis* can live in estuarine flats and brackish grazing marshes, and while it can tolerate a dilute saline solution in the soil, it seems to grow badly in standing water unless it is very shallow. *E. palustris* (Common Spike-Rush) can also live in brackish conditions, but it is more common in marshes and wet meadows, and around shallow pools where there is little competition from taller species, although it can tolerate some drying out of the soil (Walters, 1949). These may have been collected for domestic purposes, along with other rushes (*Juncus* spp.). Rush gathering is mentioned in an ordinance dated December 14th 1451. "Item that alle that bringe risshes (rushes) to towne to make the market in Wynchestrete at a place thereto ordeyned and in non' other place in peyne of forfaiture of the same" (Veale, 1933: 135).

Although Bristol is some 10km from the Severn Estuary, the effect of the tide is still very marked with a range of 10.3m at Cumberland Basin. In her paper on "The ecology of the Avon banks at Bristol", Rose Bracher (1929) looked at the vegetation on the Avon a few kilometres downstream from the Dundas Wharf excavation. She distinguishes 3 major zones of vegetation on the mud banks of the river. Zone I, the highest level, reveals a closed association consisting chiefly of grasses of coastal meadows and salt marsh; *Hordeum secalinum* (Meadow barley), *Poa rupestris* (Procumbent Poa) and *Agropyron pungens* (Sea Couch Grass). Zone II has typical salt marsh plants, including some of the grasses already mentioned, *Aster tripolium* (Sea Aster), *Glaux maritima* (Sea Milkwort) and *Plantago maritima* (Sea Plantain). Zone III is a zone of bare mud, the lowest part of which is submerged at every tide. Even during the high spring tides, when Zone I is submerged, the ebb is very rapid and the upper zones are covered for a few minutes only. The lowest part of Zone III,however, is submerged for six hours out of every twelve.

In the medieval period, areas of marsh existed along the Avon at St. Augustine's Marsh, Canons Marsh, and the area between the river Avon and the new course of the river Frome is frequently referred to as "The Marsh" (Fig. 1). Flooding of this area seems to have been a major problem, with continual repairs needed to maintain the river banks to keep the channel open. The edges of "The Marsh" along the Frome had to be made firm with stakes sunk in the mud, and gaps filled with sticks, brushwood and rubble to prevent winter floods constantly invading "The Marsh" (Vanes, 1977).

The exact character of "The Marsh" needs some clarification. It would seem likely that in the medieval period there were zones of vegetation similar to those suggested by Bracher (above) along the banks of the Avon and in the marshes, with salt marsh and brackish water species on the upper zones of the river bank. In areas only touched by the highest spring tides, non-halophytic species of pasture plants would have lived on the drier soils, with freshwater marsh plants where the mineral (inorganic) soil remained waterlogged throughout the summer, but the water supply was mostly fresh. While flooding is likely to have occurred through the winter months, both from the freshwaters of the Frome and Avon, and from the high tides, "The Marsh" could have provided suitable grazing land during the summer.

Food Plants

Fruitstones and seeds of succulent fruits are common and give the bulk of the evidence regarding the plant component of diet. The bulk samples (Table 2) show that Apples (*Malus sylvestris*), Sweet Cherries (*Prunus avium*) and Plums (*Prunus domestica*) were important. They were probably cultivated in orchards within the city. There are also a few examples from another orchard tree, the Medlar, (*Mespilus germanicus*). It was formerly much cultivated for its fruit which was not eaten until decayed and soft. It was either eaten raw or pulped into a sweet puree with honey.

Sloes (*Prunus spinosa*) are particularly common and one deposit contained over 1200 stones. They may have been used as a substitute for grapes in "verjuice", a kind of sharp vinegar used in cookery and pickling (Wilson, 1984: 294). Hazelnuts (*Corylus avellana*) were especially abundant. Nuts and such fruits as Blackberries (*Rubus fruticosus*) could have been gathered from scrub, woodland and hedgerows close to the town. Imported goods are represented by a few examples of Peach (*Prunus persica*) and Almond (*P. amygdalus*), while Grape (*Vitis vinifera*) could have been grown locally or imported as dried raisins.

Evidence for the consumption of herbs and vegetables is slight. There were single numbers of *Anethum graveolens* (Dill), *Daucus carota* (Wild Carrot) and *Brassica* sp. (Mustard, Rape, Cole etc) seeds, although the Carrot may have come from a wild variety. Vegetables and herbs present a problem, as, if they were cultivated, most species would have been picked for use before being allowed to set seed, apart from a few for the next year's crop. Documentary evidence is useful here. A charter of 1461 (Cronne, 1946)

Table 2. Results analyses of other plant materials

	Common name	Column A CCP 236 cl	CCG 237 org	BWQ 238 org	BLZ 239 org	BWB 240 cl	BWA 241 org	Column B AZO 130 org	AZS 131 gr	AZN 132 org	AYZ 133 org	AYX 134 cl	AYP 135 org	Habitat
Bryophyta	moss	+	+	+	+	+	+	+		+	+	+	+	S, W
Bud							2							
Leaf Abscission Pad					4							1		S, W
Linum sp. casule fragments	Flax		9											C
Pteridophyta leaf fragments	fern		+			+					+			
Rosaceae sp.	thorn		3	2								1		S, W
Trifolium squamosum L. Calyx tube	Sea Clover			17f			14f	2f		2f	33f	18f	5f	tidal estuaries, turf near the sea
Trifolium sp. Corolla	Clover			101f	4f		17f	13f	2f	2f	36f	16f	69f	
Trifolium sp. Pod cap	Clover			4			22	21	3	4	18	12	14	

f = fragments only; + = present.
Habitats: C = cultivated; S = scrub; W = woodland.

Fruitstones, nuts and pips recovered from bulk sieving

Cornus mas L. — Cornelian Cherry
Corylus avellana L. — Hazel
Jugulans regia L. — Walnut
Malus sylvestris Mill. — Apple
Mespilus germanicus L. — Medlar
Prunus amygdalus Batsch. — Almond
Prunus avium (L.) L. — Wild Cherry
Prunus domestica ssp *domestica* L. — Plum
Prunus persica (L.) Batsch. — Peach
Prunus spinosa L. — Sloe
Vitis vinifera L. — Grape

states that duty had to be paid on goods brought by water for sale in Bristol and these included Peas, Beans, Garlic and Onions.

Industrial plants
A crop grown possibly for industrial use is Cultivated Flax (*Linum usitatissimum*), represented by a few capsule fragments as well as some seeds suggesting that the crop was being processed in the vicinity. It may have been cultivated for the production of linen from the stem fibres, for the seeds which contain oil, and is also edible.

Moss seems to have been utilised in rope making. Hanks of plaited *Polytrichum commune* (Common Hair Moss) up to six feet in length have been recovered.

One of the most interesting discoveries was a deposit which has been interpreted as dyeing waste. This was first noticed during the excavation due to its deep red colour. Deposits of a very similar nature have been closely examined from 16-22 Coppergate, York (Hall and Tomlinson, 1983) and were seen to contain the remains of the dye plants Madder (*Rubia tinctorum*), Woad (*Isatis tinctoria*), and Dyer's Greenweed (*Genista tinctoria*). The sample from Dundas Wharf was chemically analysed by gas liquid chromatography, and was found to contain alizarin and pupurin. These both occur in Madder, the root of which contains the dye. Small fragments of both the root and the rhizome have since been identified. Medieval customs accounts show that Madder was imported to Bristol from Holland and Flanders, probably in a processed form. The Dutch were leading Madder growers and highly skilled refiners. The root would have been dried and then refined by pounding, grinding and sieving to separate the heart of the root which contained the largest quantity of pigment (Schaefer, 1941).

In the same deposit from Dundas Wharf Dyer's Greenweed was identified from fragments of stem epidermis. This plant is chiefly used to give a yellow as a basis for greens, when added to the blue dye plant, Woad. Thousands of seeds of Weld (*Reseda luteola*) were also present, and all parts of this plant went into the making of a pure lemon-yellow dye. Both these plants could have been gathered locally. Weld was formerly cultivated in England and in places was sown after the Barley was harvested to produce a crop the following year (White, 1912). Dyer's Greenweed, a plant of rough pasture, could have been gathered from common land. A 19th century account relates to its collection near Bristol by poor people to sell by the cartload to the dyer (White, 1912).

The identification of Woad has proved more difficult as the processes involved in the refining of this plant are very destructive, and so characteristic anatomical features are difficult to detect. Woad is frequently mentioned in 13th century records in Bristol, being the second major import from France after wine (Bickley, 1900). Its importance is also inferred from the fact that it was a form of property often bequeathed by Bristol merchants. Ludovic Mars, for example, left a quarter of Woad to pay for his burial in St. Mary Redcliffe (Carus-Wilson, 1954).

A selection of wool fibres have also been examined from this sample showing a varied range of colours from reds to blues, browns and blacks. Some white fibres may have originally been yellow but have since faded as yellow dyes are unstable, and so are more difficult to detect chemically than reds and blues. There is at present no information on the mordants used, but future chemical analysis may give more details on this important part of the dyeing process.

Comparison with other sites of the medieval period
Comparison with other waterfront excavations of this period is useful. Construction of the waterfront structures themselves, and associated warehouses, domestic dwellings and industrial complexes vary enormously, and deposits behind the revetments range from domestic rubbish to sand, gravel and redeposited clays and silts. Trig Lane in London (Milne and Milne, 1982) and sites in King's Lynn (Clarke and Carter, 1977) are a good comparison with Dundas Wharf. At these sites rubbish was dumped, eventually resulting in

the encroachment of the waterfront. Unfortunately in these cases little environmental evidence was forthcoming. Samples were examined from the excavation at Bristol Bridge (Greig, 1984), which is about 100 metres upstream from Dundas Wharf. Not surprisingly, a very similar plant assemblage was found. Two interesting additions are *Ficus carica* (Fig) and *Fragaria vesca* (Strawberry), which may indicate the presence of cess material. As at Dundas Wharf, hay is represented, possibly in the form of dung, by *Ranunculus acris/repens/bulbosus*, *Leontodon* sp., *Centaurea nigra*, *Torilis japonica* and *Prunella vulgaris*.

Other waterfronts where plant remains have been studied, vary from the complicated sequence of structures associated with the development of the sea port at Bergen (Krzywinski *et al*, 1983), to the simple utilisation of the waterfront at Norwich (Ayers and Murphy, 1983) where construction of brushwood matting allowed for the beaching of river craft. The organic soils at Bergen originate from latrines, cattle manure and stable deposits. Food remains are similar to those found in Britain, and include *Rubus* sp., *Prunus* sp., *Malus* sp. and *Corylus avellana*, although carbonised grains from fire layers are better represented. Deposits at Norwich are also of a mixed nature, mostly domestic rubbish, upon which transient ruderal and wetland vegetation developed. As in Bristol, the river was an important influence and may have transported some of the aquatics from further upstream. It is interesting that *Suaeda maritima*, present in the Bristol deposits, also occurred at Norwich. It is interpreted there as arriving with stock fattened from grazing in salt marsh areas in the lower part of the River Yare valley.

Other non-waterlogged deposits contemporary with Dundas Wharf reflect the underlying uniformity in the local economy of medieval towns. These range from occupation levels at York (Hall *et al*, 1983), and three Saxo-Norman craftsmen's tenements at Durham (Carver, 1979), to deposits interpreted as a local rubbish dump at Highgate in Beverley (Hall and Kenward, 1980). The contents of a 15th century barrel latrine from Worcester (Greig, 1981) included twenty edible plant taxa, mainly fruit, and allows for detailed comment on aspects of the diet and living conditions at the time.

Vertebrate Remains

A major component of the urban refuse comes from the vertebrate remains. A bone assemblage exceeding 100,000 has been recovered, and identification and recording has been completed. Although analysis is still in its early stages it was felt that the economy of the site would be misrepresented if the bone evidence was omitted. Bones from one tenement (127) were therefore extracted for initial assessment and to provide an idea of the range and quality of evidence (Shackleton and Douglass, 1983).

The largest component of the assemblage comprises the major food domesticates: sheep, cattle and pig, and there is a large number of species represented: 20 mammals, 26 birds and 16 fish (Table 3). Sheep are the most common mammal (45.2%) followed by cattle (37.6%) and pig (14.2%), the remaining mammals comprise only 3.2% of the mammal total. Bird proportions are calculated separately, and fowl and goose are nearly equally as common (48 and 49% respectively).

A number of perinatal sheep and pig partial skeletons are present which may represent breeding losses and, at least in the case of sheep, indicate close links between Redcliffe and its rural surroundings. Analysis of age structures can indicate (if somewhat simplistically), (Ryder, 1983)) whether the animals were reared simply for meat or whether other products were important. Cattle were slaughtered mainly as adults, at over 5 years old (Fig. 7) and appear to have been used as draught animals since many lower limb epiphyses showed excesssive wear on the articulation surfaces and had arthrithic-type pathology. Slaughter ages for sheep were generally earlier than for cattle, but spread over more than a year (Fig. 7). An early peak occurs at around 18 months and a second, major peak at about 3 years. This might indicate a mixture of sheep

Table 3. Species present from Tenement 127

MAMMALS

Taxon	*No.*	*%*	*Taxon*	*No.*	*%*
Cattle	1052	37.6	Fallow Deer	1	+
Sheep & Goat	1266	45.2	Weasel	14	0.5
Pig	400	14.2	Brown Rat	1	+
Horse	3	0.1	Black Rat	2	+
Dog	1	+	House Mouse	3	0.1
Cat	36	1.3	Harvest Mouse	1	+
Hare	5	0.2	Vole sp.	1	+
Rabbit	7	0.2	Shrew sp.	1	+
Red Deer	2	+	Mustelid sp.	1	+
Roe Deer	1	+			

Total fragments: 1298 (67.5 %)

BIRDS

Taxon	*No.*	*%*	*Taxon*	*No.*	*%*
Fowl	549	47.6	Snipe	1	+
Goose	561	48.6	?Common Gull	1	+
Duck	7	0.6	Herrring Gull	1	+
Heron	1	+	Lesser Black Backed Gull	1	+
Mallard	1	+	Pigeon	2	0.2
Teal	8	0.7	Skylark	1	+
Partridge	1	+	Thrush sp.	1	+
Crane	1	+	Blackbird	1	+
Moorhen	1	+	Redwing	1	+
Lapwing	1	+	Song Thrush	2	0.2
Dunlin	1	+	Finch sp.	1	+
Curlew	1	+	House Sparrow	2	0.2
Woodcock	3	0.3	Starling	3	0.3

Total fragments: 1154 (27.8 %)

FISH

Taxon	*No.*	*%*	*Taxon*	*No.*	*%*
Roker	56	28.9	Hake	12	6.2
Pike	5	2.6	Ling	7	3.6
Eel	7	3.6	Bass	5	2.6
Conger Eel	51	26.3	Herring	15	7.7
Whiting	13	6.7	Bream sp.	1	0.5
?Pollack	1	0.5	Atlantic Mackerel	9	4.6
Cod	8	4.1	Gurnard sp.	1	0.5
Haddock	2	1.0	Plaice	1	0.5

Total fragments: 194 (4.7 %)

+ = <0.1%

bred specifically for meat and sheep which were important for other products, primarily wool. The majority of pigs were killed young, 95% having been slaughtered by 3 years old (Fig. 7). Documentary evidence suggests that many were kept in the city where they would have converted refuse into a home grown supply of meat (Ralph, 1981).

Analysis of butchery marks and anatomical representation indicates a number of categories of refuse, ranging from domestic waste through slaughter and butchery waste to industrial processes. The latter are evidenced by groups of horncores of cattle and sheep chopped or sawn from the skull (Fig. 8). These probably relate to hornworking and/or tanning (often both occur together). A large number of long bone fragments broken into 3-5cm long slivers is present. This may be raw material for bone working (mainly pins) or for glue manufacture. There are also numerous worked bone and finished objects including bone handles (often sheep tibiae) toggles and pins, and many goose radii needles. Figure 8 illustrates the main concentrations of butchery which can be seen to be remarkably consistent, indicating similar butchery practices in the district. Evidence of butchery on cattle is greater than on sheep, and this can be related to the intensity of preparation: sheep joints more often being left bone-in, and cattle more often boned-out. The butchery from Dundas Wharf can be contrasted with that from the later quay-side site of Narrow Quay, Bristol where cattle butchery is much more formalised and even less varied (Levitan, pers comm).

In addition to the domesticates, a number of wild animals were exploited for food and these include deer, hare and rabbit, and birds such as pigeon, skylark, mallard, woodcock etc. The waterfowl and wetland birds (eg heron, crane, moorhen, lapwing, woodcock, snipe etc) may have been caught on "The Marsh", on the river (in more rural districts) and at the Severn Estuary or Somerset Levels. The presence of the species may indicate an autumn wildfowling market. Birds such as pigeon, skylark, blackbird, starling etc were often exploited in the past but may equally have been chance occurrences.

The fish are another important source of food, and a large range of species was caught, all of which could be found in coastal and inland waters of the South West. Conger eel, ling and herring appear to have been favourites, but simple bone counts are not ideal for assessing relative importance of fish, and analysis of these bones is not detailed enough to attempt this here.

The presence of domestic animals such as dog and cat indicates that these animals were thrown in with the rubbish and may not represent pets, but more likely scavengers, and a few cat skeletons may represent individuals deliberately drowned here.

A final group of animals are those which may not have been exploited, but which either exploited human habitats, or are chance occurrences. These are mainly the small mammals such as house mouse and both species of rat, and birds such as the gulls and corvids. The presence of voles, shrew and harvest mouse is interesting and supports the earlier assertion of close links with more rural areas, also borne out by the evidence from the plant remains.

CONCLUDING REMARKS

Milne (1981) has argued that it is important to identify whether waterfront properties were extended to gain land, create deep water berths, to overcome silting or to consolidate the frontage. At Dundas Wharf it would appear that the 13th century occupation was designed primarily to make the best use of the tides, to create deep water berths for local trading ships. This activity led to increasing bank indentation, and subtle changes in the hydraulic regime. As a result of this the rate of sedimentation greatly increased and estuarine clays rapidly accumulated against the quay walls. The quay extensions, therefore, must also be seen as an attempt to overcome this

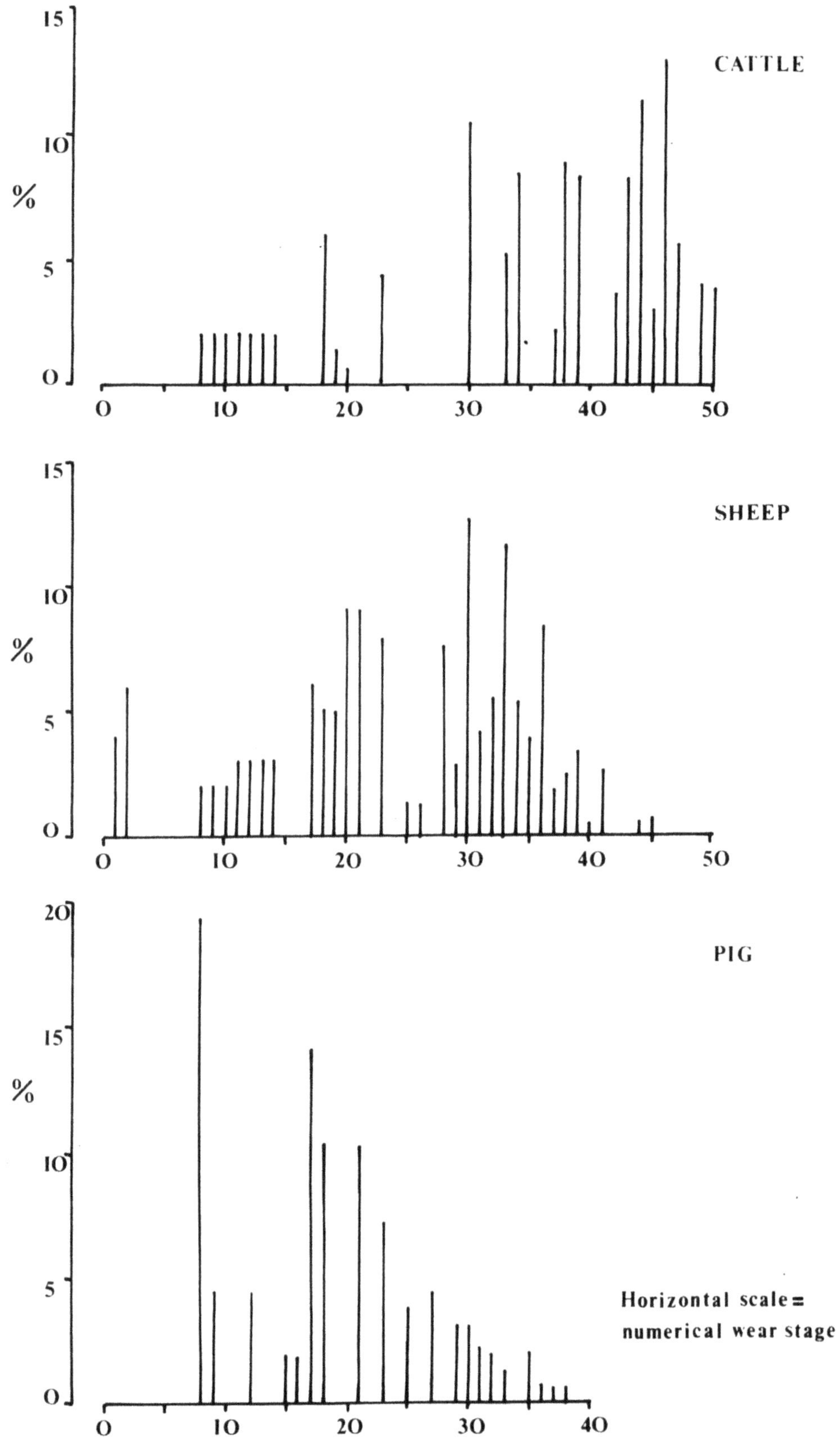

Figure 7. Diagram showing ageing evidence for cattle, sheep and pig from mandibles using the method of Grant (1975)

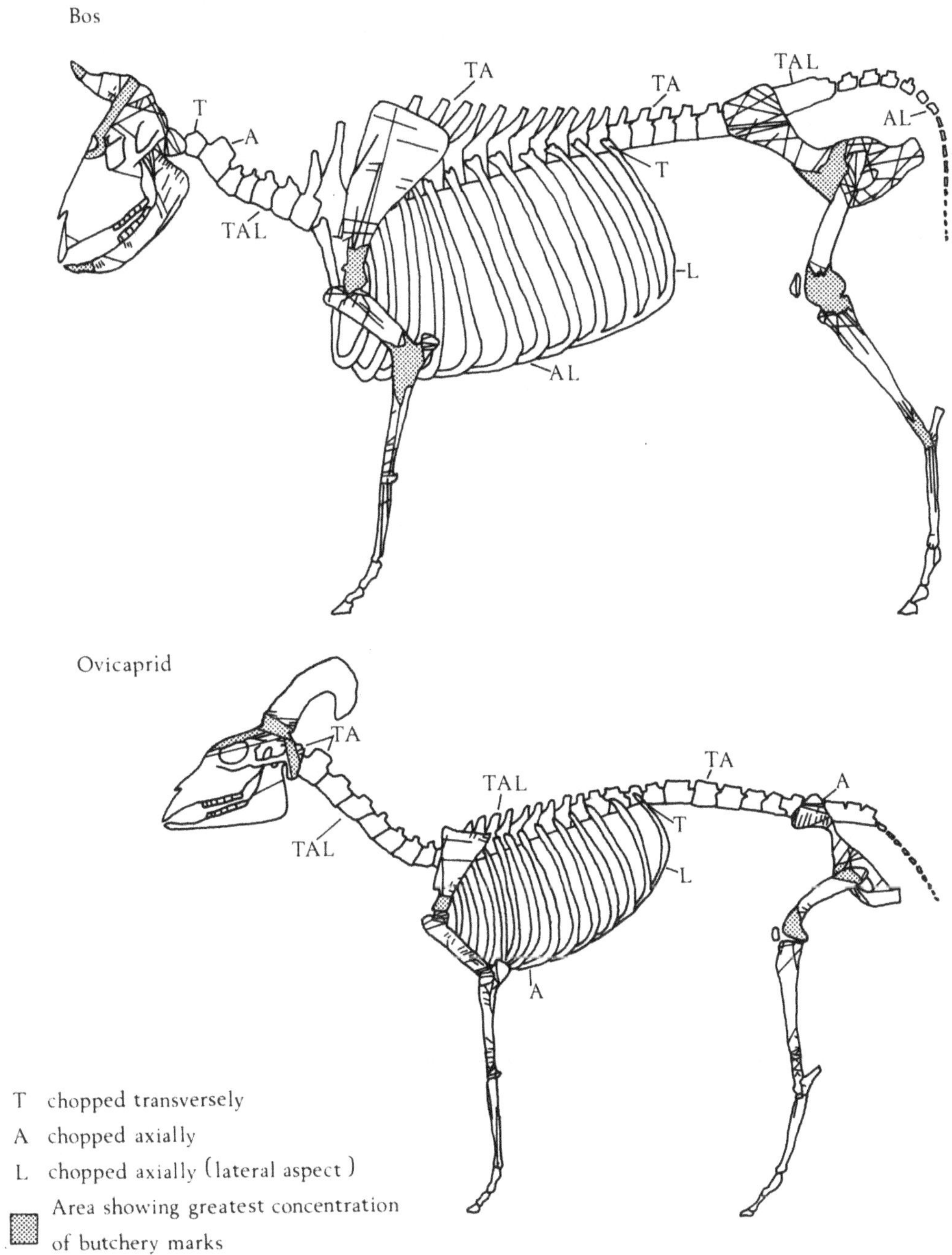

Figure 8. Diagrammatic representation of butchery marks on cattle and sheep.

problem. By the 14th century these early advances upon the river were consolidated and major quays were constructed within the neap tide range.

During the construction of these quays, vast amounts of organic refuse were disposed of into the river. Although the tidal action was probably sufficient to scour away much of this material and maintain a thoroughly oxygenated body of water, approximately three metres of organic and clay deposits built up between the quay walls. The organic remains are seen to be the result of the deposition of both plant and animal material and to represent the large input into the town of foodstuffs and raw materials for domestic and industrial purposes.

Rubbish accumulation and disposal in medieval towns is known to have been a serious problem, and some of the earliest municipal records concern efforts to control the disposal of human and animal waste (Keene, 1982). In Bristol, the Quay, The Back and the "commune ground" by the Quay were often cluttered by dumps of goods and, judging by the repetition of ordinances against this, it seems the council had some difficulty in enforcing its will. Out of concern for cleanliness, rubble and dung could only be thrown into the marsh or at fixed places along the Quay (Bickley, 1900; Veale, 1933). However it seems householders threw their rubbish into the streets despite the fines which were imposed on them.

The plant and animal remains tell us much about the diet of the inhabitants of early medieval Bristol, and show the wide range of resources which were exploited. It seems that the bulk of the bone is food refuse and represents both domestic and butchery waste. Sheep and cattle are the most commonly represented species, followed by pigs, which are known from documentary evidence to have been kept within the city. Marine resources clearly provided an important part of the diet with many species present which could have been caught in local waters. Fruit stones and seeds of succulent fruits give the bulk of evidence for the plant component of the diet, and there is slight evidence for the consumption of herbs and vegetables.

In common with other medieval towns, the plant remains illustrate the wide range of wild and cultivated plants which were brought into Bristol. Some provided the raw material for industry and the importance of dyeing in Redcliffe is indicated by the presence of dye plants in one deposit. Animal bone also provided the raw material for horn working and glue manufacture which may have been carried out locally.

While the studies of the sediments and biological remains have provided information on the the environment and activities at Dundas Wharf, it is felt that more general conclusions about the economy of medieval Bristol must await further analysis. The deposits described in this paper are not restricted to Dundas Wharf. At Bristol Bridge 2 metres of clay and organic sediment buried a 13th century wooden revetment and similar structures have been excavated at Canynges House. A comparison between the rubbish deposits from the three excavations may reveal if there are any significant differences in the activities along the medieval waterfront.

ACKNOWLEDGEMENTS

The excavation was funded by the Manpower Services Commission and the post-excavation by Bristol City Museum and Art Gallery. We are indebted to the many people who have helped during the preparation of this paper. We would like to thank the following for their specialist advice and comments: L. Good (excavation director) and P. O'Connor (excavation supervisor) have provided invaluable help in the interpretation of the stratigraphy; M. Robinson (University Museum, Oxford) and J. Greig (Birmingham University) for their comments on the plant remains; D. Hill (Bristol University) for his help with the dye plants and Professor M. Whiting (Bristol University) for carrying out the chemical analysis on the madder sample; K. Crabtree (Bristol University) and S.

Juggins (University College London) for their help with the diatoms; M. Allen (Ancient Monuments Laboratory) for the archaeomagnetic dating; A. Philpott (Bristol University) for the photography. We are grateful to the Geography Department, Bristol University for housing us. Finally we would to thank J. Douglass and B. Levitan for the bone report, and, in particular V. Straker for her continued support throughout the project.

REFERENCES

Amspoker, M.C. and McIntire, C.D. 1978. Distribution of intertidal diatoms associated with sediments in Yaquina Estuary, *Oregon J. Phycology* **14**, 387-395.

Ayers, B. and Murphy, P. 1983. Waterfront Excavation and Thetford Ware Production, *East Anglian Archaeol.* **17**.

Battarbee, R.W. 1983. *Diatom analysis of River Thames foreshore deposits exposed during the excavation of a Roman waterfront site at Pudding Lane, London,* London: University College, Department of Geography (Working Papers in Palaeoecology No. 2).

Bickley F.B. (ed) 1900. *The Little Red Book of Bristol,* Bristol: W. Crofton Hemmons.

Boyd, P.D.A. 1981. The micropalaeontology and palaeoecology of medieval sediments from the Fleet and Thames in London *in* J.W. Neale and Brasier (eds) *Microfossils from recent and fossil shelf seas,* Chichester: Ellis Horwood, 273-292.

Bracher, R. 1929. The Ecology of the Avon Banks at Bristol, *J. Ecol.* **17**, 35-81.

Carus-Wilson, E.M. 1954. *Medieval Merchant Venturers*, London: Methuen.

Carver, M.O.H. 1979. Three Saxo-Norman Tenements in Durham City, *Medieval Archaeol.* **23**, 55-9.

Clapham, A.R., Tutin, T.G. and Warburg, E.F. 1962. *Flora of the British Isles,* Cambridge: University Press.

Clarke, H. and Carter, A. 1977. *Excavations in King's Lynn, 1963-70,* LondoSociety for Medieval Archaeology (Monograph Series No. 7).

Cronne, H.A. (ed). 1946. Bristol Charters 1378-1499, *Bristol Record Society*, Publication **11**.

Fitzherbert, A. 1979. *Booke of Husbandry*, Amsterdam: Theatrum Orbis Terrarum.

Grant, A. 1975. The use of tooth ware as a guide to the age of domestic animals - a brief explanation *in* B. Cunliffe (ed) *Excavations at Porchester Castle 1: Roman,* London: Society of Antiquaries of London (Research Report No. **32**), 437-450.

Greig, J. 1981. The investigation of a medieval barrel latrine from Worcester, *J. Archaeol. Sci.* **8**, 265-82.

Greig, J. 1984. The palaeoecology of some British hay meadow types *in* W. van Zeist and W.A.Casparie (eds) *Plants and Ancient Man,* Rotterdam: A.A. Balkema, 213-226.

Hall, A.R. and Kenward, H.K. 1980. An interpretation of the Biological Remains from Highate, Beverley, *J. Archaeol. Sci.* **17**, 33-51.

Hall, A.R. Kenward, H.K., Williams, D. and Greig, J.R.A. 1983. *Environment and Living Conditions at Two Anglo-Scandinavian Sites,* (The Archaeology of York, **14/4**), London: Council for British Archaeology.

Hall, A.R. and Tomlinson, P.R. 1983. Dyeplants from Viking York, *Antiquity* **58**, 58-60.

Harding, N.D. (ed) 1930. Bristol Charters 1155-1377, *Bristol Record Society*, Publication **1**

Hendey, N.I. 1964. *An introductory account of the smaller algae of the British coastal waters,* London: HMSO.

Heyworth, A. and Kidson, C. 1982. Sea-level changes in southwest England and Wales, *Proc. Geol. Assoc.* **93**, 91-111.

Hustedt, F. 1930. *Die Susswasser flora Mitteleuropas, Heft 10 Bacillariophyta (Diatomeae),* Jena: Fischer.

Hustedt, F. 1957. Die Diatomeenflora des Fluss-systems der Weserim Gebiet der Hansestadt Bremen, *Abhandlungen Naturwissenschaftlichen Verein Bremen* **34**, 181-440.

Kay, Q.O.N. 1971. *Anthemis cotula* L. *J. Ecol.* **59**, 623-36.

Keene, D.J. 1982. Rubbish in medieval towns *in* A.R. Hall and H.K. Kenward (eds) *Environmental archaeology in the urban context*, London: Council for British Archaeology (Research Report No. **43**).
Krzywinski, K., Fjelldal, S., and Soltvedt, E.C. 1983. Recent palaeoethnobotanical work at the medieval excavations at Bryggen, Bergen, Norway *in* B. Proudfoot (ed) *Site, Environment and Economy* (Symposia of the Association for Environmental Archaeology No. 2), Oxford: British Archaeological Reports (International Series **173**), 145-169.
Levitan, B. 1982. *Excavations at West Hill Uley: 1979 the sieving and sampling programme*, Bristol: Western Archaeological Trust (Occasional Paper No.**10**).
Lord, J. and Southam, J. 1983. *The Floating Harbour, A Landscape History of Bristol City Docks*, Bristol: Redcliffe Press.
Milne, G. 1981. Medieval riverfront reclamation in London *in* G. Milne and B. Hobley (eds) *Waterfront Archaeology in Britain and Northern Europe*, London: Council for British Archaeology (Research Report No. **41**), 32-36.
Milne, G. and Milne, C. 1982. *Medieval waterfront development at Trig Lane, London*, London: London and Middlesex Archaeological Society (Special Paper No. **5**).
Moss, B. 1983. Environmental evidence: Diatoms *in* B. Ayers and P.Murphy (eds) *op cit*.
Ralph, E. 1981. *The Streets of Bristol*, Bristol: Bristol Branch of the Historical Association (Pamphlet **49**).
Robinson, M. 1979. Plants and invertebrates:methods and results *in* G. Lambrick and M. Robinson *Iron Age and Roman riverside settlements at Farmoor, Oxfordshire* (Oxfordshire Archaeological Unit Report No. 2), London: Council for British Archaeology (Research Report No. **32**), 77-106.
Ryder, M.L. 1983. *Sheep and Man*, London: Duckworth.
Salisbury, E. 1961. *Weeds and Aliens*, London: Collins.
Schaefer, G. 1941. The cultivation of madder, *CIBA Rev*. **39**(1), 398-406.
Shackleton, J. and Douglass J. 1983. Environmental sampling in Redcliffe Street 1982-3 *Bristol Avon Archaeol*. **11**, 35-6.
Sherbourne, J.W. 1965. *The Port of Bristol in the middle ages*, Bristol: Bristol Branch of the Historical Association (Pamphlet **13**).
Tansley, A.G. 1939. *The British Islands and their vegetation*, Cambridge: University Press.
van der Werff, A. and Hulls, H. 1958-66. *Diatomeenflora van Nederland*, (Eight parts published privately) The Netherlands: van der Werff.
Vanes, J. 1977. *The Port of Bristol in the Sixteenth Century*, Bristol: Bristol Branch of the Historical Association (Pamphlet **39**).
Veale, E.W.W. 1933. *The Great Red Book of Bristol*, Bristol: Bristol Record Society.
Walters, S.M. 1949. *Eleocharis J. Ecol*. 37, 192-206.
White, J.W. 1912. *The Flora of Bristol*, Bristol: John Wright and Sons.
Wilson, C.A. 1984. *Food and Drink in Britain*, Harmondsworth: Penguin.

PREHISTORIC AND ROMANO-BRITISH SITES AT WESTWARD HO!, DEVON ARCHAEOLOGICAL AND PALAEO-ENVIRONMENTAL SURVEYS 1983 AND 1984

N.D. Balaam, M.G. Bell, A.E.U. David, B. Levitan
R.I. Macphail, M. Robinson and R.G. Scaife

with contributions by
M.A. Girling, A.J. Clark, A. Locker
D. Vaughan and B.J. Wilkinson

INTRODUCTION

Evidence of prehistoric human activity preserved beneath the beach deposits at Westward Ho! Devon has been observed and recorded over more than 100 years. Details of finds recovered from the area have been published sporadically during this period but information relating to the exact nature and location of the contexts has been lacking. Despite the variety of associated biological evidence reputedly available at the site there has been no recent extensive scientific study of the remains.

The earliest accounts of prehistoric deposits at Westward Ho! were principally concerned with the "submerged forest" which was occasionally revealed when a combination of storms and spring tides scoured the covering sand from the beach. This seems first to have been reported extensively as a result of the exposure during the winter of 1863/4 (eg Pengelly, 1868) although there is a suggestion from an earlier account that the presence of the submerged forest may have been known to the local inhabitants before this date (Risdon, 1811). In 1866 record was made of flint artefacts recovered from the layer of "blue clay" which was noted as underlying the peat or "forest bed" (Ellis, 1866; Bate, 1866). This record was later augmented by the recognition of both a midden beneath the peat and lines of wooden stakes which were driven into the peat (Hall, 1870). Unfortunately, the records of these discoveries do not document the extent of the midden, nor do they plan or describe any of the stake structures in detail.

The first comprehensive account of the area and the first in which any attempt was made to illustrate and accurately to locate the site was that published by E.H. Rogers (1946). Rogers' paper draws heavily on previous work (particularly that of Inkerman Rogers, an eminent local palaeontologist of the early years of this century), but also includes his own observations, photographs and a location plan. E.H. Rogers' work was principally concerned with making a record of the midden and the area of "submerged forest" and "outer peat" although he did roughly map the area of "inner peat" as it was then exposed. In comparing his own work with that of the earlier observers he made particular note of the absence of any sign of the stake features recorded by Hall (1870; 1879) and of the much reduced size of the midden compared with the dimensions recorded by Ellis (1867).

The most substantial of the more recent reports is that of Churchill (1965). This account, whilst dealing with the archaeology to some extent (it provides the first radiocarbon date, from the upper levels of the peat), is principally concerned with the evidence for sea-level change that is provided by the sediments associated with the midden. Some analytical work on pollen and other micro- and macro-fossils was undertaken but this was limited in extent and no detailed environmental assessment was undertaken. Papers by other workers whose interests are mainly geomorphological include those by eg Stephens (1970).

Other recent papers have featured short accounts of material from the site but have not discussed or described the location and extent of the deposits in any detail. Two radiocarbon dates obtained from material collected from the midden area (by Mr and Mrs T. Jenkyn) were published by Jacobi (1979) in his general assessment of Mesolithic material from the South West.

In 1980 Dr Jacobi brought to the attention of the Inspectorate of Ancient Monuments the fact that what little remained of this important site was being actively eroded by a combination of the sea and souvenir hunters. The remains of the midden in the 1980s were reported to be significantly less substantial than those observed by Churchill. Indeed, even by the time of Churchill's visit to the site in 1962 little remained of the original deposit as seen by Ellis (1867), Hall (1879) and their contemporaries when the midden had been recorded as being up to two feet (0.6m) thick.

In view of the threat to the site through erosion it was decided that the Central Excavation Unit (CEU) should undertake excavation and survey at the site. The object of this work was to define the extent of the midden and associated peat deposits and to record accurately the relationships between the midden and the surrounding "natural" deposits (ie the peat and "clay" layers). Additional objectives were to obtain well provenanced samples for scientific dating of both the archaeological and "natural" sediments and to carry out scientific analyses on the material from the midden and associated levels.

This paper reports the results of the analyses undertaken on the material and samples retrieved from the site in 1983 and 1984. The first part of the paper is essentially a synthesis of the individual analytical reports and the records made on site. The second part consists of the principal reports from which the synthesis was made, giving full details of analytical methods and data. Additional data on the site is preserved in the project archive which will be deposited in the Royal Albert Memorial Museum, Exeter. A microfiche copy of the text and illustrations from the archive will be lodged with the National Monuments Register.

PART I - SYNTHESIS

GEOLOGICAL SETTING AND SITE LOCATION

The village of Westward Ho! lies at the southern end of Bideford Bay (otherwise known as Barnstaple Bay) (Fig. 1), where it is sited partly on outcrops of Upper Carboniferous rocks (Crackington Formation, Geological Survey, 1977) and partly on more recent sediments.

Pleistocene sediments in the area include raised beach and solifluction deposits. The raised beach is probably of Ipswichian age (Stephens, 1974) although the possibility of a Hoxnian date has been suggested (Kidson and Wood, 1974). The area was not directly affected by Devensian ice (Geological Survey, 1977) but the periglacial features (patterned ground and solifluction deposits) found to the west of the town probably date to this period (Stephens, 1974) (Part II.i).

Sediments of Holocene date include the widespread grey/blue silts and clays (of probable estuarine origin) and peats. Immediately to the north of the village these deposits are covered by blown sands (Northam Burrows).

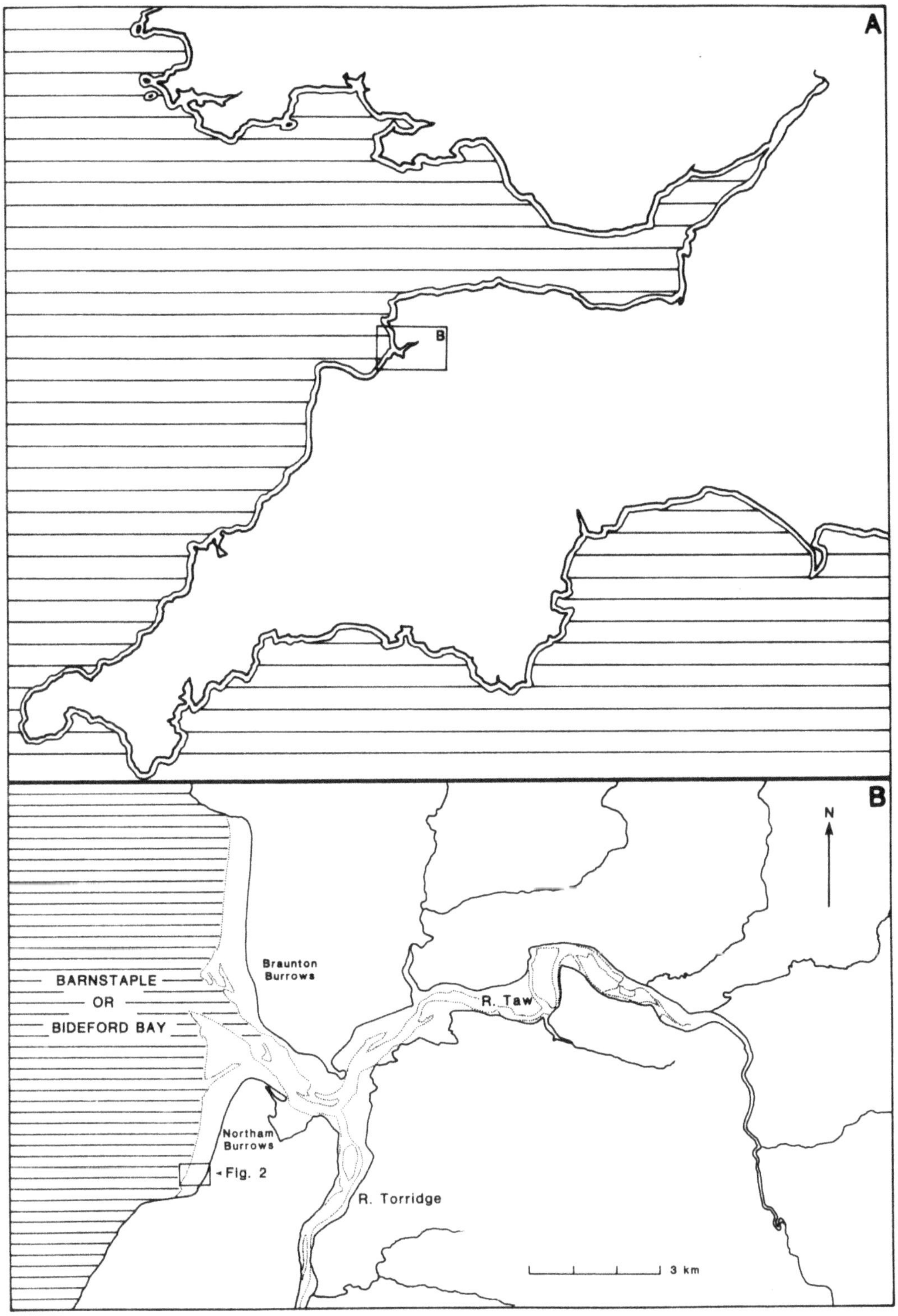

Figure 1. Location maps. South West England and Barnstaple Bay.

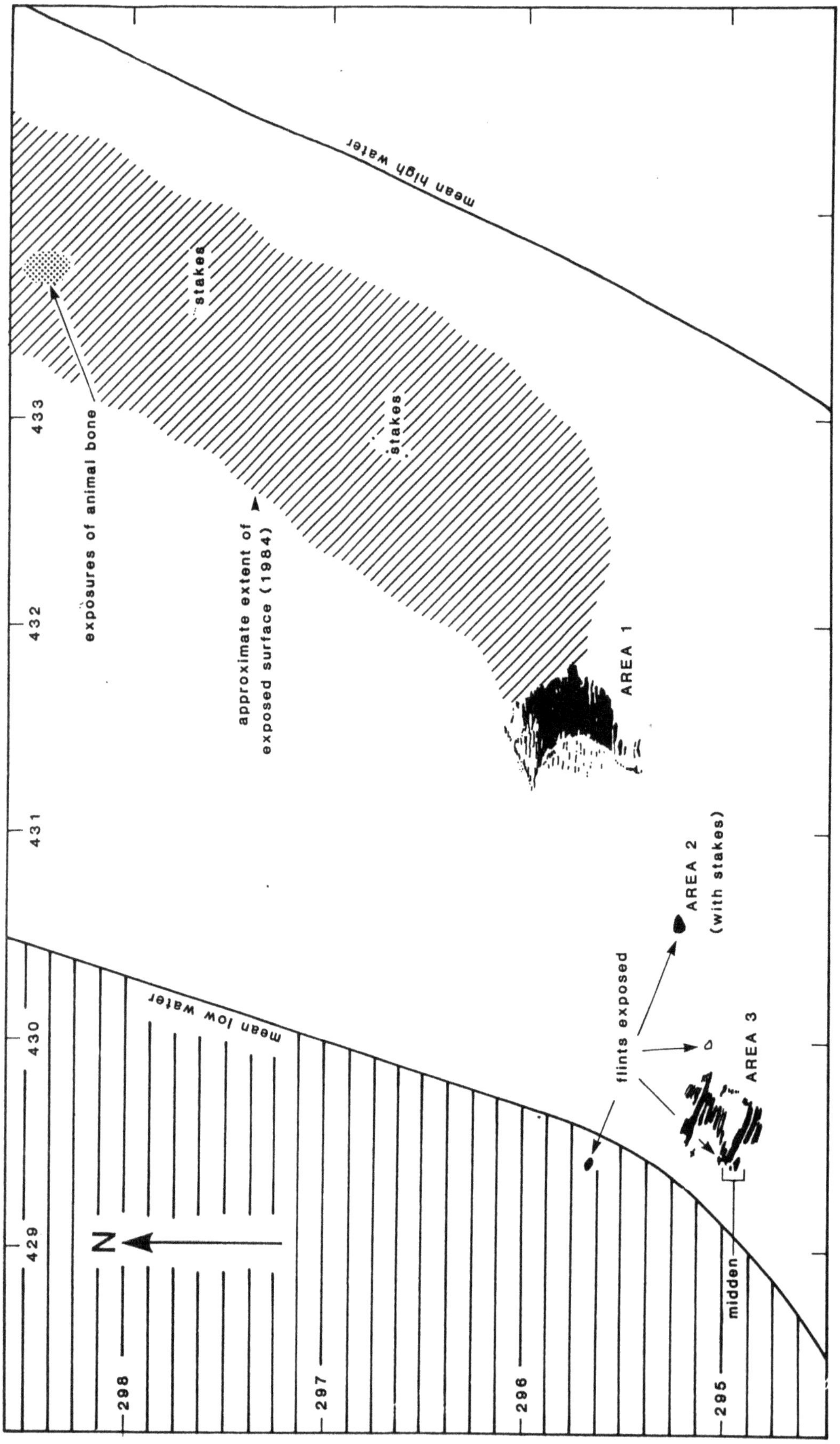

Figure 2. Plan of the beach with areas referred to in the text indicated.

Deposits exposed on the beach

The area with which this project is concerned lies slightly to the north and west of the modern slip-way (Fig. 2) and is contained within a strip of beach approximately 300-400 m wide which runs from the present pebble ridge to the limit of low spring tides. The shore exposed between high and low water marks comprises a rock cut platform to the south, with a variety of superficial deposits elsewhere.

For ease of reference during the survey and in this report the different parts of the beach with which the survey is concerned have been arbitrarily and loosely defined and labelled as Areas 1, 2 and 3 (Fig. 2, Plate 1).

Area 1 is associated with an extensive surface of peat (usually referred to as the "inner peat") that is commonly exposed by relatively minor episodes of tidal scouring. To the north of the peat itself the area extends to include a variety of estuarine deposits which are only rarely exposed in any detail. Extensive parts of this area were laid bare in the late winter/early spring of 1984.

Area 2 is very vaguely defined as being roughly midway between high and low water spring tides and is associated with small areas of peat which overlie grey-blue silty clay.

Area 3 is that area nearest to low water spring tide and includes the deposit traditionally known as the "outer peat". It is within this area that the major collections of Mesolithic material have been found.

SURVEY AND SAMPLING

Any survey of the peat or midden deposits at Westward Ho! is hampered by the position of the site. The archaeological deposits are well below high tide level and they are normally covered and completely obscured by beach sand. It is only at the times when much of this sand has been scoured from the beach by storms that the site is exposed.

Exposure of peat surfaces in some parts of the beach are not uncommon. However, the "midden" site itself lies at about -2.20m OD, NGR 242939 129493 and is only exposed when scouring coincides with spring tides. Such exposures usually take place once every one or two years. At these times the site is accessible for about two or three hours each day and the tides are only low enough for a period of about four or five days. Any detailed excavation of the deposits *in situ* is therefore impracticable. Also, the availability of the site is unpredictable - the scouring of the beach sand being dependant upon a combination of both tide and weather - it was thus not possible to organise an excavation or survey for any specific date. The work was therefore planned to involve minimal excavation on site and to concentrate on describing and surveying key areas already exposed by erosion. Detailed excavation was to be undertaken on blocks cut from the midden and removed to the laboratory.

Following storms at the end of January 1983 the deposits and archaeological features were extensively exposed. Survey and sampling took place on February 1st and 2nd, after which the site was no longer accessible. Throughout the first day the operation was hindered by a westerly gale of force 8, this made verbal communication on site extremely difficult and impossible at ranges greater than about two metres; conditions for surveying and recording were very poor. The weather on the next day was greatly improved and much more recording and sampling was undertaken. Two further short visits to the site were made in March of the same year. In March 1984 areas of the beach were again extensively exposed and the site was visited once more to clarify some points and to assess the degree to which the deposits had suffered from erosion over the intervening year.

Plate 1. Aerial photograph of the beach showing exposed areas of peat.

The survey

The objects of the survey were to identify and describe all extant major archaeological features and to plot them accurately on a large scale plan of the area so that the information could be used by future researchers (these records form part of the project archive).

One of the problems that has bedevilled the study of the Westward Ho! "midden" deposits is the confusion over the exact extent and location of the site. The possibility of the presence of more than one midden has been postulated (Rogers, 1946) as a cause of some of the confusion. The lack of accurate and detailed plans is no doubt due in part to the distance of these deposits from any permanent fixed points and the consequent difficulty in ascertaining their position with any accuracy. It was therefore a priority of this exercise to plot accurately the extent and position of any deposits exposed.

Attempts to produce a detailed plan of the peat exposed in Area 3 by a measured survey had to be abandoned because of the complexity of the outline of the peat and the inclement weather. An accurate plan of the outer peat at a scale of 1:100 was, however, made possible by the use of near vertical aerial photographs taken from an helicopter hovering at a height of *c* 300m (Fig. 3). Reference points for the aerial photograph were fixed using an Electronic Distance Measuring theodolite (EDM) which was sited on the promenade.

The small "island" of peat and clay in which traces of the midden survived (see below) was also planned in greater detail (1:10) by traditional methods. Exposed sections through the midden deposits were drawn on site (Fig. 4).

During the planning and section drawing all exposed surfaces and sections were examined for any archaeological evidence. Artefacts and features located were surveyed in, either directly by EDM, or by measuring from reference points fixed by EDM.

Once the area of the surviving midden was adequately recorded a section of the deposit was selected for removal to the laboratory for examination. Three large (50x80x30cm) blocks were cut out and lifted. Each block comprised a complete sequence of the deposits from the silts at the bottom to the peat at the top. The blocks were supported on boards and wrapped in polythene sheeting to prevent drying out.

In the laboratory the blocks were drawn in plan and section at 1:1 and were excavated and sampled in detail. Additional plans and sections were drawn during the excavation/sampling as new sections were exposed (eg Fig. 5: all other detailed plans and sections are housed in the excavation archive).

The first block to be examined (context 28) was excavated in 1cm horizontal spits (but respecting stratigraphic boundaries) and all material from each spit was kept separately. During this process it soon became apparent that the midden deposits were so extensively disturbed and contorted by prehistoric faunal and (probably) human activity and by later rooting that there was little apparent value in preserving such fine distinctions. Subsequent sampling/excavation employed coarser units (*c* 5cm) and it was felt that these preserved sufficient stratigraphic detail.

During the excavation some material was reserved for specific analyses (entomological, molluscan, botanical), all other material was sieved to 500 microns for the retrieval of macroscopic faunal and botanical remains and artefacts, and for the recovery of charcoal for radiocarbon dating.

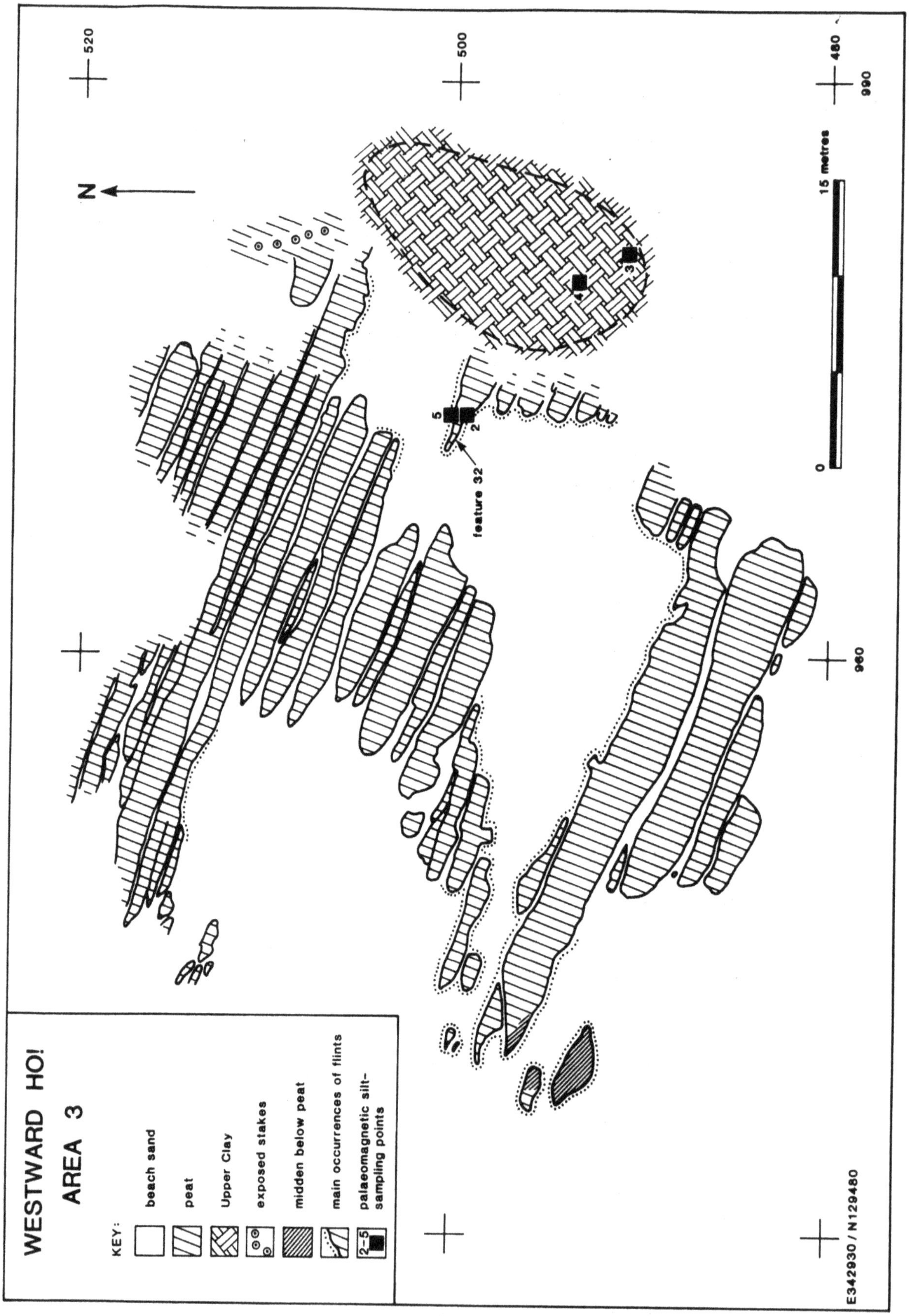

Figure 3. Plan of Outer Peat (Area 3).

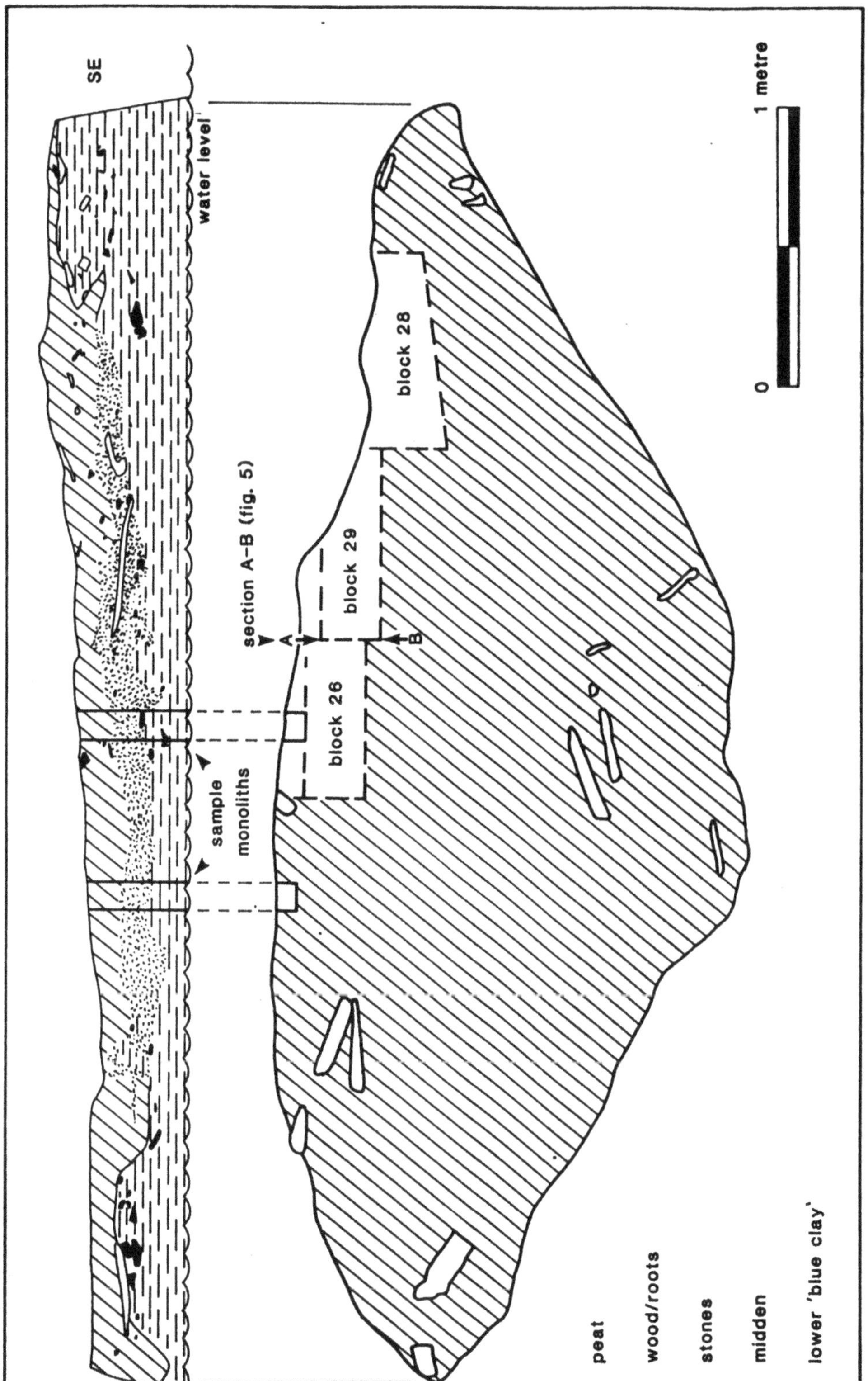

Figure 4. Plan and section of midden "island".

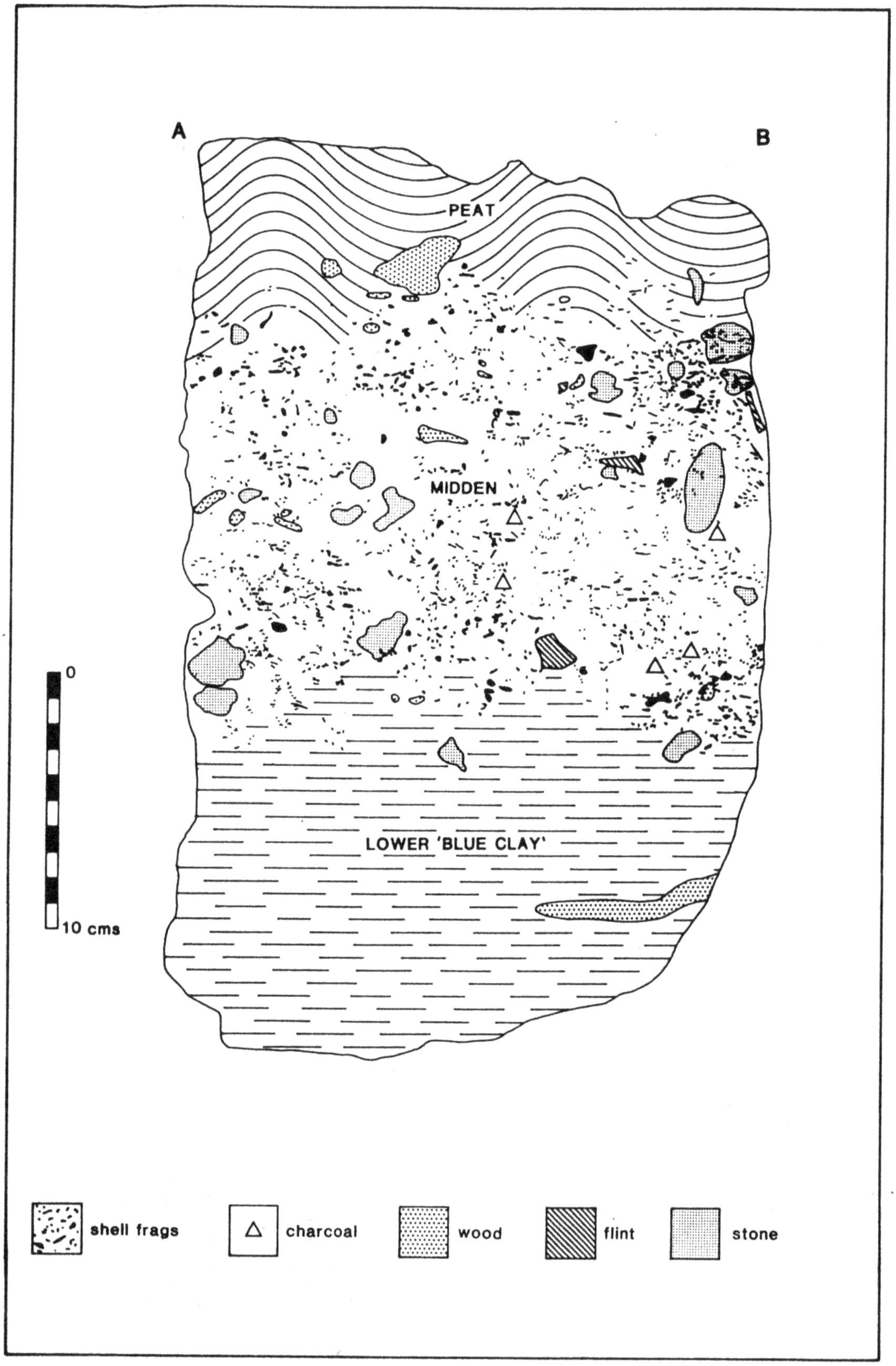

Figure 5. Section through Block 29.

DESCRIPTION OF FEATURES AND HORIZONS

Features and sedimentary horizons present a complex and enigmatic pattern at Westward Ho! and because only small and eroded sections of stratigraphy are exposed for just brief periods at various unconnected locations across the beach, correlation between them is unreliable. Although the series of radiocarbon dates obtained during this project provides an outline chronology for the features and sedimentary episodes, a more comprehensive series of dates and longer periods of observation of fresh exposures would be more satisfactory.

The features and stratigraphy recorded during these surveys are outlined below and are described in greater detail in the succeeding sections. The order in which these are presented is essentially chronological although there are some instances where the relationship or dating is less secure (eg the relative dating of the Neolithic stakes and the "upper clay" sequence).

For each of the major stratigraphic units discussed below there are descriptions of the physical characteristics of the deposit, the dating, the environmental conditions at the time of formation and an account of any evidence that there might be of human activity at the time of deposition. The details of all the radiocarbon dates obtained for material from the site are presented in Table 1 and Figure 6. The details relating to the archaeomagnetic dates are presented in Part II.vii.

The earliest deposits that may be observed on the beach are Ipswichian alluvium and subsequent Devensian drift, the surfaces of which have been subjected to cryoturbation. These stratigraphic units are best seen in Area 1 where extensive areas are frequently exposed. These deposits are overlain by up to one metre of a deposit known as the "lower blue clay".

This "clay" horizon (which may have been truncated in some places) is itself capped by a woody fen peat which survives in varying thicknesses on different parts of the beach. In many areas the peat has been completely removed by marine erosion, where both peat and "clay" survive they have been subjected to scour erosion which has produced elongated "islands" (Plate 2, Fig. 3) which are separated by gulleys of up to 0.6m in depth. The interface between the "lower blue clay" and the peat is distinct and marked in places by abundant evidence of human activity in the form of flint working debris, charcoal fragments and, in one area, the remains of a poorly preserved shell and bone midden of Mesolithic date.

The end of the period of peat formation is marked by further episodes of alluviation of various dates. The latest of these deposits is the best preserved and features a pattern of silt filled channels which are rich in molluscan remains typical of an estuarine environment.

The most recent features of the beach are the sand with which all the above deposits are usually covered and the pebble ridge which is in a constant state of change and movement and whose retreat eastwards has resulted in the exposure of the earlier deposits by marine erosion.

The Lower Blue Clay (contexts 2, 19)

This is the earliest deposit examined in Area 3. In the past it has been referred to as a "blue clay" (eg Rogers, 1946) and although this description is technically inaccurate (see below and Part II.i) it will be used here to avoid confusion with the earlier reports. The archaeological evidence from this deposit is confined to its uppermost 10cm and it is with this upper zone that the following account is principally concerned.

Table 1. Radiocarbon dates

CONTEXT MATERIAL		LAB REF	PROVENANCE	RESULT (BP)	HEIGHT OD (IN METRES)	NGR
AREA 1						
8311718	Wood	HAR-6440	Stake (*Corylus*) from line of close-set stakes	1600±80	0	24335/12977
8311353	Bone	HAR-6513	A group of animal bones from the fill of old estuarine channel	1560±80	0	24337/12984
PEAT (AREAS 2 & 3)						
8311004	Wood	HAR-5642	Two stakes (*Corylus*) from the row of stakes in Area 2	4840±70	-1.0	24305/12952
8311701	Peat	HAR-6363	Top 3cm of peat where sealed beneath upper "blue clay"	5190±80	----	242976/129495
8311028	Peat	HAR-5640	Top 2cm of peat immediately overlying the midden	5200±120	-2.1	242938/129494
8311023	Wood	HAR-5631	Outer rings (*c* 20) of trunk of *Quercus* lying on the surface of the peat	6100±100	-2.0	242960/129487
8311024	Wood	HAR-5630	Tree stool (*Salix*) on surface of peat	5630±80	-2.0	242945/129494
8311029	Peat	HAR-5641	Base of peat immediately overlying the midden	5740±100	-2.2	242938/129494
-------	Wood	Q-672	Top of peat (Churchill, 1965)	6585±130		
-------	Wood	Q-1249	Peat (Jacobi, 1975)	6680±120		
-------	Peat	IGS-42	Roots within peat (Welin *et al*, 1971)	5004±105		
MIDDEN						
8311026	Charcoal	HAR-5632	Combined samples from upper part of midden	6100±200	-2.3	242938/129494
8311027	Charcoal	HAR-5645	Combined samples from lower part of midden	6320±90	-2.4	242938/129494
-------	Bone	Q-1212	Midden (Jacobi, 1979)	6810±140		
-------	Charcoal	Q-1211	"adjacent to midden" (Jacobi, 1979)	6955±140		
LOWER CLAY						
8311705	Charcoal	HAR-6215	Upper 10cm of "lower blue clay"	6250±110	-2.5	242957/129491
8311025	Charcoal	HAR-5644	Upper 10cm of "lower blue clay"	6770±120	-2.5	242938/129494
8311019	Charcoal	HAR-5643	Gulley cut into "lower blue clay"	8180±150	-2.5	242971/129500

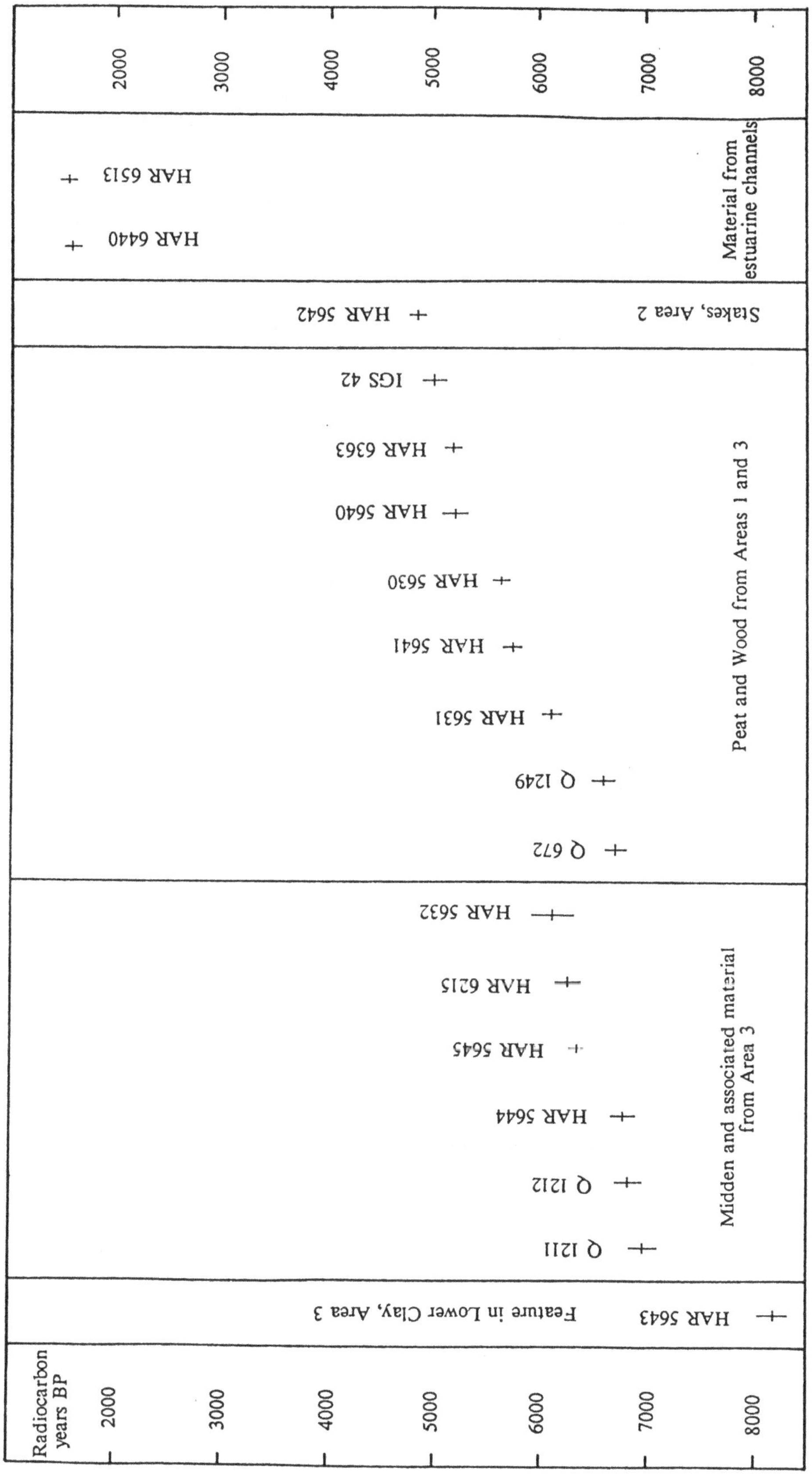

Figure 6. Radiocarbon dates.

a) Description

As it survives today the "blue clay" deposit is of varying depths up to at least 1.1m (demonstrated by augering in Area 3) and it rests variously on drift material, pebble horizons (Stuart and Hookway, 1954) or directly on the rock cut platform. It is not strictly a clay but is rather a uniform (fine) silty clay loam. It is generally dark grey (2.5YR4/0) to grey (2.5YR5/0) in colour (as opposed to blue), but sometimes with faint mottles. The deposit contains a varying, but generally small, amount of organic matter. Where the organic matter is present it typically consists of vertical (root) and horizontal (stem) monocotyledonous plant material. The sediment has very high acidity (pH 2.6) due its sulphide content (see Part II.i)

The uppermost part of this horizon (context 19) is characterised by the presence of flint artefacts and charcoal fragments. The even distribution of this material through this zone is thought to be a result of incorporation during the disturbance of an unstable surface (eg through trampling). There is no evidence of stratification within the zone and no implication that the artefacts and charcoal need be contemporary with the deposition of the clay.

To the north of the main area of peat (Fig. 3) there was an isolated example of a feature cut into the "clay". This was a "V" shaped gulley (context 32), the fill of which was almost indistinguishable from the upper horizon of the "clay". The gulley was only preserved as a small section at the very end of a peat "island".

Micromorphological studies of the sediments (Part II.i) showed them to have no pedological features other than soil ripening. The sediments also demonstrate a lack of porosity which is strong evidence for their never having undergone sustained drying out.

b) Dating

The use of both archaeomagnetic and radiocarbon dating has enabled the approximate dating of the deposition of this horizon (Table 1 and Part II.vii). The archaeomagnetic samples suggest a depositional date for the upper levels of the "clay" in the region of 8400-7800 cal BP (but also allows the possibility that it could be very much older than this (see Part II.vii for detailed discussion). Two samples of charcoal from the upper zone of this horizon were submitted for radiocarbon dating (contexts 8311705 and 8311025). Additionaly, one sample of charcoal from the "V" shaped gulley was submitted. Of these three, the two from the upper 10cm of the horizon produced dates of 6250±110 BP (HAR-6215) and 6770±120 BP (HAR-5644). The sample from the gulley was dated to 8180±120 BP (HAR-5643). In the absence of any further information, and in the light of the other radiocarbon and archaeomagnetic dates, this last seems unnacceptably early as date for the creation of the feature (context 32). It is possible that the early date may be a result of the inclusion of unnoticed residual material or some other factor. For the time being the evidence of this particular date must be overlooked. It is unfortunate that the radiocarbon dates for this deposit are beyond the range of currently available calibration curves (eg Pearson *et al*, 1986) and it is therefore not possible to render them readily comparable with the calendrical result suggested by the archaeomagnetic dating.

The two later dates (HAR-5644 and HAR-6215) are not significantly different from the dates obtained for the midden (see below) and it must remain a possibility that the material on which they were based has become incorporated in the upper part of the "lower clay" significantly after its deposition.

The question of the age of the "lower clay" must remain somewhat enigmatic for the time being. It cannot be presumed that it immediately predates the midden and its associated activity, although evidence for lack of weathering and soil development would suggest that the formation of the two deposits are not greatly separated in time.

Plate 2. Aerial photograph of Outer Peat (Area 3).

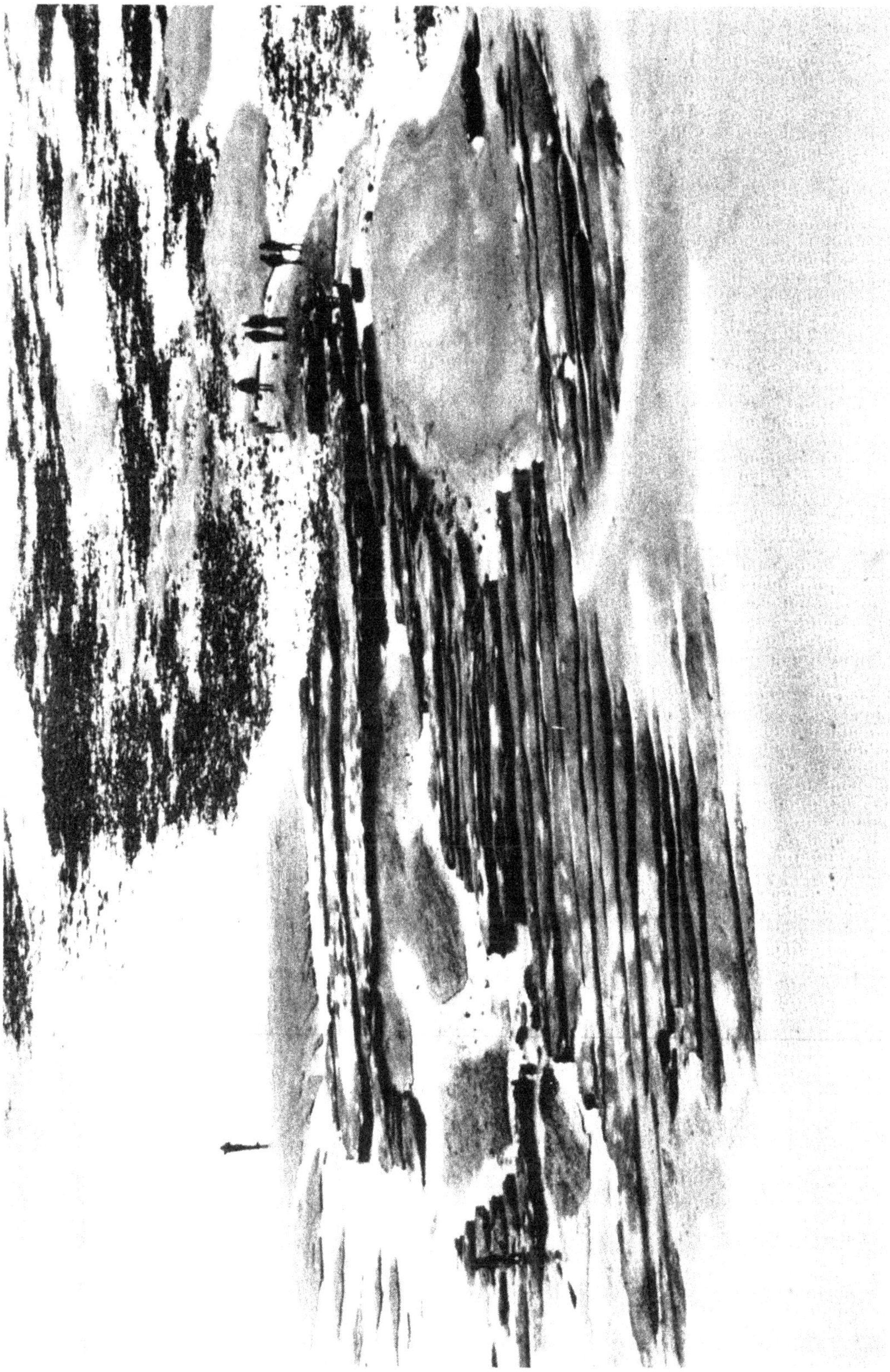

c) The Environment
There is little evidence for the character of the environment of this area during and immediately following the deposition of these "clays", consequently the genesis of these deposits is somewhat in doubt. Analysis of the sediment for both micro- and macro-fossils proved fruitless, there being a near total absence of all categories of evidence such as Mollusca, Foraminifera (J.W. Murray, pers comm), diatoms (R.W. Battarbee, pers comm) and ostracods. The field appearance of this deposit is, however, comparable to that of sediments which underlie foreshore peats elsewhere in the Bristol Channel and in some cases, such as Porlock Weir, Somerset, these contain estuarine Mollusca such as *Scrobicularia plana.*

Some indirect evidence for the environment of formation of this deposit may, however, be seen in the extreme levels of acidity recorded (Part II.i). Estuarine sediments have very marked powers of reduction (Duchaufour, 1982) which result in the formation of sulphides and the reduction of free iron (giving a bluish grey colour). If the saline water table in such sediments is lowered then sulphides oxidise and hydrolyse to produce sulphuric acid and in places the iron may also be oxidised to produce rusty patches (Part II.i). This cycle of processes may well have taken place here and in sediments such as these where there are no carbonates to buffer the acid, very low pH values can be produced and, in such conditions, microfossils such as foraminifera are destroyed.

In the absence of other evidence it seems reasonable, therefore, to adopt the traditional view (eg Churchill, 1965) and to interpret the "lower blue clay" as an estuarine sediment. Nothing more can be said of the environment, the only biological remains to survive in any quantity are charcoal fragments and these being artefactual are of little use for environmental interpretation.

d) Human activity
There is no secure evidence of any human activity contemporaneous with the deposition of the "lower clay", although its surface contains artefacts and occupation debris from Mesolithic activity this is most probably related to the episode of midden formation. Flints occurred commonly both in this interface zone (context 19) and within the midden itself.

The Midden (Context 3)

a) Description
The deposits which represent the Mesolithic shell midden are found only in Area 3 where they are largely confined to a small eroded remnant or "island" to the south west of the main area of the "outer peat". Outlying small patches of midden material were also found in two of the adjacent peat "islands" (contexts 39 and 40, Fig. 3).

What now remains of the midden appears to occupy a slight hollow in the surface of the blue clay. Sediment micromorphological evidence indicates that there may have been a certain amount of truncation of the clays - suggesting the possibility of an artificial hollow - although no evidence of this was observed during the sampling (in view of the disturbance of these horizons (see below) this is hardly surprising). The fabric of the midden was examined in the field and has been the subject of micromorphological and sedimentological examination (Part II.i). The midden comprises fragmented shell, mineral material (mainly silts), humified organic matter and occasional substantial roots which are relicts of trees and shrubs that grew in the overlying peat forming community but which rooted in this more stable substrate. There are also small amounts of charcoal, animal bone and a number of flint artefacts.

Both the micromorphological study and detailed macroscopic examination show the midden to have extensive evidence of bioturbation, which contrasts with the undisturbed and well laminated peat which overlies it (see Part II.i). It is probable that the midden has been disturbed to a significant extent by the rooting of the oak, willow and hazel woodland which grew during the development of the overlying peat. During the laboratory sampling and excavation it was apparent that clear boundaries between clay and midden or midden and peat could not be defined. Much of this disruption and blurring of horizons may be attributed to the phase of peat growth but it is likely that some of the bioturbation and micro-gulleying, especially of the lower horizon boundaries, may have been contemporary with the formation of the deposit and may have been due to animal (including human) disturbance.

b) Dating

Two samples of charcoal from the midden were submitted for radiocarbon dating (Table 1). There are, in addition, two radiocarbon dates from earlier samples which were taken from, or adjacent to, the midden by Mr and Mrs T. Jenkyn (Jacobi, 1979). There was surprisingly little charcoal in the midden, and it was therefore necessary to combine material from a number of small samples to provide sufficient for two dates. Of these, one date represents the material combined from the upper half of the midden, the other is from contexts in the lower half. The two dates from upper and lower parts of the midden are 6100±200 BP (HAR-5632) and 6320±90 BP (HAR-5645) respectively. Both these dates fit well within the chronological sequence for material from the 1983 survey (Table 1, Fig. 6), however, they are significantly later than those derived from previous investigations and which have been associated with the midden. These dates, 6810±140 BP (Q-1212, bone from the midden) and 6955±140 BP (Q-1211, "burnt wood adjacent to the midden")(Jacobi, 1979) might be better associated with the material from the upper levels of the "blue clay" (see above).

c) The environment

Multidisciplinary study of the biological remains has enabled us to develop a fairly thorough picture of the environment in which the midden was formed and of the micro-environment of the midden itself. Some evidence of environmental change during the development of the midden is demonstrated in the molluscan and pollen records although for the most part there is little distinguishable variation through time and thus the midden has been treated as a single environmental phase.

The environmental analyses undertaken present a picture of a midden set within fairly dense fen carr. The character of this fen carr is best reflected in the pollen analysis (Part II.iv), but is also confirmed by the character of the insect fauna (Part II.vi). The earliest community suggested is that of a mixed oak woodland with a few herbs, of which grasses and sedges were dominant. The presence of oak on, or near the site is also attested by the evidence of oak buds and bud scales in the midden. From the analysis of the molluscan and insect remains it can be suggested that there were pools of stagnant (but not brackish) water within the carr, indeed it seems possible that the midden itself developed from the accumulation of domestic refuse within one such temporary pool. There is certainly evidence from the character of the insect fauna for considerable quantities of damp decaying wood and other vegetable material on and around the site.

Towards the upper levels of the midden there is palynological evidence for the greater importance of willow (*Salix*), birch (*Betula*) and ivy (*Hedera*) on or around the site. The high percentages (23% of arboreal pollen [AP]) of willow pollen suggest that it must have been dominant in the immediate area, and, in the levels of peat overlying the midden there are considerable numbers of macroscopic remains of willow (leaves and wood) which support this. Throughout the part of the pollen sequence represented by the midden there is little evidence of human influence on the vegetation.

The picture of a relatively closed, damp woodland is further suggested by the remains of slow worm (*Anguis fragilis*) scales and bones of frog (*Rana* sp) and bank vole (*Clethrionomys glareolus*). Freshwater bivalves of the genus *Pisidium* suggest the presence of stagnant water, puddles or temporary ponds (as does the evidence of the caddis larvae) whilst the land molluscs such as *Discus* and *Carychium* are common inhabitants of woodland ground litter. There is no evidence from the analyses of any substantial area of open ground in the immediate vicinity of the midden. There is, however, a suggestion from the insect remains of open country and sand dunes beyond the woodland which surrounded the site itself.

The results of the various analyses confirm the status of this deposit as a Mesolithic shell midden. It is not, however, possible to confirm the suggestions of Churchill (1965) and Rogers (1946) that the midden was formed at or near the strand line. No evidence could be found for horizons of marine silt deposition within the midden, nor were the biota a characteristic strandline assemblage, indeed there was very little sign of proximity to the shore.

d) Human activity
The bulk of the material that constitutes the midden comprises fragmented marine bivalve shell. There were few bones of the larger mammals present in the surviving midden although early collections (see Part II.iii) indicate that these, in particular those of red deer (*Cervus elaphus*), may previously have been considerably more abundant. From samples taken in 1983 bones of both red deer and roe deer (*Capreolus capreolus*) were identified as well as poorly preserved fragments of a bovid radius and a calcaneum which are of a size similar to those previously recorded for female aurochsen (*Bos primigenius*). Although all the material from the midden was sieved in detail to recover small fragments and fish bone only three small fish vertebrae were found (possibly all from the tail of one small individual). These vertebrae could not be precisely identified but might represent one of the goby family (Gobiidae) (Part II.iii).

The marine Mollusca provide the major part of the evidence from the midden itself, they do, however, present considerable identification problems since the bulk of the material consists of very comminuted fragments (identification details are given in Part II.ii). Figures 9 and 10 present the results of the analysis of the Mollusca and it can be seen that the assemblage is overwhelmingly composed of two species - mussels (*Mytilus edulis*) and peppery furrow shell (*Scrobicularia plana*). There are also a number of minor species represented.

At first sight it seems strange for a presumably domestic midden to be principally composed of *Scrobicularia*. However, although *Scrobicularia* is not eaten today it does occur with edible molluscs in Mesolithic and later contexts elsewhere (see below Part II.ii) and was therefore presumably eaten.

The marine molluscan assemblage, the charcoal (including carbonised hazelnuts) and flint artefacts are thus the only firm indications for a human origin for this deposit (none of the insect taxa are particularly indicative of human activity). A total of 1074 flints were recovered from Area 3 (both from the upper zone of the "lower clay" and from the midden) and in the absence of any evidence to the contrary it is believed that the presence of all these is due to activity related to the midden. The assemblage is discussed in more detail in Part II.viii. From the analysis of these artefacts it appears that the users of the site were knapping quantities of local flint beach pebbles in order to manufacture microliths for the arming of hunting equipment. The occasional presence of scrapers and utilised flakes and bladelets suggests that at least some other, possibly more domestic, activities were also taking place on the site. The lithic inventory, however, is too incomplete to draw more than these very general conclusions.

The Outer Peat (context 1)

a) Description
The "outer peat" overlies both midden (context 3) and "lower blue clay" (contexts 2, 19). It is fairly uniform in character throughout its depth and is comprised of humified monocotyledonous material with wood and occasional hazel nuts and leaves of common sallow (*Salix cinerea*). Substantial remains of trunks and stools of oak and willow survive in places on the surface of the peat and the roots of these, and other now

vanished trees, are found throughout the peat. Micromorphological study of the deposit indicates the presence of fine laminations of organic and inorganic material within the body of the peat which suggests that the area was subject to episodes of minor (probably seasonal) flooding (Part II.i). Few molluscs are found within the peat and those that are are thought to be derived from the midden by subsequent bioturbation. The insect remains from the peat are well preserved and relatively abundant.

b) Dating

Despite their apparent similarity, the "inner" (Area 1, see below) and "outer" (Area 3) peat deposits at Westward Ho! appear to be of two distinct ages. It is possible that the discrete areas of peat which survive today represent different elements of a continuum of peat development or they may relate to temporally differing sequences.

Due to the extensive erosion the sequence of peat development is nowhere preserved *in toto*. In the immediate area of the midden only a few centimetres survive. A longer sequence is, however, preserved a few metres away where the surface of the peat is overlain and sealed by a deposit of "upper clay". The samples dated for this project were taken from the top and bottom of the peat which overlies the midden, from tree trunks and stools on the surface of this peat and from the uppermost level of peat where it was sealed by the "upper blue clay". On the basis of the radiocarbon dates obtained from material sampled by the CEU it can be estimated that the peat sequence which survives spans a period of 500 to 800 radiocarbon years (Table 1). Of the two dates obtained from earlier samples, that obtained by the Institute of Geological Sciences (Welin *et al*, 1971) fits in quite well with the sequence of CEU samples. However, the material retrieved by Churchill (1965) has produced dates very substantially earlier than any of the other peat samples and it is not clear to what this discrepancy can be attributed. The provenance of these earlier samples is too loosely described to allow any archaeological conclusions to be drawn from the differences in dating results.

c) The environment

The radiocarbon dates obtained indicate that the peat is of Atlantic (Godwin Zone VIIa, Flandrian II) date and the flora represented in the pollen analysis is in agreement with this dating. There are elements of a number of different vegetational communities present and these include the thermophilous elements such as lime (*Tilia* cf *cordata*), (probably of regional or extra-regional origin), ivy (*Hedera helix*) and mistletoe (*Viscum album*), this latter being strongly characteristic of pollen spectra of Atlantic date in lowland Britain. Other tree taxa represented are typical of those genera present during the Atlantic period and it is likely that the vegetation of the surrounding area was one in which oak, elm (*Ulmus*), ash (*Fraxinus*), hazel and willow were strongly represented. The presence of oak, willow and hazel on or near the site is confirmed by the presence of pollen clusters from partially dehisced anthers and by the appearance of macrofossils such as leaves, bud scales and, in the case of hazel, nuts. The insect remains also back-up the botanical evidence. There are clear indications of the presence of oak, willow and hazel as well as suggestions of the presence of rosaceous shrubs and elm.

The abundance of the evidence for the presence of oak and hazel at the site indicates that at least some areas in the immediate vicinity were not subjected to waterlogging. This characteristic of the environment is also reflected in the entomological evidence which is suggestive of an area of deciduous woodland with only localised swampy conditions and small streams.

Unlike many sites of comparable date, alder was not a major component of the vegetational community here. The vegetation represented in these peats was that of a willow dominated fen carr (Salicetum). Willow pollen is present in relatively high frequencies even though willows have generally poor pollen production and dispersal characteristics. Other taxa characteristic of such a fen carr community that are represented in the pollen record are, for example, privet (*Ligustrum*), yew (*Taxus*) and alder buckthorn (*Frangula*).

The evidence of the insect fauna also provides indications of the presence of grassy, open areas beyond the woodland. These may, in part, be associated with an area of sand dunes nearer the shore. The evidence for open ground is greatest in the uppermost sample of the peat and, perhaps, echos the changes in the pollen spectra that might be associated with the "elm decline".

d) Evidence for human activity
There is (surprisingly) little evidence of substantial human activity in the peat which overlies the midden. The presence of hazel nuts throughout the deposit might at first sight be attributed to human agency, however, close examination of the nuts in the macro-botanical analysis has shown that their preponderance is due to natural accumulation; there is no evidence of any human attempts to open or crack them and they appear randomly dispersed in the deposit. Similarly, the preponderance of hazel pollen throughout the pollen record attests to its growth on the site.

The existing evidence for human activity in the vicinity of the site during this phase is indirect and can only be interpolated from the evidence of the pollen and microfaunal analyses. These suggest slight vegetational changes in the upper levels of the peat which overlies the midden (Profile I, levels 1-3, Fig. 12). In the pollen record these changes are seen in the decline in the representation of elm and ash and there is also some indication of increased wetness and waterlogging. This latter effect may have been due to rises in the groundwater table prior to the marine transgression that was responsible for the deposition of the "upper blue clays" (see below) or it may have been a result of human activity within the fluvial catchment which reduced evapo-transpiration and increased ground-surface flow. Further possible evidence for anthropogeny comes from the appearance of pollen of goosefoot (Chenopodiaceae) and ribwort plantain (*Plantago lanceolata*) in pollen zone WH:IV (Fig. 12) although their presence in the pollen record does not necessarily relate to the immediate environment of the site.

Additional slight evidence of human activity *may* be seen in the presence of burnt fragments of the rib of an unidentified Bovid or Cervid and the presence of fish bones which may represent food remains although these may have been derived from the midden itself and have been incorporated in the peat by faunal or rooting activity. Similarly, bioturbation may be responsible for the presence of the few flints, including a microburin, found in the peat and perhaps having their origin with the midden activity. However, Rogers also reports the finding of a small group of flints from the peat (1946: 129, Plate XXI), and it must certainly remain a possibility that there was some human presence during this latter phase.

Structures

Early records of material found at Westward Ho! made mention of wooden structures that were to be seen amongst the remains of tree stumps etc (eg Hall, 1870). Since the reports of the latter part of the nineteenth century there have been no accounts of any such observations. Rogers (1946) notes that he could find no trace of the features referred to and concludes that large parts of the deposit, including those areas which contained the stake structures, might have been removed by marine action.

During the survey of January 1983 Dr Jacobi observed what appeared to be the remains of a stake in the peat in Area 2. This stake seemed to have been broken off flush with the surface of the peat and could easily have been mistaken for a root of one of the trees that had grown on the peat. The stake seemed to be in isolation but cleaning (by washing down with water thrown from a bucket) of the surface of the peat revealed ten others within a small area (Fig. 7). It became clear that these important features within the peat were easily obscured by the thin wash of sand that usually covered all surfaces even when most had been removed by the scouring tide.

Within the few square metres of peat that were exposed at Area 2 it was revealed that there was a pattern of two converging lines of stakes (context 8311011) each stake set approximately one metre (range from 0.9 to 1.1m) from the next (Fig. 6). All stakes were approximately 30mm in diameter and from the evidence of two stakes removed for dating and the remains of others exposed in eroded sections the surviving length varied between 20mm and 40mm.

In 1984 when the site was next visited a petrol driven water pump was used to clear the sand covering from the peat in Area 3. A further line of stakes was found at the northern edge of this area (context 24) (Fig. 3). This line appeared similar in character to the two lines discovered in Area 2 and consisted of stakes of similar size.

Both stakes removed from the lines in Area 2 show signs of pressure bends which result either from being driven into resistant ground or from subsequent stress such as carrying a heavy load or the accumulation of weighty overlying deposits. Both stakes had been sharpened to points but the working details are not diagnostic of a particular tool or period (B. Coles, project archive). The stakes were submitted for radiocarbon dating (material from both was combined) and yielded a result of 4840±70 BP (HAR-5642).

As found today, all the stakes have been eroded flush with the surface of the peat and there is no indication of the original depth of the peat deposit or of the height of the stakes. It is, however, clear that they would have been driven into a considerably thicker layer than the peat that exists today. They may have been driven in at a time when these particular peat areas were submerged or they may instead date from a period contemporary with peat formation.

Stakes from Area 3 (contexts 25-29) were lifted in blocks of clay and peat for detailed excavation in the laboratory to attempt to detect evidence of different horizons through which they had been driven. This evidence was not, however, forthcoming and the character of the surface from which they were inserted remains uncertain.

All the stakes were sampled for wood identification and all were found to be of hazel (*Corylus avellana*) with the exception of one (number 8311008) which is of alder (*Alnus glutinosa*).

With such short, and we must presume, incomplete lines of stakes and an absence of any superstructure, the function of these features can only be guessed at. Initial thoughts focussed on their possible use as part of a system of fish trapping such as still exists elsewhere in the Bristol channel. There is, however, no direct evidence to confirm that the features were constructed at a time when this area formed part of the Taw/Torridge estuary and the wide spacing of the insubstantial stakes (diameter <50mm) would probably have been inadequate for such a function. It is as likely that they constitute some sort of woodland structure as an estuarine one, a possible function that might relate to the fen peat environment is that of a trackway. The stakes could represent the substructure of trackway such as "Tinney's A" in the Somerset Levels (Coles and Coles, 1986: Plate 65).

The Upper Clay (context 23)

a) Description

This deposit is present only within an isolated hollow in the surface of the peat in Area 3. The clay survives to a depth of up to 0.7m and it seals a part of the outer peat which is up to 0.8m deep.

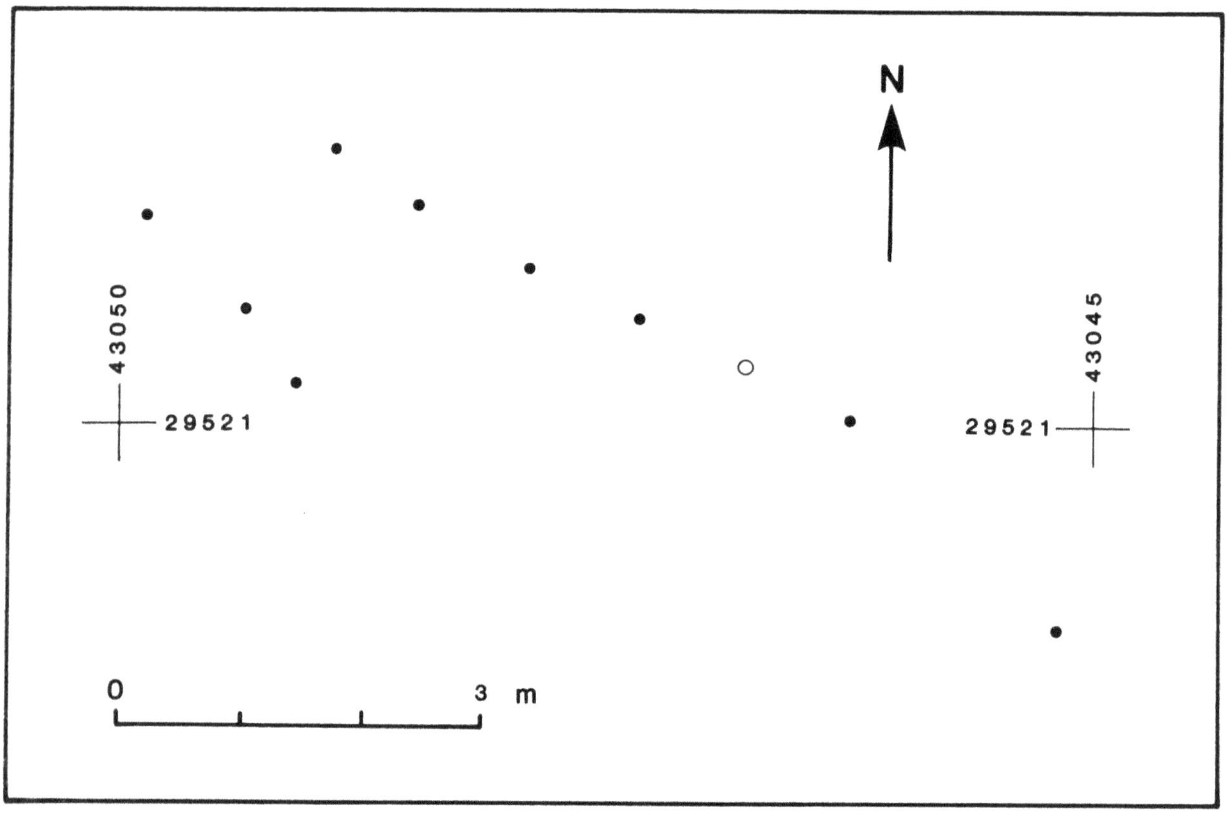

Figure 7. Plan of stakes in Area 2.

The "upper clay" is similar in character to the "lower clay" although it can properly be described as a fine clay. It can be distinguished from the lower deposit in that it contains rather less silt and more sand. The organic content is low and is similar to that of the "lower clay". The "upper clays" were not sampled extensively for faunal or botanical analysis as there was no exposed section available and insufficient time between tides to do more than auger and roughly survey the deposit.

b) Dating
The "upper clay" can only be dated to a period after or towards the end of the formation of the underlying peat and is therefore unlikely to pre-date 5190±80 BP (HAR-6363).

Measurement of magnetic intensity and susceptibility for samples taken in a column through the "upper clay" demonstrates a dramatic increase in these values towards the middle of the deposit, and this is taken to indicate a considerable input of magnetically enhanced sediments from within the local catchment area. These measurements and those relating to magnetic directions are discussed more fully in Part II.vii).

c) Environment
As with the "lower blue clay", the origin of this deposit is uncertain. It has been assumed to be of estuarine origin (Rogers, 1946; Churchill, 1965) and although this is still considered the most likely genesis it was not possible within this project to test this assumption. The clay rests immediately on top of the peat deposit (context 1) but it was not clear from observation in the field whether there had been any erosion or truncation of the peat preceding the deposition of the clay.

d) Human activity
No evidence for any human activity was found that could be associated with this horizon.

The Inner Peat (context 51)

a) Description
This peat horizon occurs extensively in Area 1. Some parts of it are quite commonly exposed although the bulk of the deposit remains covered by sand for most of the time. It varies in depth from a maximum of about 40cm at its southern end and tapers out to nothing to the north. The peat resembles the outer peats described above, being a monocotyledenous fen peat of moderately humified character and containing a substantial organic detrital component. It differs from the outer peat in that macroscopic wood remains are infrequent.

b) Dating
The peat formation in this area can be estimated as broadly contemporary with the Romano-British period. This estimate is based on the results of archaeomagnetic measurements of samples from the silts below the peat. These samples indicate a depositional magnetic remanence consistent with a date during the first century AD (see Part II.vii). Indications that these peats are substantially later than the outer peat are also to be seen in the pollen spectra (Part II.iv).

c) Environment
Pollen analysis of this deposit indicates that a much more open environment prevailed in the area at the time of peat formation than was the case during the formation of the outer peats. The vegetation comprised similar elements to those recorded in the analyses of the outer peats although there were differences in the relative abundance of the different groups. The vegetation of the inner peat is that of a Cyperaceae and Gramineae fen community in contrast to the Salicetum of the outer peat. Pollen of trees and shrubs is substantially less than in the outer peat and here the taxa such as oak, hazel, ash and holly were probably growing on better soils inland. Alder pollen is also present in relatively low numbers (*c* 15%) and thus probably represents the growth of alder carr elsewhere in the area (Part II.iv).

The upper levels of the peat show a significant increase in numbers of *Plantago lanceolata* pollen. This, and the presence of cereal pollen through the upper levels, can be taken as indicative of anthropogenic activity within the catchment area at this time.

d) Human activity
There is no direct evidence of human activity within the area of the inner peat. The evidence of the pollen analysis, however, suggests that cereals were cultivated within the vicinity of the site during the formation of the peat.

Romano-British Period Estuarine Deposits

a) Description
To the north of the inner peat of Area 1 there is an extensive and complex network of silt-filled channels which are incised into an horizon of grey silts which overlies the inner peat in places. These deposits yielded a variety of wooden features and were in places rich in animal bone and marine mollusc shells.

The deposits in this area are, perhaps, the most difficult to understand stratigraphically. Although some individual relationships are quite clear, any overall interpretation of the sequence of development of the sediments is hampered by the extensive erosion in many areas, which has resulted in the loss of much evidence of crucial relationships. Within the time available it was not possible to study adequately the deposits represented, or to obtain any coherent picture of the sequence of deposition. The detailed study of this area will require frequent observation and detailed mapping coupled with selective excavation. None of which was possible as part of this project.

b) Dating
Although it is possible that a number of different phases of channel development and infilling are represented it was not feasible to deduce a satisfactory chronological sequence for this area in the time that was available between the tides.

It was immediately clear from the range of mammal taxa represented in the deposits that these channels constitute a substantially later phase than that of the Mesolithic deposits of Area 3 (Part II.iii). This was confirmed by radiocarbon dating of bone (context 8311353) from one of the channel fills which yielded a date of 1560±80 BP (HAR-6513). A sample of wood from an adjacent group of stakes (context 57, sample 8311718) which was overlain by channel silts was dated to 1600±80 BP (HAR-6440).

c) Environment
The silts and channel fills of Area 1 are quite clearly of estuarine origin. The contents of a number of the channels include substantial molluscan assemblages which are typical of such an environment. These principally comprise species such as *Scrobicularia plana* and *Hydrobia ulvae.* The age profile of the assemblage and the presence of large numbers paired valves indicate that this deposit is a natural death assemblage. A study of the botanical component of a sample of these silts produced a number of seeds of annual sea blite (*Suaeda maritima*) whose habitat is salt marsh or sea shore (usually below HWST) and this gives further confirmation of the estuarine origin of these particular silts (V. Straker, pers comm).

d) Human activity
There is abundant, if fragmentary, evidence for human activity in and around Area 1. The nature of this activity is unfortunately rather enigmatic. The most frequently recognised evidence is that of accumulations of animal bones within some of the channels. As far as could be seen, these accumulations do not relate to any particular features within the channels nor do they appear as discrete concentrations but as generalised scatters. No excavation could be undertaken and it was only possible to retrieve a very few bones (22) on this occasion these included a number from domestic

species (cattle, sheep/goat, dog). The bones were very abraided but there was some evidence of butchery and a sheep/goat tibia also shows evidence of canid chewing (Part II.iii).

The other demonstration of human activity in this area is the presence of a number of features constructed of stakes and/or brushwood. These are found driven into and/or buried within the channel silts. It is difficult to establish any precise relationship for any of these features. Isolated examples such as the layer of brushwood (context 56, Plate 3) can be seen to be sealed by particular silt layers but the relationship of these to the channels and peat is very uncertain. For the other features it is not clear whether the stakes have been driven into the silts or whether the silts have been deposited around them once in position. In no case is there any indication of the level from which the stakes were driven in. The possible functions of these various features are discussed below.

Other than the wooden features mentioned above no other artefacts were recorded from this area by the CEU team. Numbers of flints have been collected from this part of the beach over the years (D. Trapnell, pers comm) and although many of these are apparently natural at least one artificial flake has been recognised. There is unfortunately no evidence for the exact provenance of lithic material from this general area, and although artefacts are almost certainly present, their context is unknown.

Large stakes (contexts 52, 53, 54)
One of the most obvious features located in this area were three large (*c* 10cm diameter) stakes that were driven into the silts of one of the channels of the area. The stakes were not directly associated one with another but their position within and to one side of one of the channels suggests a common function, possibly as part of a bank revetment or as mooring posts.

Stake line (context 57)
The best preserved of the wooden features/structures at the site was found to the north of the main area of estuarine deposits in Area 1. This feature consisted of a short line (*c* 50cm long) of close set small stakes (Plate 4). As found, the stakes projected between 5cm and 15cm above the brown clay (Part II.i) into which they had been driven to a depth of at least 15cm. They were almost all contiguous with no gap between any two greater than about 5mm.

A sample of one of the stakes was removed for identification and dating (it was not possible to remove the stake *in toto*). The sample (which was identified as *Corylus avellana*) yielded a date of 1600±80 BP (HAR-6440), this agrees well with the date derived from the animal bone from the adjacent deposits.

Although the evidence is slight, it is very tempting to associate this feature with some form of estuarine fish trapping. Archaeological examples of such fish traps are comparatively rare in this country but they are very well documented from historical sources and recent studies and observations in the British Isles and elsewhere (eg Went, 1966; 1969). Unfortunately, the short surviving length of the feature and the lack of other evidence precludes any serious consideration of its function.

Brushwood layer (context 56)
A layer of twigs/brushwood was noted (Plate 3) adjacent to the stake line (57) but it could not be stratigraphically linked to it. This layer was sealed beneath a thin horizon (*c* 3 cm) of estuarine silts. The layer was only exposed in section and little idea could be gained of its extent or function (if any).

Plate 3. Brushwood (Context 56, Scale in 10cm intervals).

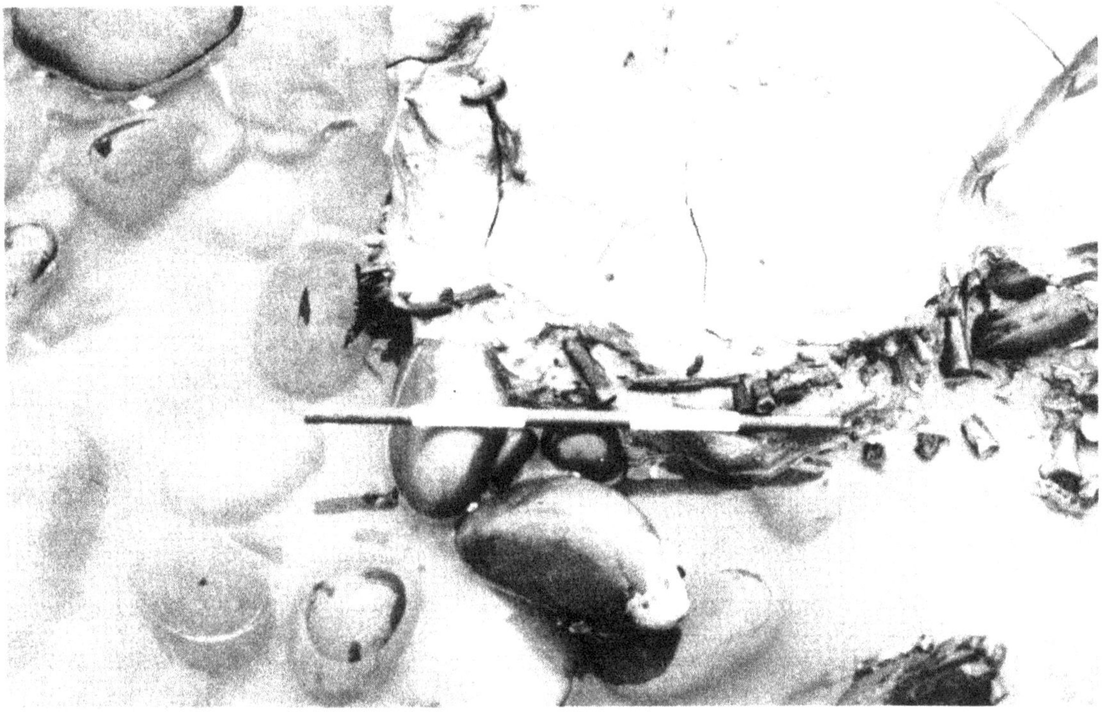

Plate 4. Stake line (Context 57, Scale in 10cm intervals)

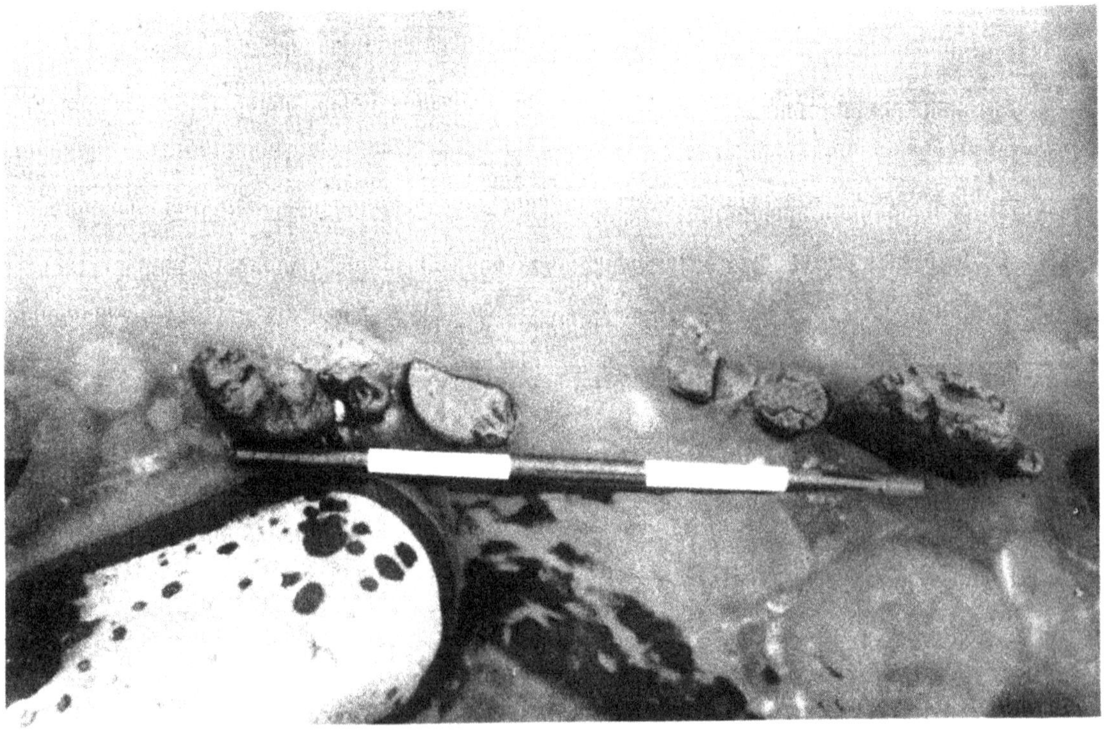

CONCLUSION

Identification of sites and the attribution of material

One of the problems that arose in dealing with the material from Westward Ho! was the integration of the information derived from the current surveys with that published by earlier workers. The variety of independent workers who have researched and reported on the material and the absence of plans, sections and proper documentation of the location of the site has lead to assumptions being made about the nature and source of some of the material which cannot now be validated.

E.H. Rogers noted the discrepancy between his own observations and those of others (Rogers, 1946, and eg Hall, 1870; Ellis, 1867) and suggested that this discrepancy might be due to the presence of more than one midden in the area. This early suggestion has not been taken up or considered to any extent by later writers and reference to all work in the area has continued to assume a single source for all the material recovered.

The situation was substantially clarified for this particular project by the very extensive exposures of foreshore deposits in 1984 which enabled the recognition of a number of discrete areas of human activity. The simultaneous exposure of a number of these "sites" encouraged the critical reassessment of the materials and records available.

It is clear that the earlier published and manuscript accounts lack sufficient detail to enable the accurate attribution of material to any particular site. This state of affairs is due in part to the difficulty of surveying the features accurately (in the short time available between tides) without specialist modern distance measurement equipment. The work of E.H. Rogers is an exception to the general rule and his location plan proved to be accurate if lacking in detail. Another hindrance and complication to the location of sites reported over the last century is the continual state of change of the beach and the pebble ridge. Accounts and photographic records demonstrate the degree of eastward movement of the pebble ridge over the last century (Rogers, 1946; Stuart and Hookway, 1954) and it is this retreat of the pebble ridge that has allowed the exposure of those sites which had previously been protected from erosion during periods of tidal scouring.

The problem of identifying the areas of the beach to which records of previous work relate was relieved in part by information provided by Mr David Trapnell, a resident of Westward Ho! Mr Trapnell's information related to the collection of material formed by the late Inkerman Rogers whose published account of the site (1908) prompted many questions by listing domesticates in apparent association with Mesolithic material. Mr Trapnell had visited the site as a boy with the elderly Rogers and he has continued to visit the same area of the beach ever since. He has consequently built up a substantial collection of material (principally animal bone) from the site. From Mr Trapnell's information it was clear that at least some of the material collected by Rogers was of Romano-British age (a fact that was also suggested by the reported presence of bones of domesticates in his collections) (I. Rogers, 1908; E.H. Rogers, 1946).

Prompted by the identification of the area from which Rogers collected his material it was decided that a comparison of the different groups of faunal remains in each collection would be worthwhile even though the total amount of material was quite small. The analysis of the small quantity of faunal material retrieved by the CEU survey indicated elementary differences between that from the Mesolithic midden and that from the later site. Bruce Levitan therefore undertook a study of the surviving material in the various locations in Devon (Part II.iii). Despite the small numbers of specimens involved in this study the analysis clearly demonstrates differences between the various assemblages and these differences can be attributed to the collection of bones from different sites.

Two distinct groups of collections can be demonstrated (Fig. 8), of which those of Hall, Jenkyn and probably that of Heathcote appear similar to the material collected in 1983 from the Mesolithic midden while the collections of Inkerman Rogers, Trapnell and Ellis have more in common with the material from the Romano-British period deposits. However, in view of the small sample sizes involved and the possibility of the existence of other sites elsewhere on the foreshore it is not possible to attribute these earlier collections (with the exception of those of Trapnell and Jenkyn) to any of the sites described here.

Mesolithic Midden and adjacent area

A "V" shaped feature penetrating the lower clay in Area 3 and containing a number of flint flakes is perhaps the earliest evidence for human activity at the site. It is dated by its contained charcoal to 8180±150 BP (HAR-5643). This is the earliest date yet recorded for the Post-glacial settlement of the south-western peninsula, but as an isolated determination should be treated with great caution (see above). Unfortunately, the small lithic assemblage from the feature does not provide any hint of its likely chronological affiliation which, on the radiocarbon evidence, might be expected to fall near the break in assemblage typology recognised between the earlier and later Mesolithic (Jacobi, 1973; 1979). Certainly, sea level at this time is likely to have been at least 15 m below present OD (Kidson and Heyworth, 1977) and one would have to consider a rather different genesis for the "lower blue clay" into which the feature is cut than the estuarine one proposed. The radiocarbon date also contradicts results of the archaeomagnetic determinations and further evidence must be forthcoming before any weight is placed on this date.

There is, unfortunately, no completely satisfactory evidence of the nature of the environment prior to the accumulation of the midden. In view of the absence of any significant signs of soil development in the upper levels of the clay - and the incorporation of flint and charcoal throughout the top 10-15cm of the silts - it seems likely that vegetation cover of the area was sparse or absent. During and following the development of the midden the area became largely covered by fen woodland (as evidenced by the terrestrial Mollusca, pollen and insect remains). Within this there were glades which (from the evidence of the pollen analysis of the peat adjacent to the midden) may have been quite long lived (Part II.iv).

As it survived to 1983, the Westward Ho! midden was not at all substantial and it is probable that the deposit visible today is only a fragment of a much larger deposit. It is not clear how many of the earlier published reports which describe more extensive deposits relate to this midden alone. The possibility must still exist that other Mesolithic deposits have been exposed in the past which are now lost through erosion. The midden is too disturbed by bioturbation to identify any meaningful sequence of deposition.

Earlier records (eg E.H. Rogers, 1946; I. Rogers, 1908) of the remains of many large mammals in the deposits have here been demonstrated to be unreliable and little evidence of such material was found in the deposit when it was examined in 1983 and 1984. It is, however, clear from the available reliable Mesolithic material that mammalian resources were exploited to some degree (Part II.iii) but the evidence for the taking of fish is slight.

Other evidence for Mesolithic subsistence from the midden is provided by the marine Mollusca whose remains constitute the greater part of the deposit (described in detail in Part II.ii). The overwhelmingly predominant marine species present in the midden were the mussel (*Mytilus edulis*) and the peppery furrow shell (*Scrobicularia plana*). These, and the other less well represented species, represent the exploitation of two distinct

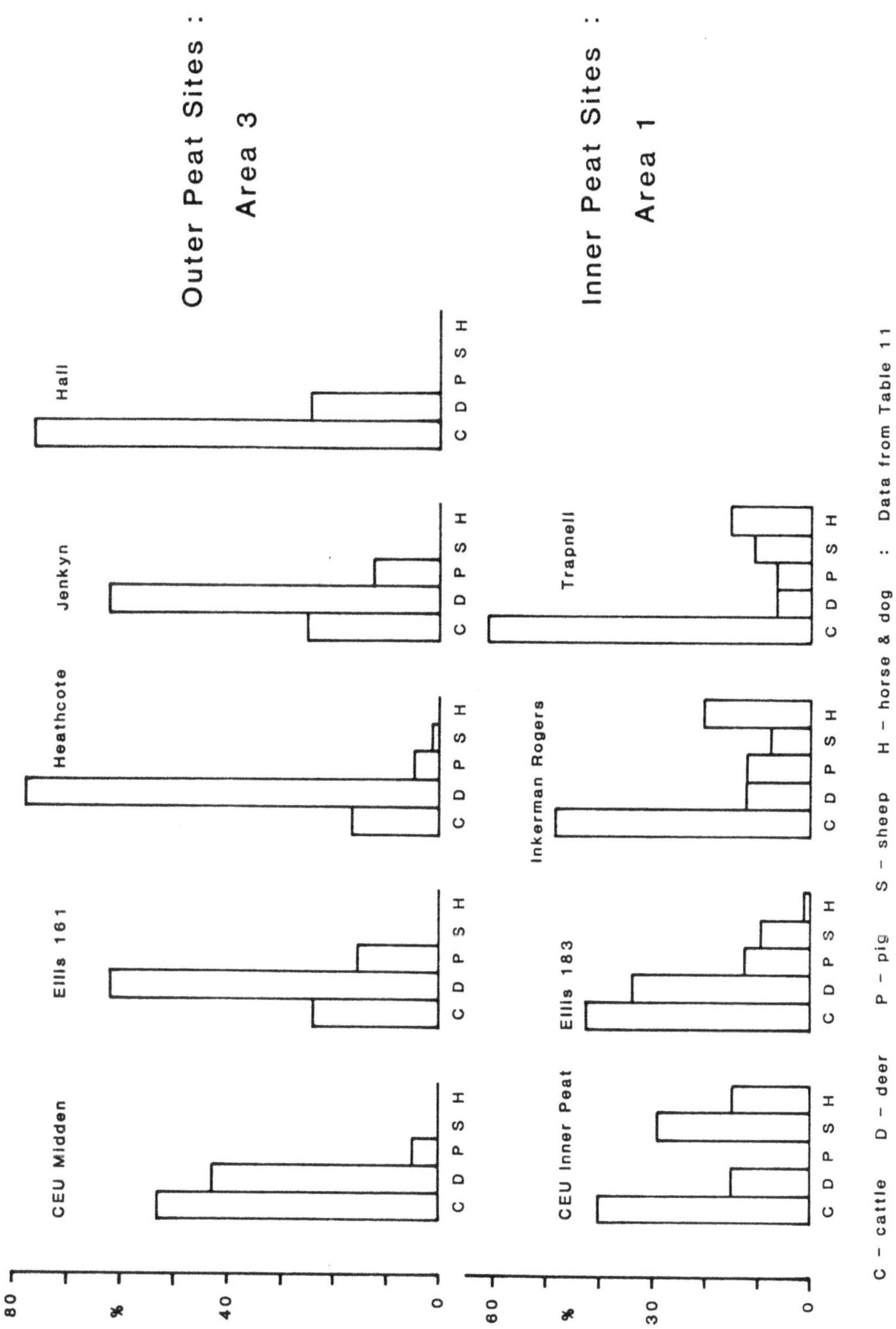

Figure 8. Comparison of bone assemblages from Inner and Outer Peat.

ecological zones, the mussels being gathered from rocky shores such as would be found to the south of the site whilst the *Scrobicularia* would have been taken from areas of muddy or sandy deposits in the estuary.

The midden is thus an accumulation of the debris of an ephemeral occupation or exploitation site which can now be shown to have been within wooded, sometimes wet, surroundings and which was a little way from the contemporary high-water mark. However, the several lines of evidence drawn upon here to understand the environment and the human activity still leave a very inadequate and piecemeal picture of the nature of human activity on the site. Consequently there is not sufficient evidence for any detailed comparison with other midden sites or with the very many other Mesolithic sites identifiable only as lithic scatters, and which are so abundant on western coasts. The presence of land mammals and also of inter-tidal resources here demonstrates the attractiveness of this coastal zone for exploitation of the various and rich resources of this environmental interface. Lithic raw material was also abundant, if only of modest quality, and at Westward Ho! this was clearly being used for the preparation of processing and especially hunting or fishing gear. A continuing puzzle must be the absence here of the pebble tools and heavier duty stone implements found commonly on other coastal sites. The suggested association of bevelled pebbles with seals (Jacobi, 1980) is perhaps supported in this case where there are only a very few bones (and none of seal) and where the site is not favourably situated for their capture.

However fragmentary the evidence from the midden at Westward Ho! it yet remains one of a very small number of sites where any evidence is preserved at all. Its low-lying estuarine position contrasts with the majority of other sites situated near cliffs and, as such, it is a rare and fortunate preservation of an element within a suite of sites that were subsequently inundated and destroyed by the eustatic sea level rise which must have swamped much of our evidence for Mesolithic coastal exploitation in the southern part of the British Isles.

Neolithic activities

Neolithic activity is only definitely represented by the stakes found within Area 2 and must remain an enigma for the time being. The extant features in Areas 2 and 3 (that might be contemporary) are particularly difficult to interpret in that we cannot identify the nature of the environment within which they existed. Without this information and in the absence of any particular diagnostic pattern in the construction of the features there seems little point in making what would only amount to crude guesses as to their function. Detailed analysis must await the discovery of more substantial Neolithic features or evidence for the Neolithic environment.

Romano-British activity

The location of deposits and features of Romano-British date was an unexpected product of this project. All material recovered and published to date had been interpreted as "prehistoric", or more particularly Mesolithic, on the strength of the flint artefacts known to have come from parts of the beach.

Evidence for Romano-British activity is concentrated in and around an area of small silt filled channels in Area 1. Within this area is an extensive deposit of grey silts which contains numerous mammalian bones and this deposit seems to have been the subject of the attention of collectors over the past century or more. The bones retrieved from this area form the basis of a number of the collections of "Mesolithic" material from Westward Ho! It is assumed that the material relates to an activity area nearby as it is confined to a relatively small section of the channels and not widely distributed throughout the system. The numbers of specimens recovered so far are too small to enable any detailed study or assessment of the character of the occupation to which they relate.

Other artefacts and features of Romano-British date are the stakes and stake lines. These features are similar in many respects to the Neolithic stakes of Area 2 although here it is possible to make some attempt at functional interpretation. The principal difference between these features and the earlier stakes is our knowledge of the nature of the environment at the time they were in use. It is clear that the features were used at a time when the channels were part of the Taw/Torridge estuary. In the case of the three large stakes (52, 53, 54) which stand within and to one side of one of the channels it could reasonably be suggested that they performed some mooring function. They seem to be too widely spaced (more than 10 m between) for them to have formed any single structure although the possibility of the remains of additional stakes at a lower level in the silts cannot be ignored. The line of close set small stakes (57) may be tentatively identified as a remnant of a fish trap although a more extensive investigation would be required to confirm this suggestion.

Evidence for sea level change

For more than a century the submerged peats of Barnstaple Bay have provided data for debate about changing sea levels in south west England. Although the research reported here was not primarily directed at questions of sea level it does provide some relevant data in the form of additional radiocarbon and archaeomagnetic dates and the biological evidence has enabled us to review previous assumptions about the relationship of the midden to the sea. It had been hoped that it would also be possible to characterise the "lower blue clay" and to determine whether or not it was an estuarine sediment. Conclusive evidence of the genesis of this deposit was not, however, forthcoming although an estuarine origin still seems the most likely (see above).

The cycle of alternation between estuarine/brackish minerogenic sediments and freshwater/terrestrial biogenic deposits at sites such as this has formerly been taken as evidence for relative rises and falls in sea-level. More recent work has shown that this need not always have been the case and that such variation in sediment type may represent the effects of quite localised factors such as change in the rates of sedimentation or changes in the coastal geomorphology (Kidson, 1982; Tooley, 1982). The effect of either type of change might be to enable the colonisation of brackish water sediments by freshwater/terrestrial communities. The major influence on the shorescape over the past 2000 years, a period during which the sea level has been relatively stable, has been the movement eastwards of the pebble ridge, a phenomenon still to be observed today whenever storms are sufficiently strong. This movement has resulted in the loss of hundreds of acres of common grazing on Northam Burrows in recent centuries.

The earliest sediments with which this aspect of the project has been concerned are the "lower blue clays". The marine/estuarine environment that these represent was succeeded by a freshwater/terrestrial (woodland) environment in which the midden deposits were formed. Churchill (1965) has argued (on the basis of observations by Rogers, 1946: 119) that the midden was formed in the zone between neap and spring high tides. However, no stratigraphic evidence for this was found in 1983 or 1984 and the biological evidence presented here does not support a strandline origin for the deposit.

The growth of peat on the site subsequent to the formation of the midden sealed both the midden and "lower blue clay" and represents a considerable period during which terrestrial conditions prevailed at the site. The existence of the peat-forming fen carr community was brought to an end in this area by a marine transgression which laid down further estuarine deposits both in Area 3 (the "upper blue clay") and, later, in Area 1.

The peats and clays in Area 1 present a much more complex picture in which it is not possible to identify the exact sequence of deposition. The phase of peat accumulation is probably terminated by the deposition of estuarine silts and clays which are seen within the channels to the north of the peat. Although the sediments within one of the channels have been dated (by the contained animal bone), deposits such as these can tell us little about the contemporary sea level (Kidson and Heyworth, 1979).

As there is now a considerable amount of data on marine and terrestrial sediments from Westward Ho! it was thought to be worthwhile comparing this information with published sea-level data from the region. Curves representing the change in relative sea-level for the Bristol Channel have been prepared by Hawkins (1971; 1973), Kidson and Heyworth (1973) and Heyworth and Kidson (1982). The earlier of these papers attempt to portray sea-level change through the use of single line curves. The most recent representation, however, has been produced after a much more critical appraisal of the evidence of the radiocarbon dating and it provides an indication of the range of uncertainty associated with these and other measurements. This last curve was therefore used for comparison with Westward Ho!

Following the recommendations of Kidson and Heyworth (1979), the radiocarbon dates for the remains of trees in Area 3 were used to provide a date for the marine transgression that halted peat growth in the vicinity of the midden. Comparison of the dates and the height of the stumps (OD) with the data published by Heyworth and Kidson (1982) shows that the Westward Ho! material falls within the height range predicted for the relevant period, that is, a sea level (however defined) approximately 7m below that of the present day. It is, however, worth noting the very significant variation in the results of the radiocarbon determinations on different materials and contexts at the site and this emphasises the need for well defined and consistent sampling and recording of materials that are to be used in any study of sea level change.

Although there is clearly additional information about post-glacial sea-level change contained within other deposits at this site, their discontinuity and the absence of radiocarbon dates from other areas of peat exposure prevent a thorough study at this stage. More detailed accounts of the sequences of transgression and regression will only be possible as a result of careful and accurate observation and measurement as different features and strata are exposed from time to time.

ACKNOWLEDGEMENTS

The work reported here would not have been possible without the help and information provided by Julie Turner and Diana Warmington, who observed the beach at Westward Ho! and kept us informed of extensive exposures of the sites. We are also indebted to Roger Jacobi for initiating and encouraging the work. Joy Ede, Mike Allen, George Smith and Richard Bartokowiak helped with the recording; Joe Dowson undertook the surveying. Various specialists provided analyses and reports which do not necessarily appear in this publication, these include Dr R.W. Battarbee, Dr B. Coles, Professor J.W. Murray, Dr Eric Robinson and Miss V. Straker to all of whom we are very grateful. Permission to work on the site was granted by Torridge District Council. Particular thanks are also due to Mr and Mrs T. Jenkyn and to Mr D. Trapnell who provided much information about the site and made their own collections freely available for study.

PART II - ANALYTICAL DETAILS

II.i - SEDIMENTS, SOILS AND GEOLOGY

R.I. Macphail

Westward Ho! lies just west of the confluence of the rivers Taw and Torridge on Upper Carboniferous rocks (Crackington Formation - shales, siltstones, turbidite sands) (Geological Survey, Sheet 292, 1977) (Fig. 1). To the north of the town these rocks are covered by blown sands (Northam Burrows) and recent marine alluvium which also occupies the river valleys.

During the Wolstonian glacial the drainage of the Taw and Torridge was disrupted by ice moving up the estuary, and till was deposited as far east as Instow (Geological Survey, 1977). Ipswichian alluviation and river gravels (Geological Survey, 1977) may probably be correlated with the old cliff line and raised beach to the south-west of the town. These features are also regarded as Ipswichian by Rogers (1946) and by Stephens (1973) but Kidson (1974) has suggested a Hoxnian date for them.

The area was not directly affected by Devensian ice (Geological Survey, 1977) but the cryoturbated patterned ground to be seen on the beach (see below) and the soliflucted deposits to the west of the town probably date to this period (Stephens, 1973).

Modern soils

Gleyed soils exist in the river valleys where clayey pelo-alluvial gley soils (Wallasea 1 Association, Findlay *et al*, 1983) are developed on the alluvium and to the west of the town where the drift deposits have a clayey, fine loamy and fine silty pelo-stagnogley soil cover (Hallsworth 2 association). The soils of the blown sand areas are typical para-rendzinas (Sandwich association). Inland, well-drained typical brown podzolic soils (Manod association) and typical brown earths (Denbigh 2 association) occur directly on the carboniferous rocks.

Deposits presently exposed on the beach

The shore exposed between high and low water marks comprises a rock cut platform to the south, with superficial deposits elsewhere. The area for which the stratigraphy has been studied lies due west of the modern recreation ground (NGR SS 294434) and is contained within a strip approximately 300-400m wide which runs from the present pebble ridge to the limit of low spring tides.

The major deposits observed in each area are descibed in summary below, the deposits for each area are listed in stratigraphic order.

Area 1 (Fig. 2)
a) polygonal patterned clay loam near the modern pebble ridge (the "brown clay" of Rogers, 1946).
b) "brown clay" containing oriented beach pebbles.
c) grey fine silty clay loam to fine clay.
d) peat, ("inner peat"; Rogers, 1946).
e) reworked "brown clay" and pebbles contained in a channel cut through the peat.

Area 2
f) grey fine silty clay loam.
g) peat - "middle peat".

Area 3
h) grey fine silty clay loam - "lower blue clay".
i) an "outer pebble ridge" 400m west of the modern pebble ridge
j) "midden"
k) peat - "outer peat" (after Rogers, 1946).
l) a grey fine silty clay loam or clay contained in a depositional hollow in the peat (*c* 10m by 18m, Fig. 3) - the "upper blue clay".

The polygonal patterned "brown clay" (a) features varying, but coarse (*c* 0.35m wide) ice wedge polygons with strong brown (7.5YR4/6) centres and light grey (5YR7/1-6/1) cracks.

The cracks commonly contain woody root fragments *in situ*. The sediment itself is a clay loam with small to large stone inclusions. The deposit is probably alluvial and may be Ipswichian in date (Geological Survey, 1977). Micromorphological analysis (Table 2) shows that the "brown clay" has been slightly weathered, probably by lower subsoil pedogenesis which mainly resulted from tree rooting along the grey polygonal cracks, giving this sediment slight soil characters (hydromorphism, *in situ* roots, other minor pedofeatures, small organic carbon content). A presumed post-glacial woodland cover (see below) rooted in the polygonal cracks produced the colour variations - the more silty cracks allow freer soil water movement while the interior remains mainly oxidised. The original soil must have been a metre or so thick and would probably have resembled the local pelo-stagnogley soils (Hallsworth 2 Association) found in the area today. This soil has since been truncated by marine erosion.

The "brown clay" with pebbles (b) may be related to that described (a) above, but in addition it has incorporated beach pebbles, possibly derived from the local raised beach. These pebbles are oriented in a compacted soil matrix, again this is probaby a result of Devensian periglacial activity. The brown colouration may again relate to its Devensian character.

The "blue clays" (c,e,f,h) are exposed in all the areas examined. Although similar in many respects they represent at least three sedimentary episodes (see also Part II.vii). The earliest of these is represented by the "lower blue clay" (h) in Area 3. Augering here showed that the horizon was up to at least 1.10m deep and rested on stoney head or the rock cut platform. It is a uniformly (fine) silty clay loam (Table 3), generally dark grey (2.5YR4/0) to grey (2.5YR5/0) in colour, sometimes faintly mottled. It contains varying quantities of organic matter, especially vertical and horizontal monocotyledonous plant material (see Profile Descriptions). They are dissimilar to the "brown clay" in texture, colour and content. The "blue clays" are more obviously post-glacial silts and have an extremely low pH (due to sulphide content). Micromorphology (Table 2) shows the "blue clays" in both horizons in Area 3 to be sediments with no pedological features other than soil ripening. This latter is mainly represented by minor pyrite framboid formation (Miedema *et al*, 1974). The lack of porosity in these sediments is also very strong evidence against the blue clays ever having undergone sustained drying out.

The "midden" (j), which rests unconformably on the "lower blue clay" (Profile 2), is a very heterogeneous deposit. It is characterised by large quantities of shell fragments (which vary in their states of preservation) and fine and coarse charcoal. Its fine matrix (Micromorphological description, Table 2) is primarily composed of organic matter and silts - in much the same way as the "outer peat" described below. It contains more sand and organic matter than the underlying "blue clay" although minor silt laminations testify to minor sedimentation. The "midden" is, however, strongly turbated with micro-gullies and in-fills and displays obvious mixing.

The "outer peat" (k, g) rests immediately on top of the "lower blue clay" - except where midden deposits are interposed. It is seen to vary considerably in thickness ranging from *c* 0.1m to 0.9m where it exists in hollows in the "blue clay". Although this deposit is undoubtedly humose it does contain large quantities of mineral material (Micromorphological descriptions; Tables 2 and 3). Texturally this mineral fraction resembles that of the "midden" being coarser (containing a greater proportion of silts and sand) and less well sorted than the "lower blue clay". There has been some pyrite replacement of organic material. Of major significance are thin horizontal laminae showing variations of mineral content within the peat. These may relate to seasonal flooding episodes.

A later episode of sedimentation is represented in Area 3 by a deposit of "upper blue clay" (l). This is only present within an hollow in the peat surface where up to 0.7m of blue clay overlies about 0.8m of peat. The "upper blue clay" differs from the "lower blue clay" by containing rather less silt but more sand. The organic content is rather low - similar to sample 2 of the "lower blue clay" .

The deposit of pebbles traditionally known as the old pebble ridge (i) is not clearly related stratigraphically to the deposits described above. It does, however, appear in some areas to be overlain by a small amount of peat and to be embedded in a matrix of "blue clay". The pebbles may be derived from a reworking of the deposit of brown clay with pebbles noted in Area 1.

PROFILE DESCRIPTIONS

Profile 1 (Fig. 4)

Horizon Depth (cm)	
Peat (Of) 0-10	Reddish brown (5YR5/3) peat, containing old roots, pieces of wood - up to tree trunks; few rounded to sub-rounded sandstone, siltstone, shale fragments.
"Lower blue clay" 1 10-19	Dark grey (2.5Y4/0) moderately sticky, moderately plastic and fine silty clay loam; massive; many old fine roots, organic matter fragments; monocotyledenous, both in vertical and horizontal plane; coarse and fine charcoal present; fine organic matter present; common rounded top sub-rounded medium to large stones; gradual wavy boundary.
"Lower Blue clay" 2 19-40+	Grey (2.5Y 5/0) Moderately sticky, moderately plastic fine silty clay loam, with few fine diffuse mottles; massive; common fine, few medium old root fragments (as above).

Profile 2 (Midden, NW face)

Peat (Of) 0-5(6.5)	Dark yellowish brown (10YR3/4) peat, containing wood fragments, hazlenut shells; *in situ* tree roots; tree trunks; mineral laminations (see Table 2, Micromorphology); narrow, irregular boundary.
Midden 5(6.5)-17(26)	Dark grey (5Y4/1) moderately weak, slightly plastic fine sandy clay loam; abundant small to medium shell fragments; much organic matter present ("peaty"); few large rounded stones at base (siltstone, quartzite pebbles); few quartz gravel; many charcoal fragments present; discontinuous silt laminae present; shelly at base; clear irregular boundary.
"Lower blue clay" 1 17(26)-31	Very dark grey (5Y3/1) moderately sticky, moderately plastic finc silty clay loam; few small stones (gravel); organic matter present; old root fragments; few fine and coarse charcoal; gradual wavy boundary.
"Lower blue clay" 2 31-49+	Dark grey to grey (5Y4/1-5/1) moderately sticky, moderately plastic, fine silty clay loam with in places common faint medium mottles; rare shell fragments; small stones; rare charcoal fragments; common root/plant fragments.

AUGER DESCRIPTION THROUGH OUTER PEAT AND "LOWER BLUE CLAY"

Depth (cm)	
0-53	Peat
53-102	Grey (5Y5/1) silty clay loam
103-108	Grey (5Y5/1) silty clay loam, with small subangular siltstone fragments.
108-132	Grey (5Y5/1) silty clay loam

132-140	Olive grey (5Y2.5/1) silty clay loam with black (5Y2.5/1) organic areas; contains small stones (<5cm), subangular to subrounded; few coarse sand, fine gravel; zones of high coarse sand content; verticcal monocotyledenous material; faint mottling.
140-156	Dark grey (5Y4/1) silty clay loam with faint mottling.
156-162	Dark grey (5Y4/1) silty clay loam with faint mottling.
162-	Either bedrock or stoney head.

MICROMORPHOLOGICAL DESCRIPTION (Bullock *et al*, 1985)

A - Polygonal patterned "Brown clay"

Homogeneous: *microstructure*; massive with well developed coarse "prisms" (ice wedge polygons, *c* 35cm across): <1% (see Table 2) total voids; discontinuous; elongate channel (40-150 microns across) at polygon boundary; rough walled; short, fine (40 microns across) smooth-walled channel within polygon:

Mineral Components Coarse/fine (60 microns) 20/80: moderately sorted: 19% *coarse mineral*, moderately sorted dominant quartz fine sand (<200 microns) sand; sub-rounded; few (1%) coarse-sand sized siltstone rock fragments: 80% *fine mineral*, well sorted, mainly quartz silt and clay; dark brown and pale grey:

Organic components (<1%); probable *in situ* dark reddish brown lignified coarse and fine root material confined to polygon boundary (coarse woody root material present in the field); few fine amorphous material (same locality):

Groundmass: mainly undifferentiated; small areas of weak parallel orientation: porphyric:

Pedofeatures: <2% non-oriented fine channel infill; rare "silt" intercalation: (19%) depleted zones; pale grey especially along polygon boundary and sporadically elsewhere: 81% diffuse, dark reddish brown nodular "amorphous iron" staining.

B - "Lower Blue clay" below midden

(Comparable to two samples of Lower "Blue clay" below peat at Profile 1)

Homogeneous: *Microstructure*: massive: <2% voids, compound packing voids (no channels), smooth walled (40 microns across) elongate; random.

Mineral components Coarse/fine (60 microns) 5/95; very well sorted: 4% *coarse mineral*; dominant quartz fine sand (<200 microns): 87% *fine mineral* well sorted; quartz silt and clay; greyish brown.

Organic components <3%; few coarse tissue fragments; mainly fine amorphous; reddish brown (pyritised in places - see below): fine charcoal present:

Groundmass undifferentiated, unoriented: porphyric.

Pedofeatures: <1% depletion: <2% amorphous diffuse dark reddish nodular iron staining along void margins: pyrite framboids present (3%); individuals 15 microns; microaggregates 30 microns; clusters up to 400 microns.

C - "Midden"

Heterogeneous: random mixing of varying microfabrics.
Microstructure: massive <2% voids; compound packing voids; channels present; smooth walled; random; some infilled; some have probable root pseudomorphs: void space within organic tissue not counted.

Mineral components: Coarse/Fine (60 microns) 25/75; poorly sorted: 24% *coarse mineral*: dominant (19%) shell fragments; aragonite; coarse sand size to fine; includes (4%) decomposing fragments, edges splaying - probable fibrous calcite: coarse sand sized sandstone (2%) and siltstone (0.3%) rock fragments; moderately rounded: 70% *fine mineral*: dominant (68%) mixed low proportions with fine organic; generally random distribution of mineral/organic components; clumping; few laminae of increased mineral content: porosity (micro-gulleying) infills (2.4%) of mainly fine mineral; mineral interleaving of organic tissues: greyish brown to reddish brown.

Organic components: 73%; dominant (68%) fine amorphous material mixed in varying proportions with fine mineral; reddish brown: few (1%) organic fragments; (1.5%) brown tissues; (2.4%) black tissues: (1.5%) coarse (>60 microns) charcoal; dominant fine charcoal throughout: probable pyritised *in situ* fine root material; with possible silt in fill relating to root channel (*c* 400 microns across).

Groundmass: undifferentiated, unoriented:porphyric.

Pedofeatures: pyritised plant tissues present; few pyrite framboids (as Silt): textural: silt in fills of "micro-gullies" and of possible "root channel": probable turbation.

D - Peat ("Outer peat")

Homogeneous: *Microstructure*: microlaminated eg organo-silt, fine sand: <1% voids; compound packing voids; smooth walled; random; within organic tissue void space not counted.

Mineral components: Coarse/Fine (60 microns) 20/80; poorly sorted: 16% *coarse mineral*: dominant, fine-sized sand; quartz; moderately rounded: 76% *Fine mineral* low density fine mineral in fine organic groundmass (as midden) eg silt interleaving organic tissues: microlaminations (as above)(200-300 microns).

Organic components: 82% 76% amorphous fine organic mixed with fine mineral: few (1.6%) brown and (2.2%) black tissue material: few (1.9%) organic fragments: few (1.0%), coarse (>60 microns) charcoal fragments: fine charcoal present.

Groundmass: undifferentiated, unoriented: porphyric.

Pedofeatures: few pyrite framboids; pyritised organic tissues.

Table 2. Approximate percentages micromorphological features (1100 point counts).

Sample	void	Coarse Mineral FS	MS	CS	Rock	Shell	Fine Mineral Silt+Clay	Infill	Inter-calation	Coarse Charcoals	Organic components Organic	Fine Tissue	Amorphous	Pedofeatures Pyrite framboids	Amorphous nodule
Outer Peat	0.3	16	*	*	*	*	(76)	*	-	1.0	1.9	3.8	76	**	-
Midden	1.1	2.4	*	*	2.7	18.7	(68)	2.4	-	1.5	1.0	3.9	68	*	-
"Lower Blue clay" (below midden)	1.4	3.9	-	-	-	-	87	-	-	*	-	-	2.8	3.2	< 2
Polygonal patterned "brown clay"	< 1	17	0.5	0.5	1	-	77	1.7	1.5	-	0.5	-	-	-	(81)

Table 3. Analytical data: Pedology.

Sample	pH	% Loss on ignition	% Organic carbon	Clay	FZ	MZ	CZ	Silt	FS	MS	CS	Sand	Texture
"Upper Blue clay"	3.2	2.98	0.51	36	12	16	14	42	21	1	< 1	22	Fine clay
Outer peat	1.8	42.56	16.46	13	2	6	27	35	23	12	16	51	Humose fine sandy clay loam
Midden	6.3	10.04	9.23	14	5	19	13	37	23	14	11	48	Humose fine sandy clay loam
"Lower Blue clay" (1)	2.6	9.15	2.85	31	14	18	21	53	15	2	< 1	17	Fine silty clay loam
"Lower Blue clay" (2)	4.3	3.27	0.49	34	18	17	15	50	13	2	1	16	Fine silty clay loam
Polygonal Patterned "Brown clay"	6.6	2.90	0.12	29	9	16	9	34	28	7	3	38	Fine clay loam

II.ii - THE MOLLUSCS

Martin Bell

INTRODUCTION

This analysis was concentrated on one of the monoliths cut from the midden (context 29). Work began in the laboratory by making traced drawings of two faces of the block. A column of sediment (Column A) 20cm wide by 10cm thick by 35cm deep was then carefully excavated in depth units of *c* 5cm taking care to respect stratigraphic boundaries. Plate 5 shows the face of the block once this series of samples had been removed. Subsequently the remainder of the block was divided to form a second series of samples (Column B). Results from the two columns have been kept separate to facilitate strict comparison with other biological data from each column and so that the results from the two adjacent columns can be compared. The stratigraphy of the columns was as follows:-

Column A	Column B
0-5cm peat	0-4cm peat
5-20cm shell midden	4-7cm peat/midden interface
20-35cm "blue clay"	7-26cm midden
	26cm+ "blue clay"

In addition to the columns a number of other samples were examined. Samples 8311730-32 were pieces of the block which had become detached in transit. Sample 8311218 was a spot sample of midden taken in the field immediately adjacent to the monolith sampled in the laboratory. Sample 8311234 was collected by Dr Jacobi for preliminary analysis during the planning stages of this project. A number of individual shells recovered in the field or during laboratory dissection of the monolith were also identified. Finally, an unrelated sample from a different site, the "inner clay", is considered separately at the end of this report.

Repeated freeze-thaw cycles were found to be the most efficient way of disaggregating the sediment. Following this some shells floated and were decanted, the remainder of the sediment was decanted onto a nest of sieves of sizes 6mm, 2mm and 0.5mm. Material retained was dried and sorted under a binocular microscope. In addition to the molluscs, flint artefacts, seeds, charcoal, beetle fragments and ostracods were separated out. A single foram from samples 8311724 has not been identified.

Nearly all molluscs came from the midden; those few from the peat and blue clay in Column A are considered to be due to small-scale mixing which is shown by the sections (Plate 5) to be partly the result of subsequent tree rooting. Thus it is concluded that the marine, land and freshwater molluscan assemblages relate to a single stratigraphic horizon.

MARINE MOLLUSCS

These presented considerable identification problems since virtually no whole shells were present, most being very comminuted. It is unclear to what extent this resulted from physical damage before deposition of the peat and to what extent it results from readily identifiable secondary processes such as the effects of rooting. Scanning electron micrographs taken by Dr Maureen Girling show that some shells are breaking down into individual fibres of calcite and aragonite suggesting that the organic shell matrix has been lost.

Standard texts on marine mollusc identification are concerned with complete shells. This material has to be identified on the basis of detailed comparison with reference material of hinge line and apical fragments or pieces with distinctive surface sculpture or shell structure. Details of the species present are given in Tables 4 and 5 but it is desirable to begin by describing each species in turn drawing attention to some features employed in identification, and to the main areas of uncertainty as well as some points of nomenclature:-

Plate 5. Block 29 after removal of mollusc column.

Bivalves

Anomia sp., Saddle oyster. Complete tiny juvenile shell from peat, probably intrusive.

Mytilus edulis (L.), Common mussel. Grey blue in colour with surface sculpture of rather irregular concentric lines. They have a markedly fibrous macrostructure and tend to break into pieces which in cross-section are parallelogram shaped in contrast to the more rectangular sections of other species. There were 1371 hinge line fragments and even tiny ones tended to preserve the crenulations and, within the shell, slightly etched triangular depressions within the anterior adductor scar. Other *Mytilus* species could also be represented but the reasonably preserved examples compare closely with *Mytilus edulis.*

cf. *Cerastoderma edule* (L.), Common edible cockle. White/yellow in colour with exterior sculpture of strong radiating ribs and interior crenulate margin. Weathered specimens revealed an internal macrostructure of rayed corrugations. Seventeen hinge lines were present. Certain identification of Cardiacea on fragmentary material is not possible but Dr John Taylor considers they are more likely to be *C. edule* than any other species.

Venerupis spp. Yellow/white in colour, generally well preserved with sculpture of numerous concentric ridges and grooves with fine radiating ribs. Twenty-six hinge lines are present. Nine were identified as *Venerupis decussata* (L.) (carpet shell) on the basis of size and the characteristic decussate sculpture. Three were identified as *Venerupis pullastra* (Montagu) (pullet carpet shell) and the remainder could not be specifically identified.

Scrobicularia plana (da Costa), Peppery furrow shell. White/light grey in colour usually quite well preserved sculpture of fine, rather irregular, concentric lines and ridges. The 747 hinge line fragments were readily identified on the basis of the large half-moon shaped internal ligament pit (chondrophore) and associated cardinal teeth.

Ensis sp., Razor shell. Two hinge line fragments.

Solen sp. Single hinge line fragment identified by Dr John Taylor.

Pholadidae, Piddocks. Intrusive into the peat and not specifically identified.

Gastropods

Patella sp., Limpets. Colour variable. Some fragments retain the distinctive radiating and concentric surface sculpture. Other fragments have been identified on the basis of a similar concentrically arranged cross foliated structure. No apices were found and specific identification was not possible.

Littorina spp. Colour grey, thick and generally well preserved. *Littorina littorea* (L.) Edible winkle was identified on the basis of two complete shells and one apex. *Littorina "saxatilis"* is now considered an aggregate of several species which it is not possible, in this instance, to separate. It was identified on the basis of three complete shells, five apices and fragments with distinctive ribbing, wide grooves and a slightly keeled whorl. As regards most of the fragments it was not possible to distinguish between juvenile *L. littorea* and *L. "saxatilis"* so this material is listed as *Littorina* spp. *Littorina obtusata/mariae* (formerly considered as a single species - *Littorina littoralis*) was represented only by fragments and identified on the basis of its smooth surface, much depressed spire and shallow suture.

Nucella lapillus (L.) Dog whelk. Single thick fragment of the junction between two whorls showing flanges and irregular growth.

Extraction and quantification procedures differed slightly between the two columns:

Column A An attempt was made to divide all shell fragments larger than 2mm into groups on the basis of colour, surface sculpture, fracture patterns, etc. These groups were then identified on the basis of the larger fragments with diagnostic features. On Table 4 the results are quantified in two ways for comparison: firstly the total weight

Table 4. Marine mollusca from Column A.

Sample number	8311202-6 8311220		8311207-11 8311221		8311212 8311222		8311213 8311223		8311214 8311224		8311215 8311225		8311216 8311226		8311218		8311234			
Sample depth	0-5		5-10		10-15		15-20		20-25		25-30		30-35				Sample A Jacobi		Individual shells	
Context	Peat		Top of Midden		Midden		Bottom Midden		Top of clay		Clay		Clay		Midden		Midden		Midden	
Weight of sediment (g)	1269		1950		1993		2059		2313		1932		1849		1099		300			
BIVALVES	Wt	no	Wt	no	Wt	no	Wt	no	Wt	no	Wt	no	Wt	no	Wt	no	Wt	no	Wt	no
Anomia spp. juvenile (?modern)	0.01	1																		
Mytilus edulis L.	4.81	5	81.48	166	101.09	150	83	72	14.96	9	0.56	2	0.14	-	132.4	130	1.61	34		1
Mytilus edulis L. juvenile																				1
Mytilus edulis L. juvenile complete (?modern)	0.01	5													0.1	2				
cf *Cerastoderma edule* (L.)	0.01		0.48	-	0.39	2	0.19		0.01	-			0.01	-	0.3	-	0.2	-		2
cf *Cerastoderma edule* (L.) juvenile	0.01	1			0.01	-	0.01	1												
Venerupis pullastra (Montagu)			0.19	1					0.3	1										
Venerupis decussata (L.)			0.91	1																
Venerupis decussata (L.) juvenile			0.02	1			0.3	2												
Venerupis spp.			0.65	1	0.36	-	0.24	9	0.1	-			0.11	-						
Scrobicularia plana (da Costa)	0.41	2	11.05	51	31.4	134	74.55	191	14.78	46	0.18	-	0.05	1	1.9	6	2	47		
Scrobicularia plana juvenile	0.01	1	0.01	1							0.01	-								
?*Ensis* spp.																	0.1	1		
Pholadidae (?modern)	0.12	1	0.05	-	0.02	-			0.01	-										
GASTROPODS																				
Patella spp.			0.12	-	0.08	-	0.1	-							0.1	-	0.1	-		
Littorina littorea (L.)					1.64	1			0.1	-							0.1	-		
Littorina obtusata/mariae (formerly *L. littoralis*)			0.89	-											0.1	-				
Littorina "*saxatilis*" agg.					0.05	-											0.1	1		
Littorina spp. "*saxatilis*" & *littorea*	0.03		0.05	-			0.35	-	0.06	-					0.3	-				
Nucella lapillus (L.)					0.33	-														
Gastropod fragments, unidentified			0.18	-	0.16	-	0.52	-	0.05	-	0.06	-	0.03	-						
WEIGHT OF SHELL + HINGE LINE NUMBERS	5.40	16	96.08	222	135.53	287	159.26	275	30.31	56	0.81	2	0.34	1	135.4	139	5.2	83		4
CRUSTACEANS																				
Cirripedia: Barnacles: no of plates		5		4	0.01	7	0.02	22	0.01	1					0.2	7		9		
Crabs/lobsters: no of fragments			0.02	2	0.04	5	0.04	21	0.01	1										

Table 5. Marine mollusca from Column B.

Sample number	8311723	8311724	8311725	8311726	8311727	8311728	8311729	8311730	8311732	8311731	8311702
Sample depth	0–4	4–7	7–12	12–17	17–19	19–26	26 +				
Context	Peat	Peat / Midden	Top of Midden	Midden	Middle Midden	Bottom Midden	Clay	Midden	Top of Midden	Midden (29)	Inner Clay
Weight of sediment (g)	1570	1300	2900	3030	2220	2420	4000	2100	620	1460	1606
BIVALVES											
Anomia spp. juvenile (?modern)											8
Mytilus edulis L.		86	42	211	244	62		156			5
Mytilus edulis L. juvenile											15
Mytilus edulis L. juvenile (?intrusive) paired valves		2									
Musculus sp											1
cf *Cerastoderma edule* (L.)		1 (22)	1 (7)	3 (15)	(20)	1 (5)		4 (25)			
cf *Cerastoderma edule* (L.) juvenile					1						16
Venerupis pullastra (Montagu)			1								
Venerupis decussata (L.)		1		3							
Venerupis decussata (L.) juvenile					1						
Venerupis spp.		(8)	(3)	1 (11)	(12)	(4)		(3)			
Venerupis spp juvenile								2			
Veneracea											1
Mactracea											1
Scrobicularia plana (da Costa) adult pairs											4
Scorbicularia plana (da Costa) umbo		65	17	46	38	74		25			15
Scrobicularia plana juvenile					2						33
?Ensis spp.			1								
?Solen				1							
Pholadidae			1 (2)		1			(2)			
GASTROPODS											
Littorina littorea (L.)			1					1			
Littorina "saxatilis" agg.			(1)	4	3 (2)	(1)		(1)			(4)
Littorina spp. *"saxatilis"& littorea*				(8)	(2)	(1)					
Hinge line/apical numbers	0	155	64	269	290	137	0	188	0	0	99
CRUSTACEANS											
Cirripedia: Barnacles: no of plates				3	5	1		5			3
Crabs/lobsters: no of fragments				1		1					33

of all the fragments larger than 2mm in each taxa, secondly the number of hinge line fragments/apices present.

Column B With experience it was only thought necessary to remove the apical and umbonal fragments together with certain pieces showing characteristic sculpture (ie, from species other than *Scrobicularia* and *Mytilus*). Thus Table 5 gives only the number of umbonal/apical fragments with the number of sculptured pieces (where known) in brackets. Graphs of results from both columns (Figs. 8 and 9) are based on the number of hinge line fragments/apices as percentages of the total marine assemblage. It should be noted that the number of hinge line fragments cannot be taken strictly as the minimum number of valves because some hinge lines will have broken into more than one diagnostic fragment.

To an extent these problems are academic because, however we quantify the assemblage it is quite clear that it consists overwhelmingly of two species *Mytilus edulis* (63% of the total hinge line fragments/apices) and *Scrobicularia plana* (34%). Figure 8 shows that in Column A *Mytilus* was much more abundant at the top, *Scrobicularia* at the base. This pattern is not so clear in Column B, (Fig. 9) and it may simply reflect variability within the midden, which is also suggested by the spot samples eg 8311218 with 93% *Mytilus* and 4% *Scrobicularia.* Species present in much smaller proportions are *Venerupis decussata* and *Venerupis pullastra* (together 1%), cockle cf *Cerastoderma edule* (0.8%) and tiny proportions of *Littorina littorea*, *Littorina "saxatilis"*, *Littorina obtusata/mariae*, *Patella*, *Ensis*, *Solen* and *Nucella lapillus.* Additionally there were a few *Mytilus* juveniles, one *Anomia* juvenile and some piddock fragments which are so well preserved that they are suspected to be modern contamination.

This list expands on, and quantifies, those species noted by Churchill (1965) from Westward Ho! but curiously does not include *Ostrea* sp. which Ellis (1867) noted as a major constituent of the midden. This discrepancy no doubt arises from confusion with some of the later deposits at Westward Ho! Studies of Ellis' animal bone collections (Part II.iii) have shown that they bear great similarity to those collections from the Romano-British deposits and Ellis' mention of *Ostrea* may point to the existence of a separate Romano-British shell midden.

We can divide the present assemblage into broad ecological groups. Firstly, species which are likely to have lived on the exposed rocky shore adjoining the site to the south: *M. edulis*; most *Littorina* spp., *Patella* sp. and *Nucella lapillus.* Today the rocks support very large populations of *M. edulis* with much smaller percentages of *Patella*, *Nucella* and members of the Trochidae (top shells), the latter unrepresented from the midden. The second ecological group comprises species derived from soft muddy and sandy bottom deposits of the Taw/Torridge estuary which, at that time of lower sea level, would have run to the north of the site. *Scrobicularia* is the chief among these species; it can occur in very large numbers in estuaries where conditions of varying salinity are present with abundant organic debris. Ancient *Scrobicularia* clays are present at the inner clay site on the foreshore (see below) and such clays can be seen today forming further up the estuary, eg, the mud flats at Bideford. Other species of sandy or muddy/sand bottom deposits are the cockles, *Venerupis* spp., *Ensis* sp. and *Solen* sp.

It is interesting that *Scrobicularia*, the second most abundant species, is not eaten today. However, it does occur with edible molluscs in Mesolithic terrestrial contexts at Blashenwell, Dorset (Reid, 1896: 71; Preece, 1980: 355) and Nannas Cave, Pembrokeshire (Lacaille and Grimes, 1955) and in the midden at Morton, Fife (Coles, 1971). It is also the major constituent of a Romano-British midden at Stoke Gabriel on the edge of the Dart estuary, Devon (Woolner and Woolner, 1966); this evidence, together with the hint provided by its common name, peppery furrow shell, leaves little doubt that it was once eaten. Others of the inedible species such as *Littorina "saxatilis"*, *Littorina obtusata/mariae* and the other juvenile *Littorina* sp. are present in tiny proportions and could have been brought to the site with other shells, stones or seaweed.

Freshwater Molluscs

On Figures 8 and 9 freshwater species are plotted as percentages over and above the land mollusc total. Numbers given for bivalves are individuals (ie, valves divided by two). Bivalves are abundant and well preserved, the periostracum often being present and a number of the valves still paired. It is remarkable that all the bivalves are of a single species, *Pisidium personatum*, a species of stagnant water, puddles and temporary ponds. Similar conditions are suggested by the only other freshwater species: a few examples of *Lymnaea truncatula* and *Anisus leucostoma*. This very restricted freshwater assemblage points to specialised conditions and accords well with the insect evidence that there were only very small areas of standing water, which the mollusca suggest were subject to periodic drying out. Certainly we do not seem to be able to account for the better preservation of freshwater as compared to marine molluscs by invoking a major episode of fluvial reworking because at no stage in the midden is there any evidence of running water. Interestingly the freshwater molluscs decline at the top of the midden and this hints at a trend towards drier conditions.

No examples were found of the brackish water species *Hydrobia ulvae* which Rogers (1908: 253) reports as being abundant in the blue clay. However, it is very abundant in the clay of Area 1 and it is demonstrated elsewhere (Part II.iii) that much, if not all, of Rogers' material came from that area.

Land Molluscs

Preservation was variable, the periostracum on some shells was preserved while others were etched and fragile. Identifications are presented as summary diagrams (Figs. 8 and 9) and in tabular form (Tables 6 and 7). The summary diagrams follow recent work of Kerney and Preece (eg, Kerney *et al*, 1980; Preece, 1980), on mollusc assemblages of partly comparable date, in grouping certain terrestrial species thus:-

Terrestrial Group A. Have wide ecological tolerances and are found in open ground, marshes and coniferous and deciduous woodland: *Cochlicopa*, *Columella ?aspera*; *Punctum pygmaeum*; *Vitrea*; *Nesovitrea*; Limacidae; *Euconulus fulvus*; *Arianta* and *Cepaea*.

Terrestrial Group B. Species more critical in their requirements than Group A being most frequent in deciduous woods and similar well shaded places: *Carychium tridentatum*; *Acanthinula*; *Ena*; *Aegopinella* and Clausiliidae.

Both columns show similar assemblages right through the mollusc bearing midden deposits without any hint of ecological change. Four species predominate: *Discus rotundatus* and *Carychium tridentatum* are common in woodland ground litter, *Vitrea contracta* and *Oxychilus cellarius* are also common in shaded places. If we consider the total land assemblage in terms of the ecological groups defined by Evans (1972) then 92% belong to his shade loving group. Terrestrial Group B (a smaller range of species than Evans' group) makes up 30% or more of all samples. This, together with the ecological preferences of nearly all the other species, implies very strongly that this is a woodland/scrub assemblage. Species included by Evans (1972) in his open country group make up less than 1% of the total assemblage. Even this low figure is made up entirely of Vallonias which do live in woodland but, like the so-called intermediate species *Trichia hispida* (eight apices), become very much more abundant with anthropogenic clearances in the late Flandrian. Thus there seems to be no clear molluscan evidence for any sort of open or cleared area associated with the coast (eg, the often quoted exposure factor) or with the midden itself. Two of the species present in small numbers, *Carychium minimum* and *Zonitoides nitidus*, are regarded as obligatory swamp species. It should also be mentioned that *Oxychilus* and *Vitrea* are known to be flesh eaters, and the same is suspected of *Discus* (Evans and Jones, 1973: 121), so it is possible that these three abundant species may have been attracted by rotting food debris within the midden itself.

Westward Ho! provides welcome sub-fossil molluscan records for the south west peninsula where assemblages of this date are seldom preserved. In some cases there is an interesting comparison with the present geographical distribution of species as shown by Kerney (1976a). Distributions have often become more restricted, particularly as a result of human impact through clearance, drainage etc. *Vertigo pusilla*, for instance, seems to be totally extinct in the south west peninsula but interestingly the only old non-subfossil record is from this area. *Leiostyla anglica* is restricted to a single site in south Devon and occurs on three Cornish sites, the nearest being at Bude with an old record from Lundy. Evans (1972) notes that this species is strongly anthropophobic and consequently rare on archaeological sites. *Zonitoides excavatus*, though still fairly extensively distributed in the South West is a calcifuge species and is rarely found as a post-glacial fossil. *Cochlodina laminata* is conversely much more common on calcareous deposits to the east and reaches its most western British limits in this area.

Table 6. Land and freshwater mollusca from Column A.

Sample number	8311202–6 8311220	8311207–11 8311221	8311212 8311222	8311213 8311223	8311214 8311224	8311215 8311225	8311216 8311226	8311218	8311234	
Sample depth	0–5	5–10	10–15	15–20	20–25	25–30	30–35		Sample A	Individual
Context	Peat	Top of Midden	Midden	Bottom Midden	Top of clay	Clay	Clay	Midden	Jacobi Midden	shells Midden
Weight of sediment (g)	1269	1950	1993	2059	2313	1932	1849	1099	300	
LAND MOLLUSCS										
Carychium minimum Müller		2	9	3						
Carychium tridentatum (Risso)		30	33	26	2			15	7	
Carychium spp		1	7							2
Cochlicopa spp		2		+						
Columella ?aspera Waldén			1							
Lauria cylindracea (da Costa)		1	3	2						
Vallonia costata (Müller)			2	1						1
Vallonia pulchella (Müller)			1							
Vallonia excentrica/pulchella				2						
Acanthinula aculeata (Müller)		2	10	5				1		
Ena obscura (Müller)		1	5	5	1			1	2	
Punctum pygmaeum (Draparnaud)		1								
Discus rotundatus (Müller)	1	35	80	60	7			11	6	14
Vitrea crystallina (Müller)		4	1	3				1		
Vitrea contracta (Westerlund)		30	34	21	3			4	2	1
Vitrea spp		2								2
Nesovitrea hammonis (Ström)		1		+				+		
Aegopinella pura (Alder)		3	2	4	1			2		
Aegopinella nitidula (Draparnaud)		16	13	5				+		1
Oxychilus cellarius (Müller)	3	29	33	17	4		2	6	5	7
Oxychilus alliarius (Miller)		1		2						
Zonitoides excavatus (Alder)		1	3							
Zonitoides nitidus (Müller)			4							
Limacidae	2	3		2				2		
Euconulus fulvus (Müller)			2	1						
Cochlodina laminata (Montagu)		2			1					1
Clausilia bidentata (Ström)	1	1	4	6						1
Balea perversa (L.)		1						1		
Clausiliidae		+					+			
Trichia hispida (L.)		+	2	2	+			+		
Arianta arbustorum (L.)				+						
?Helicigona lapicida (L.)			1	1						
Cepaea nemoralis (L.)		1								1
Cepaea hortensis (Müller)				1						1
Cepaea / Arianta	1	3	2	9	+		+	2	2	+
FRESHWATER SPECIES										
Lymnaea truncatula (Müller)			6	15						
Anisus leucostoma (Millet)			1	2						
Pisidium personatum Malm (single valves/2)		6	100	45.5				2	1	
Pisidium personatum (conjoined valves)		8	8	7				1		
LAND MOLLUSC TOTAL	8	173	252	178	19	0	2	46	24	32
FRESHWATER MOLLUSC TOTAL	0	14	115	69.5	0	0	0	3	1	0

Table 7. Land and freshwater Mollusca from Column B (Area 3) and fresh and brackish water Mollusca from Area 1.

Sample number	8311723	8311724	8311725	8311726	8311727	8311728	8311729	8311730	8311732	8311731	8311702
Sample depth	0-4	4-7	7-12	12-17	17-19	19-26	26 +				
Context	Peat	Peat / Midden	Top of Midden	Midden	Middle Midden	Bottom Midden	Clay	Midden	Top of Midden	Midden (29)	Inner Clay
Weight of sediment (g)	1570	1300	2900	3030	2220	2420	4000	2100	620	1460	1606
Carychium minimum Müller		3		2	2			1			
Carychium tridentatum (Risso)		30	22	70	14	11		10	6	8	
Carychium spp					13	12		12	2	4	
Cochlicopa lubricella (Stabile)				1		+					
Cochlicopa spp				3		4		2			
Vertigo pusilla Müller				1							
Leiostyla anglica Wood										1	
Lauria cylindracea (da Costa)		1									
Vallonia costata (Müller)		1			+	1		2		1	
Vallonia pulchella (Müller)				1							
Acanthinula aculeata (Müller)		2		4	1	3				1	
Ena obscura (Müller)		5	1	3	1						
Punctum pygmaeum (Draparnaud)		2			1						
Discus rotundatus (Müller)		51	24	69	42	39		28	5	15	
Vitrea crystallina (Müller)		1	2					1	1		
Vitrea contracta (Westerlund)		33	21	40	12	18		15	11	10	
Nesovitrea hammonis (Ström)		1	+								
Aegopinella pura (Alder)		3	1	5	4	6				2	
Aegopinella nitidula (Draparnaud)		15	9	11	5	13		9	3	4	
Oxychilus cellarius (Müller)		18	9	36	8	9		13	12	7	
Oxychilus alliarius (Miller)		3									
Zonitoides excavatus (Alder)		1		1							
Zonitoides nitidus (Müller)				1							
Limacidae				1							
Euconulus fulvus (Müller)				2	1						
Cochlodina laminata (Montagu)		1	2	2							
Clausilia bidentata (Ström)		2	1	4	3	2					
Trichia hispida (L.)		1		2		1		+			
Cepaea / Arianta		10	5	6	5	5		1			
FRESH AND BRACKISH-WATER SPECIES											
Valvata piscinalis (Müller)											1
Hydrobia ulvae (Pennant)											1143
Ovatella myosotis (Draparnaud)											2
Leucophytia bidentata (Montagu)											4
Lymnaea truncatula (Müller)		1	1		1			1			
Anisus leucostoma (Millet)			1	1	2			3			
Pisidium personatum Malm (single valves/2)		7	4	64	18	1		17			
Pisidium personatum (conjoined valves)				22	6			1			
LAND MOLLUSC TOTAL	0	184	97	265	112	124	0	94	40	53	0
FRESH AND BRACKISH TOTAL		8	6	87	27	1	0	22	0	0	1150

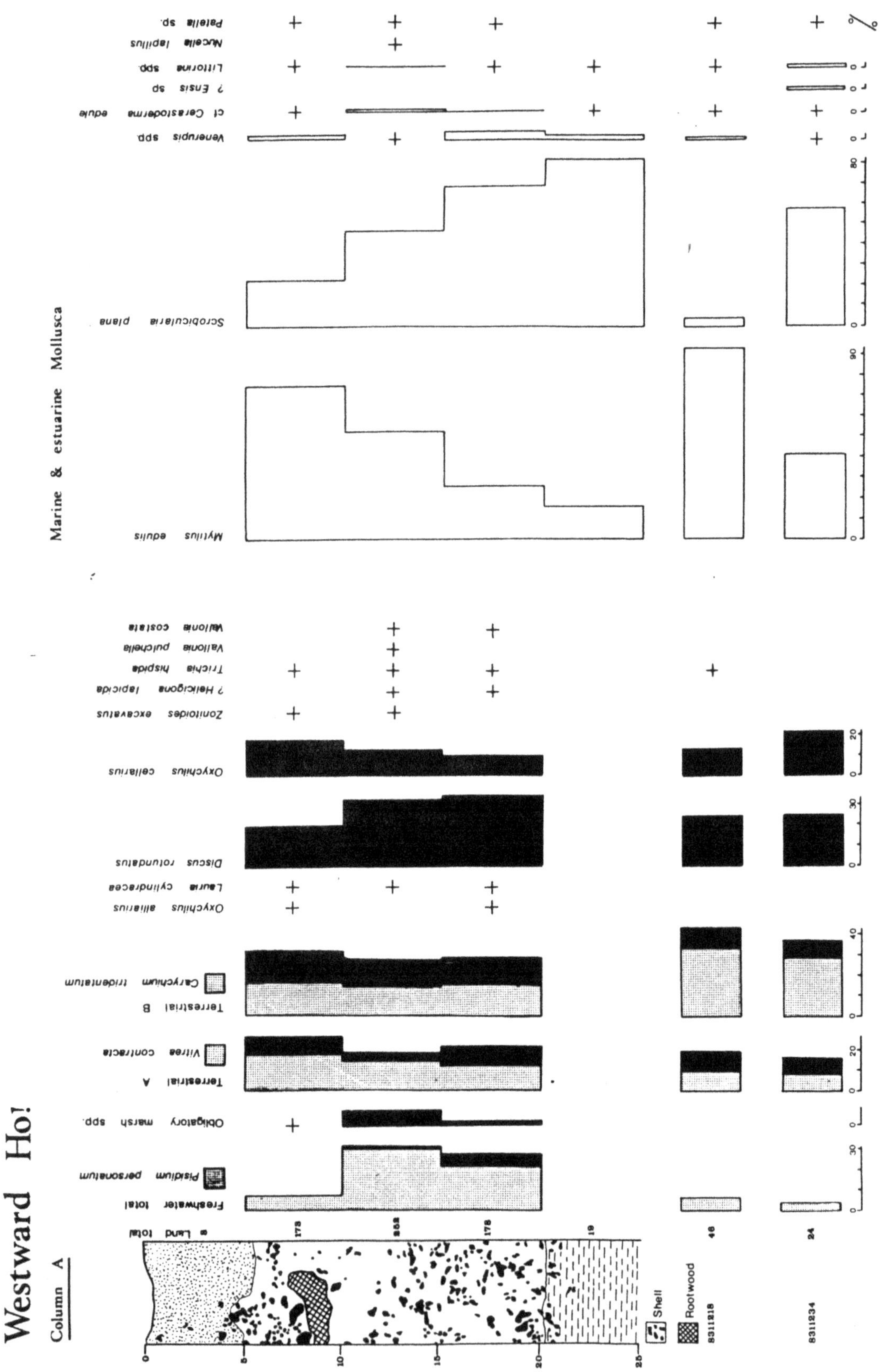

Figure 9. Mollusca from Column A.

The assemblage compares quite closely with that reported by Kerney (1976b) from organic debris peat dated 6730±150 BP (Q-1343) exposed on the modern foreshore at Blue Anchor Bay near Minehead. Westward Ho! clearly belongs to Kerney's molluscan biozone *d* (Kerney *et al*, 1980). The radiocarbon dates suggest it is contemporary with zone *d2* as defined in the South East by the occurrence of three species of which only *Leiostyla anglica* is represented here by a single shell. Of the other two, *Spermodea lamellata* is present at Blue Anchor Bay but not here and *Acicula fusca* is absent at both sites. Possible reasons for these absences may include the factors already suggested in relation to the comparatively low species diversity. Kerney (1977: 388) has in fact suggested that the subdivision and dating of zone *d2* may to some extent be of local value only. Compared to the climatic optimum assemblages from the South East, the Westward Ho! range of species is in any case a little restricted. This may to some extent reflect the non-calcareous nature of the surrounding geology, the nearest limestone being 40 miles to the east, and the possibility that we are dealing with a site not long colonised from alluvium, as hinted at by the absence of pedological development on the top of the underlying "blue clay".

Crustaceans (Tables 4 and 5)

Many of the samples contained barnacle plates, which probably arrived attached to mollusc shells, and small pieces of crab or lobster shell and nippers.

CONCLUSIONS

A second column of samples was examined because the writer was puzzled by the absence of certain rarities found in other contemporary sites, particularly those characteristic of zone *d2*. In the event this exercise did produce a couple of new species but otherwise the results are very closely comparable, which is at least reassuring.

The mollusc evidence all relates to one brief period when deposition of the midden created conditions sufficiently calcareous for the land and freshwater Mollusca to be preserved. At that time it is probable that the site was in the later scrub/fen woodland stages of a succession which began with alluvium and continued with the woodland conditions of the overlying peat. Swamp species were present but pools were small and subject to periodic drying out; the Mollusca do not suggest running water.

We now need to consider whether the marine shells comprise a Mesolithic midden or a natural death assemblage. The fragmentary material and the importance of a species not eaten today might be thought to hint at the latter. Neither piece of evidence is, however, convincing since fragmentation is at least partly secondary and *Scrobicularia* was certainly collected, and presumably eaten at other sites. Had this been a natural strandline death assemblage then we might have expected a much larger number of juvenile shells, such as we find in the clays in Area 1 (see below), in fact the vast majority were adult (Tables 4 and 5). We might also have expected an altogether broader range of species. The mollusc evidence therefore, together with the density of flints and charcoal, persuasively argues for human activity and it seems that molluscs were brought from a nearby shore for consumption in a basically terrestrial situation. This is suggested because neither molluscs, insects nor the other evidence indicates that the midden formed on the strandline. This new evidence does not support Churchill's (1965: 81) premise that the midden necessarily stood at "sea level" between the tides.

It is most unfortunate that the fragmentary nature of the marine assemblage has greatly limited research possibilities. The key issue of seasonality cannot be confronted because oxygen isotope work or the analysis of growth line patterns would require at least some whole shells or, at least, large fragments (Margaret Deith, pers comm).

The composition of the mollusc assemblage illustrates very clearly the exploitation of an ecotonal situation. The site seems to have lain more or less at the junction of three ecosystems: the rocky *M. edulis*-rich coast, the biologically very rich and productive environment of the neighbouring estuary, and on the site itself and inland the resources of the fen wildwood. In this respect Westward Ho! may offer something of a contrast to the much better preserved Oronsay middens where there is a suspicion (Mellars, 1978, 394) that emphasis on the resources of the seashore was an economic necessity engendered by a paucity of other food sources on a small island.

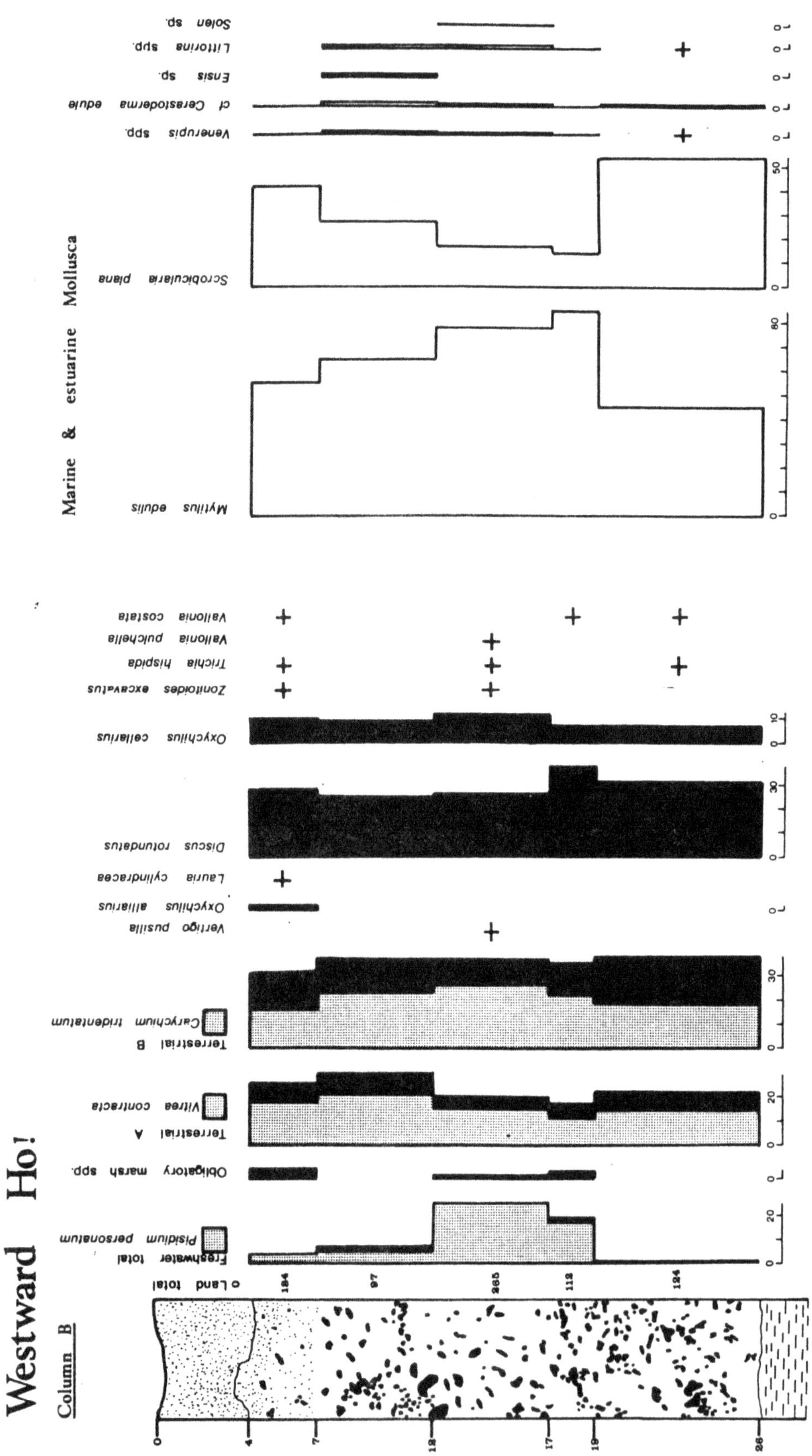

Figure 10. Mollusca from Column B.

SAMPLE 8311702 - CHANNELS IN AREA 1

This sample is discussed separately because it comes from a context completely unrelated to those associated with the midden. The shells are much better preserved than those from the midden. Among the marine species (Table 5) *Scrobicularia plana* is predominant, at least four adult valves were still paired. Other taxa are *M. edulis*, cockles, *Anomia*, *Littorina "saxatilis"* aggregate, *Veneracea*, *Mactracea* and *Musculus*. In contrast to the midden assemblage it is interesting that the majority of *Scrobicularia*, *Mytilus* and cockles are tiny juvenile shells.

As regards the non-marine Mollusca, no land species were present and only one freshwater shell of *Valvata piscinalis*. A group of three brackish water species is overwhelmingly dominated by large numbers of *Hydrobia ulvae*, a species of estuarine mud flats, particularly the mid to upper part of the tidal zone. Also present are *Ovatella myosotis* which lives on rotting rubbish near high water in estuaries and *Leucophytia bidentata*, another estuarine species. Some barnacle plates and fragments of tiny immature crab/lobster were found in the same sample.

The general impression from the paired *Scrobicularia* valves and the predominance of tiny juveniles is of a natural death assemblage. Its location seems to have been the tidal zone of estuarine mud flats.

ACKNOWLEDGEMENTS

I am grateful to Nick Watson and Joy Ede who helped with the sorting and preliminary identification. Advice on specific identifications and interpretation was kindly given by Dr Michael Kerney, Imperial College, London and Dr John Taylor, British Museum (Natural History).

II.iii - THE VERTEBRATE REMAINS

Bruce Levitan and Alison Locker

INTRODUCTION

Subsequent to the sampling by the Central Excavation Unit, bones recovered by hand and from sieving were identified by A. Locker who prepared a preliminary report on these remains. B. Levitan then visited museums at Exeter (Royal Albert Memorial Museum) and Barnstaple (North Devon Athenaeum) to record bones from the site which had been collected in the 19th century, and also identified and recorded bones from collections made in the 20th century, mainly by Messrs. Trapnell and Jenkyn, plus a small amount of material from Dr R. Jacobi. The analysis of the collections predating the recent CEU excavation was undertaken for two major reasons: to provide a detailed record of all (or most) of the available material collected to date, and to compare the earlier collections with the provenanced CEU material. The CEU investigations revealed two sites which yielded animal bone (see Part I), one in Area 1 and the other, the Mesolithic midden in Area 3. A third aim of the bones study was to find out whether the two sites were similar or dissimilar. The following report tackles each aspect in turn, and authorship is given in parentheses: AL= Alison Locker, BL= Bruce Levitan. The first section contains the analysis of the CEU material (AL/BL) and the analyses of the earlier collections (BL). The second section compares the two groups of material (BL) and the third section discusses the material in relation to the two sites (BL). The report is concluded with a general discussion of the material (BL).

CENTRAL UNIT COLLECTIONS

Large fragments of bone were recovered by hand, but the greatest number of fragments were recovered from sieving of the column samples taken for mollusca. Much of this material was highly fragmented, very small, and in many cases unidentifiable (79.7% of the bones are unidentified: Table 8). The majority of the bones are from the midden and overlying peat in Area 3 and only 22 (7.1%) come from Area 1. Both sites will be considered here but possible differences between the sites are considered below. The measurements were taken according to Jones *et al* (1981), and are summarised in Table 9.

Area 3 (AL)

From the peat the slow worm (*Anguis fragilis* L.) was identified by the characteristic bony scales that cover the body. This species prefers well vegetated habitats, with extensive ground cover, often in damp situations (Arnold and Burton, 1978: 175). It is frequently found in certain coastal areas in the south and west of England (Smith, 1969: 173).

Larger mammals from the peat include a fragment of cervid/bovid sized rib, and two burnt fragments. At the transition between the peat and the clay was the lower third molar of pig (*Sus scrofa* L.). The enamel is burnt; the length of this tooth (Table 9) indicates this animal was wild (Clason, 1967). Also present was the proximal end of a right radius of red deer (*Cervus elaphus* L.).

In the midden were more scales of slow worm. Frog (*Rana* sp.) was identified from a maxilla, and bank vole (*Clethrionomys glareolus* Schreber) from two molars. This latter animal is typically found in woods, at their edges and in clearings (Hanak, 1977: 108). Other small mammal fragments were not identified to species. The distal end of a red deer right humerus and a cervical vertebra were found, and other fragments of (?) red deer are suggested by the proximal posterior surface of a metatarsal shaft, the eroded proximal end of a humerus and a lunate. A few pieces of cervid/bovid bone from within the midden are not specifically identifiable, some of which are lightly burnt. Roe deer (*Capreolus capreolus* Lonnberg) was identified from the distal end of left metacarpal and the proximal end of right humerus.

The two most interesting finds from the midden are the broken tuber area of a bovid calcaneum, the measurements of which fall within the lower (female) range for aurochs (*Bos primigenius* Bojanus), and at the boundary of the midden and clay a very eroded bovid radius, which also falls into the lower size range for aurochs (C. Grigson, pers comm). The female aurochs size range overlaps with larger domestic cattle, but given the securely stratified Mesolithic context of these finds, they may be assumed to represent female aurochsen.

Two eroded metapodial fragments were also found, one fragment of the proximal articulation and shaft of a right metatarsal, and part of the posterior surface of the midshaft. Both of these bones match red deer, but are so fragmentary that these identifications must be tentative.

Other bones from the midden include three very small fish vertebrae, possibly all from the tail of one individual. These proved impossible to identify, but most closely resemble goby-type (Gobiidae), a shallow water coastal group. If this suggestion is correct, these vertebrae may be the remains of a predator's meal, caught on the shoreline.

The clay beneath the midden contained little material except for a bovid tooth fragment, a vole incisor and one frog bone. The unstratified material includes a fragment of roe deer metatarsal, and the lightly burnt distal end of metapodial which might also be roe deer.

The species identified suggest wooded conditions, with the exception of the frog and fish bones, both which may represent food remains. The slow worm, bank vole and larger mammals are all indicative of wooded or partly wooded areas, and the burning might be evidence of man's activities.

Area 1 (BL)

The 22 bones from this area include bovid, cervid, sheep/goat and dog fragments. Identifications in the first two categories are tentative since fragments are small and often eroded. Four bones are probably bovid, and although only a right distal scapula is measurable, they are too small for aurochs so may be domestic cattle (*Bos taurus* L.).

They include a left upper first molar, a left proximal half of metacarpal and a left navicular. The last has a hole which joins the proximal and distal surfaces, but appears to be a pathological rather than artificial condition. Three antlers are from red deer, one of which had been shed (not known for others). The sheep/goat fragments include a left humerus with unfused epiphyses, a left pelvis, and a right tibia with canid chewing at the distal end. These cannot be confidently assigned to species, but probably are sheep (*Ovis aries* L.) rather than goat. The three bones of dog (*Canis familiaris* L.) are an articulated seventh lumbar vertebra and sacrum, and a left tibia.

With the exception of the deer, these are domestic animals and cannot give clues about environment; however they are direct evidence that man was active at this site. The presence of red deer antler might indicate that this had been collected by man and brought to the site, possibly for working, though no evidence for bone working was discovered.

EARLIER COLLECTIONS
(BL)

The bones from earlier collections comprise seven groups, four from the 19th and three from the 20th century. The earlier collectors were Inkerman Rogers, Townsend Hall, Heathcote and Ellis, the first two of which are presently housed in the North Devon Athenaeum, Barnstaple, and the others in the Royal Albert Memorial Museum, Exeter. The report by Churchill (1965: 77) lists bones reported by Inkerman Rogers and D. Bate, but gives no quantifications. Taxa listed include rodent, hedgehog, red deer, fallow deer, pig and wild boar. Not all these were seen by the present author, so it seems likely that there is more material, but possibly it is very widely dispersed (R. Jacobi, pers comm). E.H. Rogers (1946) also lists *Bos longifrons* but this may be incorrect, paticularly if the context is Mesolithic (but see below). The Inkerman Rogers collection seemed surprisingly small in light of his reputed collecting abilities and Arber (1982: 202) describes "More than 30 vertebrate bones and teeth" from I. Rogers' Bideford collection as well as "About 20 specimens from the Submerged Forest at Westward Ho!" at the Athenaeum (Arber, 1982: 207). The former collection is supposedly at the Royal Albert Museum (Arber, 1982: 199), but this proved to include only his fossil collection - it is possible that some of Inkerman Rogers' Westward Ho! collections are now lost. Hall's collection at the Athenaeum includes two papers in manuscript describing his finds and work at Westward Ho! and listing some of his identifications. The Heathcote collection at Exeter includes a number of interesting handwritten labels, the majority of which have unfortunately come adrift from the bones. Many are dated and range from early 1871 to mid 1872. They make interesting reading, and range from simple dated labels to one with the note "Portion of antler of Red (?) Deer Given by ___ 10/2/1872 the forest ranges about 2 miles" indicating a much larger deposit than survives today. (This label accompanies specimen 4161).

Ellis collection

This comprises 181 fragments and dates from around 1866 (Churchill, 1965: 74) and has been divided into two groups according to collector's lot numbering (183 and 161). There is no information with the collection about provenance so the material cannot be directly related to Rogers' paper (1946) which describes Ellis' activities. The two groups are rather different. Group 161 contains sixteen bones, half of which are deer (7 red deer, 1 roe deer), 7 bovid/cervid and 1 pig. The bovid fragments are too fragmented for measurement, so it is not possible to determine if they fit into the aurochs size range, but if they do they represent small (female) individuals. Three of the bones cannot be definately said to be bovid, so only three bones (18.8%) are probably bovid (right distal fragment of radius, shaft fragment of femur and shaft fragment of tibia). The two pig bones, a radius and ulna, are from young individuals so it is not possible to determine if they are wild pig. The red deer bones include a left proximal tibia, a left navicular, one right distal and three shaft fragments of metatarsal and a first phalanx. The roe deer bone is a large left shaft fragment of metatarsal, and this might alternatively be fallow deer (*Dama dama* L.).

Table 8. Species present.

	CEU-midden		CEU-inner		Ellis		Hall		Heathcote		I. Rogers		Trapnell		Jenkyn		Jacobi		TOTAL
	n	%	n	%	n	%	n	%	n	%	n	%	n	%	n	%	n	%	N
cattle	10	22.22	8	40.00	21	33.87	22	75.86	10	16.13	12	48.00	28	60.87	2	25.00	-	-	113
pig	1	2.22			9	14.52			3	4.84	3	12.00	3	6.52	1	12.50	-	-	20
red deer	5	11.11	3	15.00	24	38.71	7	24.14	48	77.42	3	12.00	3	6.52	5	62.50	1	50.00	99
roe deer	3	6.67	-	-	2	3.23	-	-	-	-	-	-	-	-	-	-	-	-	5
sheep	-	-	6	30.00	5	8.06	-	-	1	1.61	2	8.00	5	10.87	-	-	-	-	19
horse+dog	-	-	3	15.00	1	1.61	-	-	-	-	5	20.00	7	15.22	-	-	-	-	16
sm. mammal	2	4.44	-	-	-	-	-	-	-	-	-	-	-	-	-	-	-	-	2
reptile	18	40.00	-	-	-	-	-	-	-	-	-	-	-	-	-	-	-	-	18
amphibian	3	6.67	-	-	-	-	-	-	-	-	-	-	-	-	-	-	-	-	3
fish	3	6.67	-	-	-	-	-	-	-	-	-	-	-	-	-	-	1	50.00	4
Total	45	100.00	20	100.00	62	100.00	29	100.00	62	100.00	25	100.00	46	100.00	8	100.00	2	100.00	299
bovid/cervid	9		2		84		39		25		3		-		8		-		170
indet.	134		-		35		3		2		-		2		-		-		176
indet.sm.mam.	101		-		-		-		-		-		-		-		-		101
TOTAL	289	15.57	22	90.91	181	34.25	71	40.85	89	69.66	28	89.29	48	95.83	16	50.00	2	100.00	746

% in TOTAL row = % identified
bovid/cervid bones include some identified to anatomy but not species.

Group 183 contains many unidentified fragments (67.3%), and is described by the collector as from the "kitchen midden". All the above species are represented, but sheep/goat and dog are also present. The deer are less common, eighteen bones making up 33.4% of the identifiable bones. The red deer bones include a left mandible (no teeth), right lunate, six fragments of metacarpal (with one right, one left), two right tibiae, four fragments of metatarsal, one metapodial fragment and a first phalanx. The roe deer bone is a fragment of metatarsal. The 23 bovid bones (42.6%) include five which may be cervid (two vertebra and three metapodial fragments). The bovid bones are all fragmentary, but some at least are complete enough to compare with aurochs (though not measurable). A right ulna, left proximal femur and first phalanx all seem too small to be aurochs so are tentatively identified as domestic. (It should be noted that if these are early cattle they will have been much smaller than aurochs, but even so it is extremely difficult to tell with such eroded and fragmentary material: these comments also apply below). The seven pig bones include an atlas vertebra and a fragment of mandible which appear to be too small to be from wild pig, so are tentatively identified as domestic (*Sus domesticus* Erxleben); other bones are either too fragmentary or from individuals too young to tell. Five sheep/goat bones are present, four of which are sheep (left distal humerus, two left pelves [one male] and one right distal metatarsal) - the remaining bone (radius) could be sheep or goat. Dog is represented by a single right radius fragment.

The two groups are in similar condition, but the 183 group is more fragmentary and appears to be more consistently stained grey or grey/blue - presumably from clay.

Townsend Hall collection

This collection also dates from around 1866 and is also described as coming from the "kitchen midden". All the bones are heavily stained grey/blue so appear to have come from a clay, rather than peaty deposit. There are 71 bones, 37 of which are unidentified (52%) - most are bovid/cervid but a minority are from smaller animals: small cervid / sheep/goat / pig. Only two species are identifiable. Red deer bones are three metacarpals (one right proximal, one right shaft, one shaft), one left patella, a right shaft and shaft fragment of metatarsal and a proximal part of metapodial. The bovid fragments include four cervical, one thoracic and two unknown vertebrae (which are possibly cervid). Two right (one proximal, one distal) and one left distal humerus, two left radii (proximal, distal), one right distal tibia, one right calcaneum and a third phalanx are all from small aurochs/large domestic cattle, but are not measurable; other bovid fragments are too fragmentary to attempt any kind of division. The left proximal radius is broken proximal-distally in the anterior/posterior plane, and may heve been deliberately split along the length of the bone (this is not typical of canid action, and there is no evidence of chewing).

Heathcote collection

There is quite a lot of information with the collection. The bones were obtained in 1871 and 1872, some directly by C.D. Heathcote, others donated. In the latter case, labels are cryptic, eg the label quoted above, and another with "given by 'M'". The labelling is extremely haphazard, some bones having unique specimen numbers, others with lot numbers and a few with both. These were recorded where possible in order to see if they form any kind of groups (by date, content?), but do not appear to do so except that all the numbers on antlers are different to all the numbers on other bones. This may be more than coincidence since the antlers do seem to form the major part of the collection (51.7%). Some of the bones are described as from "blue clay" and one or two from the "bone bed" thus at least two sites of provenance are indicated, but such descriptions are too few to attempt grouping the bones on this basis. Many of the antlers have coronet present, and only one out of eleven is not shed (9.1%) indicating that if collected by man, they were possibly collected during the short shedding season of late February-early March (but see Levitan, *forthcoming*; Clutton-Brock, 1985 for discussions of collection of shed antlers - deer in fringe/poor environments may quickly eat their antler to regain calcium, deer in better environments may not, so the collection period may be longer). The unshed antler indicates that some hunting of deer may have taken place. One antler has a hole in the base of the beam, possibly having been drilled. The more complete antlers are illustrated in Figure 11 which shows that stage of development is not readily discernable. The (?) worked antler is the only specimen with battering around the coronet suggestive of use of a tool, and the fact that tines do not seem to have been deliberately removed might argue against manufacture into tools such as the picks from Grimes Graves (Legge, 1981: 101-103). However analysis of the antlers

Table 9. Summary of measurements (excluding antlers).

Taxon/ Anatomy	Collection	Measurement
Cattle		
scapula	CEU (Area 1)	SLC:43.5, BG:40.8, SG:47.9
	Trapnell	41.7 39.2 45.2, SLC:55.9
radius	CEU midden	Bd:88.0 (aurochs)
	Trapnell	73.5, BFd:65.9
	Trapnell	Bp:64.4, Bfp:60.7
calcaneum	CEU midden	RJ1:109.7, RJ4:45.0, RJ5:37.3 (aurochs)
metatarsal	I.Rogers	Bp:42.0
Sheep		
horncore	Trapnell	41:48.7, 42:35.3, 40:138.7
Pig		
radius	Heathcote	Bp:35.0
lower M3	CEU midden	L:44.7, B:17.9
Horse		
metatarsal	Trapnell	Bd:42.2, Dd:33.8
Dog		
humerus	I.Rogers	GLl:170.0, Bp:40.0, Dp:47.5, BT:22.0, Bd:34.5
ulna	I.Rogers	LO:21.1, BPC:15.5, GL:>164.5, SDO:21.0
tibia	I.Rogers	GL:164.0, SD:10.0
	Trapnell	173.0 11.1, Bp:27.5, Bd:21.4
	Trapnell	11.3 34.4
Red deer		
scapula	I.Rogers	SLC:44.0, BG:40.0, SG:49.5
humerus	CEU midden	BT:49.8, Bd:52.7,
	Trapnell	54.1 61.9, SD:29.5
radius	CEU midden	Bp:59.1, Bfp:56.0
metacarpal	Heathcote	Bp:41.4
Roe deer		
humerus	CEU midden	Dp:40.3
metacarpal	CEU midden	Bd:22.0, Dd:13.9, DD:12.7

All measurements described in von den Driesch (1976) except calcaneum, from Jones *et al* (1981).

from Hazleton indicates that such need not have been the manufacturing procedure in every case, the Hazleton antlers almost definitely having been used as picks, but without removal of tines (Levitan, *forthcoming*). Where possible the antlers were measured, and measurements are summarised in Table 10.

The remaining 43 bones comprise ten bovid, one sheep/goat (radius), three pig (mandible, right radius, left calcaneum), and two red deer metacarpals (left, right). Twenty-seven fragments are not identifiable. Again none of the cattle can be definitely assigned to aurochs or domestic, though the larger fragments appear more of domestic size. These include left distal humerus, left proximal tibia, left calcaneum and left navicular. The calcaneum is particularly interesting because it is the only bone with definite butchery having been chopped on the anterior of the tuber calcanei in a dorso-ventral direction.

Inkerman Rogers collection

There are no records concerning the collection at the Athenaeum, and any bones which were in the Bideford collection do not appear to have gone to Exeter. The collection dates from about the turn of the century, about which Rogers published an article describing specimens later deposited in the Bideford collection (Rogers, 1908). The 28 specimens at the Athenaeum have been numbered for display 1-24, but bear earlier specimen numbers running from 1002 to 1017. Some of the identifications listed at the Athenaeum (not made by Rogers) are incorrect. Bovid remains are the most numerous with twelve bones (42.8%), the other species being represented by four bones or less. The cattle appear to be domestic and include a left proximal metatarsal with proximal breadth 42 mm. The bones include the occipital condyles from a skull, three fragments of mandible, two (right) of humerus, three of pelvis, one left astragalus, one left metatarsal and an upper first molar. A nearly complete left mandible has no second premolar, a foramen on the lingual surface and pigmentation of the teeth. The absence of the premolar is interesting, indicating an (?) early occurrence of this phenomenon. Another (right) mandible has heavy canid chewing at the angle. The sheep/goat remains comprise two skull fragments. The pig bones include a left lower third molar, too small to be from wild pig. Horse (*Equus cabalus* L.) is represented by a left upper second molar. The deer remains comprise two scapulae (left, right), one of which is measurable. The dog bones include a complete right humerus indicating a height of 55.7 cm. An incomplete left ulna gives a height of greater than 46cm, and a weathered left tibia indicates a dog of around 49 cm (Harcourt, 1974: 154).

Trapnell collection

This recent collection comprises 48 bones from Area 1. It is divided into two groups (recorded as 1 and 2) but they are essentially similar and will be considered together. The bovids are most numerous, all fragments probably being of domestic cattle. They include the best preserved specimens, and measurements are summarised in Table 9. Bones include fragments of skull, radius (3: left), metacarpal (left), 2 pelvis (1: left), 5 tibia (1: left, 2: right), navicular (right), 3 metatarsal (2: right, 1: left), first and third phalanx. One radius has been split lengthways, probably deliberately, and one (cervid/bovid) rib has cut marks near the caput which could have been made using a flint knife. The sheep/goat bones are a right horncore of a ram, left mandible and atlas vertebra (also sheep), an upper third molar and right femur (sheep or goat). Pig (domestic) is represented by a right fourth metacarpal, right juvenile pelvis and left femur. Dog, represented by three left tibiae, include an individual of 51.5 cm; other dog bones are a right pelvis and a sacrum. The two horse bones are a left distal metatarsal and a first phalanx, the former of which has been battered in midshaft. Finally, red deer comprise a left humerus, a fragment of antler and a cervical vertebra. Several of the bones, especially bovid, have canid chewing marks.

Jenkyn collection

This collection comprises sixteen fragments including two fragments of (?) one atlas vertebra, and a cervical vertebra. One rib is cervid/bovid, and there are six unidentified cervid/bovid fragments. Pig is represented by a lumbar vertebra which has been cut laterally, removing the transverse processes. Red deer bones include three left metacarpals, one with a very smooth, worn distal end - possibly due to artificial smoothing - a left tibia and a right astragalus.

Table 10. Red deer antler measurements.

Specimen no.	Measurements 1	2	3	4
	182.0	144.5	70.2	76.1
	-	136.0	-	-
	124.5	139.4	62.5	65.0
	156.5	131.0	53.5	59.5
	-	101.2	50.1	54.2
	-	-	47.5	56.5
	111.5	90.0	44.0	51.2
	-	145.0	-	-
N	4	7	6	6
mean	143.6	126.7	54.6	60.4
sd	-	22.0	9.9	9.0
cv	-	17.4	18.1	14.9

measurement1: circumference of beam between coronet and brow tine
2: circumference of beam between brow tine and bez tine
3: least diameter coronet
4: greatest diameter coronet
[measurements taken from Clutton-Brock (1985).]

Jacobi collection

Two bones collected by Dr Jacobi from the midden were identified: a left proximal metacarpal of red deer and a fragment of hake (*Merluccius merluccius* L.) dentary.

COMPARATIVE ANALYSIS
(BL)

Together these bones total 746 fragments, an assemblage too small to support any kind of valid statistical analysis. Criticism on these grounds of any attempts to compare the collections and draw conclusions from them may be valid, but in the light of the importance of the site, this has been attempted below, and it is felt that the conclusions drawn are reasonable.

The most general of comparisons reveals a single major difference between the collections: one group contains large proportions of domestic species, characterised by high proportions of cattle and the presence of sheep, pig, dog (and horse in two examples), while the second group is dominated by deer with cattle also important, and sheep, dog and horse absent or virtually absent. Group one consists of the CEU midden site, Heathcote collection, Jenkyn collection, part of the Ellis collection (161), and, less obvoiusly, Hall collection. Group two comprises the CEU (Area 1) collection, Trapnell collection, Inkerman Rogers collection and part of the Ellis collection (183). Table 11 gives proportions of species for these groups excluding unidentified bones, and summing horse and dog, and red and roe deer. The differences are both consistent and convincing, and no doubt would stand up to statistical testing, though this has not been attempted due to small sample size.

This would seem to neatly answer one of the main questions posed at the outset, and so it would appear that a separation on the basis of location is tenable.

PROVENANCE OF THE MATERIAL
(BL)

Figure 8 is a plot of the data in Table 11, it reinforces visually the conclusion that the collections form two separate groups. Four of the collections (Jacobi's two bones are not included!) have known provenance and it seems to be stretching coincidence too far to suggest that the fact that these separate logically into the two groups could be mere chance (with the midden group being characterised by deer and cattle, and the Area

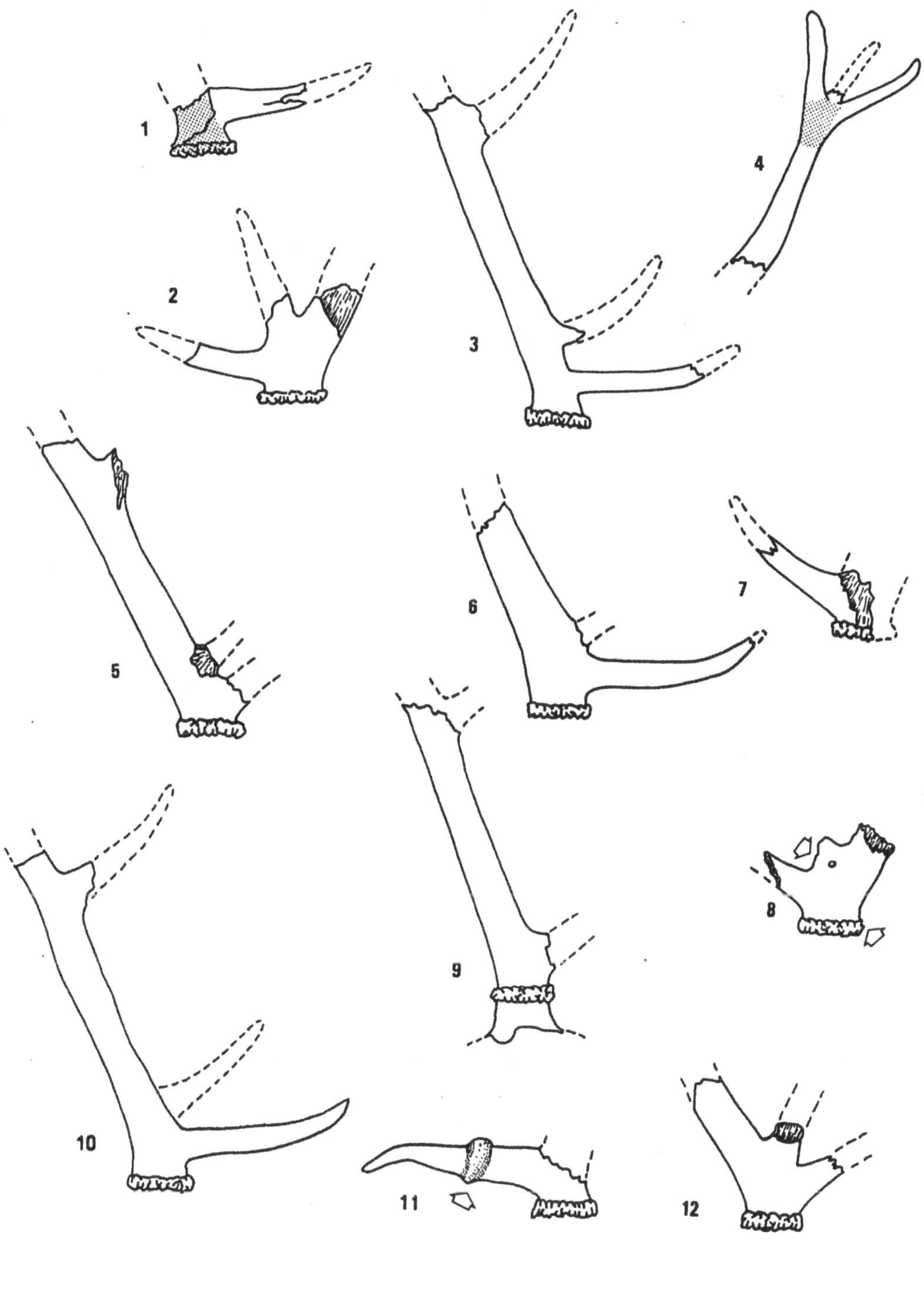

Figure 11. Red deer antlers from various collections.

Table 11. Species summary.

Species		Outer peat sites								??					Inner peat sites				
	CEU-midden		Ellis 161		Heathcote		Jenkyn		:	Hall		*	CEU inner		Ellis 183		I.Rogers		Trapnell
	n	%	n	%	n	%	n	%	:	n	%	*	n	%	n	%	n	%	n	%
cattle	10	52.63	3	23.08	10	16.13	2	25.00	:	22	75.86	*	8	40.00	23	42.59	12	48.00	28	60.87
deer	8	42.11	8	61.54	48	77.42	5	62.50	:	7	24.14	*	3	15.00	18	33.33	3	12.00	3	6.52
pig	1	5.26	2	15.38	3	4.84	1	12.50	:	-	-	*	-	-	7	12.96	3	12.00	3	6.52
sheep	-	-	-	-	1	1.61	-	-	:	-	-	*	6	30.00	5	9.26	2	8.00	5	10.87
horse+dog	-	-	-	-	-	-	-	-	:	-	-	*	3	15.00	1	1.85	5	20.00	7	15.22
Total	19	100.00	13	100.00	62	100.00	8	100.00	:	29	100.00	*	20	100.00	54	100.00	25	100.00	46	100.00

1 group by the many domesticates). The Heathcote collection contains some sheep, but possibly this collection is a mixture of material from sites in Areas 1 and 2; note that this almost certainly is the case for the Ellis collection, but luckily the two groups were labelled as separate. The apparent division into two groups for the Heathcote collection has already been noted.

CONCLUSION
(BL)

Bone collections from two sites examined by the Central Unit and seven collections of earlier date were identified and recorded. All the collections are small, and the entire assemblage numbers less than 1000 fragments. However, detailed examination of the evidence reveals that certain consistent differences occur which divide the bones into two groups on the basis of species represented and location. Bones from Area 1 are characterised by domesticated species and a small amount of deer, and must date from the Neolithic period or later; bones from the Mesolithic midden and adjacent peat, do not include sheep, horse, domestic pig or dog. They do include red and roe deer (which form the majority) and aurochs. Some bovid material is not definitely identifiable as aurochs or domestic, but the fragments are larger than the domestic cattle from the Area 1 groups, and are suggestive of small (ie, female) aurochs. The human activity is more evident in the material from Area 1, with evidence of butchery, burning (and canid chewing - an indirect indicator). Human activity is more tenuous at the midden site, but antler seems to have been exploited, and perhaps formed into tools, and deer, aurochs and wild pig may have been hunted.

ACKNOWLEDGEMENTS

Bruce Levitan would like to thank Kelvin Boot (Royal Albert Museum, Exeter) and staff at the North Devon Athenaeum, Barnstaple for arranging access to the early collections. Thanks are due to Dr Roger Jacobi, Mr D. Trapnell and Mr and Mrs T. Jenkyn who kindly lent the bones from their collections for this analysis.

II.iv - POLLEN ANALYSIS

R.G. Scaife

Pollen analysis was carried out on three profiles from Westward Ho!. Of these, Profile I (monolith 8311302) is from the midden itself and the peat which overlies it, Profile II (monolith 8311335) is from a more complete peat profile nearby but still within Area 3, Profile III (monolith 8311301) is from the inner peat (Area 1). Profile I was sampled and analysed in 1983, Profiles II and III were sampled later (1984) and were analysed to clarify some points raised by the earlier analysis. The three profiles are described and discussed separately but Profiles I and II, which are essentially contemporary, are also compared one with another.

PROFILE I

Two metal box monoliths were obtained for pollen analysis from the exposed section (Fig. 4). The total range of lithostratigraphy and biostratigraphy of the midden and its over- and underlying contexts was sampled. Sampling at 2cm intervals was carried out in the laboratory with 1-2ml of sample being taken at each level. These were subjected to standard procedures for extracting the sub-fossil pollen and spores (Moore and Webb, 1978). They were then mounted in glycerol jelly, counted and identified with the aid of phase contrast microscopy. The state of preservation throughout the profile was highly variable, ranging from moderate to good in the peats (levels 1-8). Absolute pollen frequencies (although not calculated fully) were similarly highest in these levels. By contrast, the lower silts, although sampled to a depth of 30cm below the base of the midden, were almost devoid of pollen. Pollen contained within the midden deposit fell between the two extremes. In consequence, the pollen sum counted was variable between

580 in the lower peats and 131 in the midden deposits (Zone WH:II). Results of this analysis are presented in Figure 12, further details are available in the project archive or in Scaife (1983). Arboreal pollen (AP) has been calculated as a percentage of its sum. Herbaceous and spore taxa are given as a percentage of arboreal pollen. The zone lines relate to both the lithostratigraphical and biostratigraphical divisions which, with the exception of Zone WH:V, concur.

The divisions are detailed from the base upwards:

WH:I
(15cm) The "lower blue clay": only the data from the uppermost sample is presented here. Pollen was almost totally absent in these highly inorganic sediments. A small number of highly degraded *Dryopteris* type and *Pteridium* spores were noted from a number of levels. No vegetation interpretation was possible.

WH:II
(12-14cm) The lower midden: pollen present, but relatively sparse; comprised dominant *Quercus* (73% AP), *Corylus* (230% AP), and *Ulmus* (15% AP). Relatively higher *Pinus* percentages are noted in the basal part of this zone. *Alnus* and *Hedera* occur from the centre of the zone. Herbaceous taxa are dominated by Gramineae and Cyperaceae. Spores (mostly degraded) are well represented by *Dryopteris* type and *Polypodium*. Some spores of pre-Quaternary type were found only in this zone.

WH:III
(8-11cm) The upper midden: comprises greater organic content than zone WH:II. Arboreal pollen spectra are again dominated by *Quercus*, *Corylus* and *Ulmus* but with a notable expansion of *Betula*, *Salix* and *Hedera*. Herb taxa are fewer, but with Gramineae and Cyperaceae as the primary constituents.

WH:IV
(4-7cm) The lower peat levels: comprising humified monocotyledonous peats with wood, *Corylus* nuts and leaves of *Salix* cinerea. Characterised by increased values of *Alnus*, *Salix*, *Betula* and *Fraxinus* with increasing herbaceous diversity. The spectra are dominated by *Quercus*, *Corylus*, *Salix* and *Hedera*. Gramineae, Cyperaceae and *Filipendula* are the dominant herb categories but with notable occurrences of *Plantago lanceolata*, Chenopodiaceae and fen wood taxa.

WH:V
(1-3cm) The upper peat levels: comprising the uppermost zone of the monocotyledonous peats. *Salix cinerea* leaves and *Corylus* nuts were readily identifiable. The zone is differentiated palynologically by declining *Ulmus* values (23% to 5%) from level 3 and increasing percentages of pollen of Cyperaceae and spores of *Pteridium aquilinum*, *Dryopteris* type and *Polypodium*.

Discussion

Zones WH:II and WH:III represent the levels in which midden material was present. Zone WH:II contained less organic matter and relatively sparse pollen, whereas Zone WH:III's greater quantities of organic matter and pollen possibly accumulated in damper conditions. Palynological differences are evident between the two zones although both illustrate a regional and local vegetation dominated by *Quercus*, *Ulmus* and *Corylus* woodland. Higher values of *Pinus* in levels 13 and 14 are possibly a function of over-representation due to poor pollen preserving conditions or differential fluvial transport or are due to a greater importance of this taxon in the landscape. If the latter postulation is correct, the percentage occurrence is small and its real growth at this time must have been minimal, possibly representing the final remnants of the Boreal pine woodland noted by Godwin (1940) in the Swansea Bay submerged peat sequences. Zone WH:III differs from Zone WH:II by having greater importance of *Salix* (also *Filipendula*) which along with the greater organic content may be indicative of a damper (but not flooded) environment locally. *Salix*, being entomophilous, is largely under-represented in pollen spectra and the values of 23% AP are indicative of substantial importance in the local vegetation. Evidence of Mesolithic anthropogenic influence is not immediately apparent in the pollen spectra of these two midden levels but possible elements of anthropogenic causality are discussed (see below).

Figure 12. Profile I.

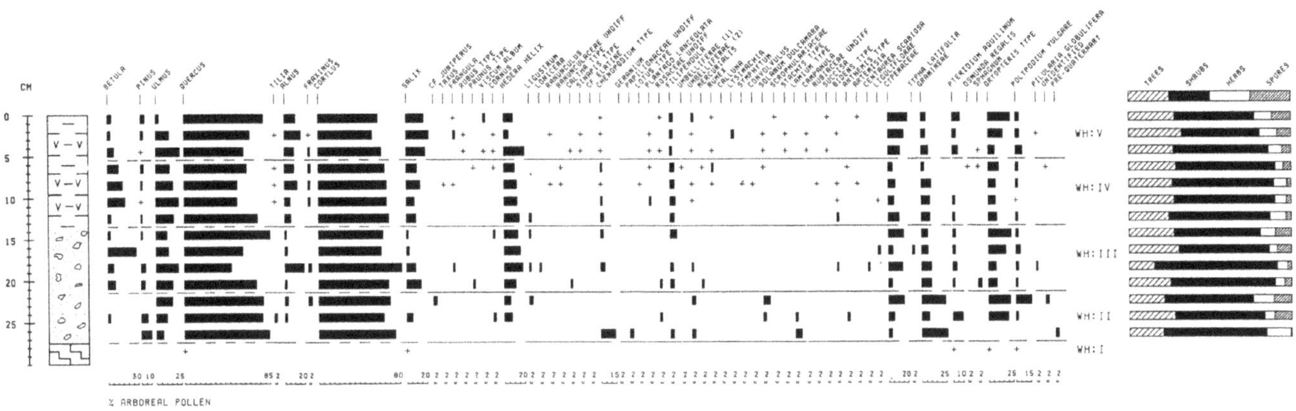

225

Please note that a full-size version of this figure is available to download from www.barpublishing.com/additional-downloads.html. The original foldout has been reduced in size to match the A4 format of this book, the image is therefore not as clear as the original foldout. The image might also be cut off or not complete in the printed book. Please refer to the original foldout via the download for the original content.

Zone WH:IV which spans the period dated by radiocarbon to 5740+100 BP (HAR-5641) to 5200+120 BP (HAR-5640) is therefore of Atlantic (Godwin's pollen zone VIIa) date. The vegetation represented in the pollen spectra of this zone is in accordance with this view, with a number of different vegetation community elements being present. These include the thermophilous elements *Tilia* cf *cordata*, *Hedera helix* and *Viscum album*. Although not present in quantity, the last of these is strongly characteristic of pollen spectra of Atlantic date from lowland Britain. *Tilia* cf. *cordata* is sporadically represented throughout this zone and is similarly a taxon becoming significant during the Atlantic period. The entomophily of this taxon results in its under-representation in pollen spectra (Andersen, 1970) and therefore in its spatial representation. Sporadic occurrences as noted here may therefore be viewed as local occurrences of this genus although long distance transportation cannot be ruled out. Immediately local soil status would not have been suited to this species and these occurrences may relate to stands growing on the surrounding better drained areas at some distance. Thus they would be of regional or extra-regional occurrence. Pollen analyses covering the Atlantic period in south-west England are few, but distributions show that *Tilia* was marginal in this area of the country (Birks *et al*, 1975), being absent in the far South West (Scaife, 1984). The remaining tree taxa are typical of those arboreal genera present during the mid-Flandrian, Atlantic period. Whilst the idea of monoclimax mixed deciduous woodland (Quercetum mixtum) spanning the country during the Flandrian chronozone II has been realistically shown as a polyclimax of dominant tree taxa in suited edaphic and lithological cicumstances (the polyclimax view: Godwin, 1975a). In these pollen spectra we see a strong representaion of *Quercus*, *Ulmus*, *Fraxinus* and *Corylus* and it is likely that the vegetation of the surrounding area was one of this character. The analysis of plant macrofossils has shown quantities of *Corylus* nuts in these peats and the presence of pollen clusters from partially dehisced anthers of *Quercus* and *Corylus* attest to its autochthonous growth and are indicative of possibly damp but not waterlogged conditions in at least part of the peat forming community.

Quantities of *Salix*, *Alnus* and other autochthonous elements of the peat forming plant community are also present and include *Taxus*, *Frangula*, *Rubus*, *Ligustrum*, *Caltha* type cf *Elatine*, *Filipendula*, *Lysimachia*, *Symphytum* type, *Solanum*, *Succisa*, Cyperaceae, *Osmunda*, *Sphagnum* and *Pilularia*. From these a picture of the environment of the peat formation may be gained. Whilst *Alnus* is a highly characteristic taxon for this period, rising to high values at *c* 7500 BP (albeit asynchronously - Smith and Pilcher, 1973), its presence in quantity at this site is not likely because its high pollen production would give rise to greater pollen percentages than the 16% maximum recorded here (see for example Scaife, 1980; 1982; Thorley, 1981) where percentages of up to 95% AP are not infrequent in lowland southern English sites). Turner (in Kerney, 1976b) at Blue Anchor Bay has recorded 56% of total pollen as being *Alnus* and in Swansea Bay (Godwin, 1940) 62% AP indicated the presence of alder carr woodland (Alnetum). At Westward Ho! the vegetation of the fen carr is seen to be *Salix* (Salicetum) which, in spite of poor pollen production and dispersal characteristics, is present in relatively high frequencies and this is further supported by the presence of plant macrofossils of *Salix cinerea*. As noted above, other fen carr vegetation was present, including *Ligustrum*, *Taxus*, (Godwin, 1975b) and *Frangula*. A fen carr environment of typical lowland character is thus postulated. Comparison with existing communities which have similar vegetation and have provided longer pollen sequences would indicate that the environment here was relatively dry. Areas of greater wetness would have supported those telmatic elements present which tolerate seasonal inundation and locally standing pools of water (with *Salix* and shade tolerant fen carr herbs eg, *Filipendula*, *Symphytum* type and Cyperaceae (Tansley, 1939)). This assessment is in agreement with the analysis of two spot samples analysed previously from the midden site (Churchill, 1965) and also from Swansea Bay (Godwin, 1940) and Minehead (Kerney, 1976b) with the difference that *Salix* not *Alnus* dominated the fen carr.

In levels 1-3 (Zone WH:V) a number of changes occur. *Ulmus* and *Fraxinus* pollen decline markedly from more than 20% to 3% AP whilst Cyperaceae, Filicales (*Dryopteris* type) and *Pteridium aquilinum* increase. Radiocarbon dating of these upper peat levels is commensurate with that of the "elm-decline" at *c* 5500 BP (Smith and Pilcher, 1973).

It is unfortunate that peat levels higher in this profile are not available to assess this phenomenon throughout (but see below, Profile II). A discussion of the nature of the "elm-decline" cannot be given here, but the possibility of natural and/or anthropogenic causality is held (see Smith, 1970; 1981). The evidence seen here is for the opening of the woodland canopy with declining *Ulmus*, increasing *Pteridium aquilinum*

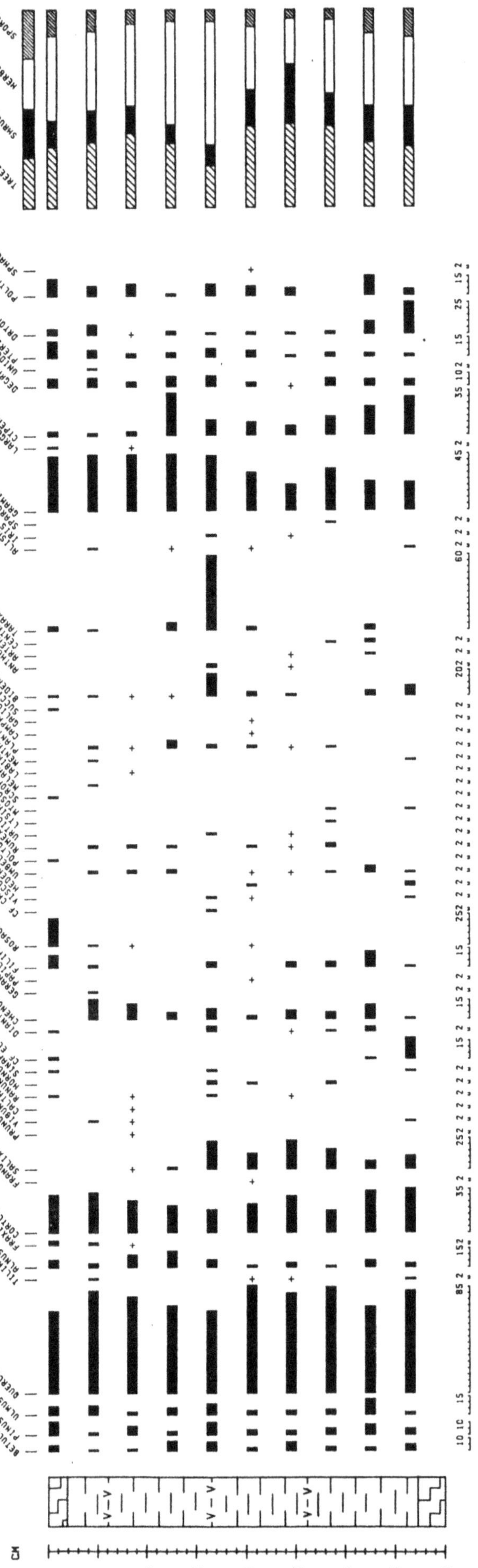

Figure 13. Profile II.

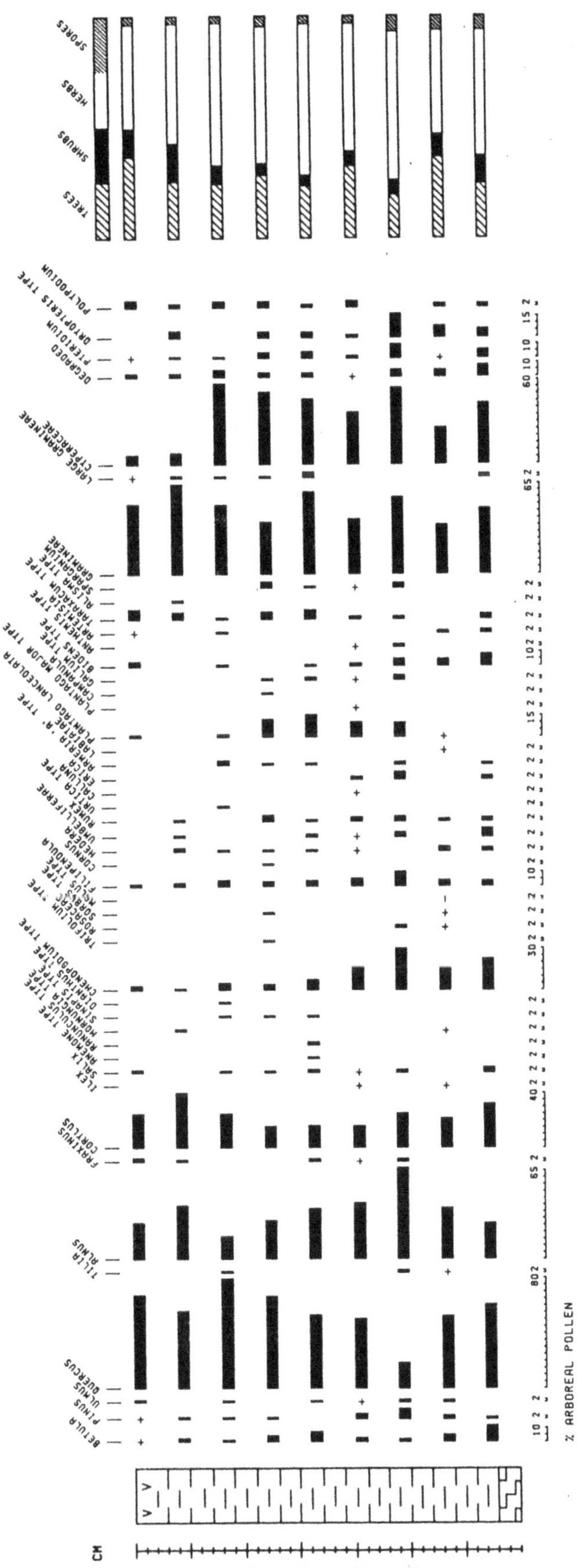

Figure 14. Profile III.

and by the slightly higher *Pinus* percentages which may be viewed as an extra-regional element penetrating and settling more readily where the canopy is more open.

Other ruderals and cultigens often associated with the "elm-decline" and Neolithic subsistence are not found, possibly because levels from immediatley above the "elm-decline" are absent. There are also indications in this zone of increasing wetness/waterlogging. This may have been a function of the ponding back of the freshwater hydrology prior to an estuarine/marine transgression referred to above and which deposited the overlying "upper blue clays". Alternatively an anthropogenic cause may be postulated whereby a greater degree of human impact on the fluvial catchment may have increased surface run-off and groundwater table by reduction in evapo-transpiration and increase in ground surface flow.

The evidence for mesolithic anthropogeny

Zones WH:II and WH:III, being the midden contexts, might be expected to show some evidence for utilization of natural resources by Mesolithic man. This was not, however, found to be the case. The vegetation illustrated in these levels bears the same general characteristics as the overlying peat (formed in the fen carr community albeit with lesser herbaceous diversity). This is not unexpected in an environment of dominant mixed oak forest and fen carr woodland. The presence of higher numbers of Gramineae and Chenopodiaceae (although this might include *Atriplex*, a saline tolerant/halophytic element) might be indicative of some localised clearance within the canopy. In the peats of Zone WH:IV, some disturbance to the woodland may also be evidenced by the herbs Chenopodiaceae (but see above) and *Plantago lanceolata*. The former may also be cited as evidence for broken ground and especially the presence of large grazing herbivores (McAndrews, 1966). The presence of *Plantago lanceolata* in pre "elm-decline" contexts has been widely discussed in terms of the possibility of ephemeral canopy openings (Smith, 1970). Dimbleby (1976) and Simmons and Dimbleby (1974) have assessed the evidence for *Hedera* often being found in abundance in association with Mesolithic contexts as being indicative of some form of stock control by Mesolithic man. Here, although *Hedera* pollen percentages are unusually high, the length of time involved (*c* 500 radiocarbon years) must indicate a natural cause. In view of its extremely poor pollen production and dispersal characteristics its extreme domination on or immediately adjacent to the site is likely. The south western oceanic character of the region must have favoured its flourishing growth. Much literature has been dedicated to the possibility of *Corylus* being of anthropogenic importance (see Smith, 1970; Jacobi, 1978) and whilst its use by Mesolithic man is unquestioned (Woodman, 1981; 1985), the domination by hazel here must be seen as part of the natural ecosystem. Higher percentages of *Betula* in levels 4, 5 and 6 (Zone WH:IV) might be evidence of forest canopy opening or clearance by fire. Numerous microscopic charcoal fragments were noted throughout all levels and in the soil micromorphology (see above). However, it is not possible to ascertain whether or not the *Betula* formed part of a natural fen carr succession.

PROFILE II

Samples of 1ml were taken at 4cm intervals through the peat profiles which were obtained in metal box monoliths. The pollen was in a large number of cases somewhat degraded but not seriously enough to negate identification. A pollen sum of 300 grains was counted for each level. The raw data from which pollen diagrams have been constructed (as percentages of various sums; see Figure 13) are in the project archive and in Scaife (1986). The peats in both monoliths were well humified dark brown to black comprising largely organic detritus with silt laminations and monocotyledonous root and stem macrofossils. Wood was present in the outer peat but not evident in the inner peat sections.

Comparing the pollen data (Fig. 13) of this peat profile with that originally constructed from the midden (Profile I, Fig. 12), it is clear that the overall vegetation elements are similar but that certain differences between the dominant elements are apparent.

Arboreal and shrub pollen are dominated by *Quercus* and *Corylus* but percentages are lower than in Profile I. This is perhaps a response to the higher values of other vegetation elements (*Salix* and Gramineae) which may also have had a depressing effect upon other percentages within the pollen sum. Inspection of the total pollen sum illustrates, however, that the arboreal pollen sum remains the same in both diagrams from the outer peat, which are contemporary. It is the shrub pollen (largely *Corylus* type) which is substantially reduced in relation to expansion of herb pollen (largely Gramineae and Cyperaceae). Other differences between the arboreal/shrub spectra of the two profiles are the minor increases in *Pinus* and the lower percentage values of *Alnus* and to some extent *Betula.* Gramineae (to 40%TP) and Cyperaceae (to 20%) are the dominant herbs. The greater importance of a number of types is notable. Compositae types have a greater representation. *Chenopodium* type (including *Atriplex*) are present throughout but show slightly higher values towards in the upper and lower levels of the diagram.

Discussion

A landscape in which *Quercus* and *Corylus* are important constituents is indicated. *Ulmus* and *Fraxinus* appear to have been lesser elements. This was not, however, the autochthonous vegetation as was postulated for Profile I. From the relatively high values of *Salix* (to 15%) in WH:I (from 16cm-38cm), it may be suggested that *Salix* carr (*S. caprea* was identified from leaves within the peats) was of local importance because *Salix* being entomophilous is usually under-represented in pollen spectra. The overall greater abundance of herbs and fewer shrubs in this profile may be suggestive of a more open area or perhaps clearing within a relatively dry open woodland community. Alternatively, the greater numbers of wetland taxa suggest that this area was of a wetter nature (see below). Evidence for local variability within the vegetation community also comes from the earlier pollen analyses of Churchill (1965, 79) who records *Alnus* as 17% of total pollen in the peats overlying the section of midden he investigated.

The pollen profiles I and II are contemporaneous, both having been taken from the continuous outer peats at a distance of some 50 metres apart and both profiles having similar radiocarbon dates for the upper levels. Analysis of this second profile was undertaken to provide data which may heve been missing through truncation of the uppermost peats above the midden section and to confirm/deny the existence of a phase of openness and possibly "elm-decline" at the top of the pollen profile. Inspection of the two profiles indicates that only some 4-8cm of additional peat was present in this second profile by virtue of protection from erosion by the overlying silts. This evidence is based upon comparison of the pollen frequencies of *Salix* and *Alnus.* It is interesting to note also that there is no further or real indication of an "elm-decline" as seemed to be suggested by Profile I. The radiocarbon dating confirmed that the uppermost levels of the peat profiles date to the latter part of the sixth millennium BP (5200±120 BP, HAR-5640 and 5190±80 BP, HAR-6363) and so the possibility remains (though is not proven) that the "elm-decline" is represented in the pollen spectra of these upper levels.

COMPARISON OF THE POLLEN PROFILES I AND II

As noted above, a second profile (II) of the outer peat was analysed to study in more detail the possibility of an open phase in the upper peat levels. It became clear, however, that although the site of the second core was in relatively close proximity to the midden site, there were some differences in the vegetation community present. These changes may be summarised briefly as being:

i) Fewer shrubs, especially *Corylus* which was shown to have been an important constituent of the vegetation during deposition of the midden and during subsequent peat accumulation.

ii) Sporadic occurrences of *Fraxinus* in the midden profile are fewer in Profile II. Since pollen of this genus is poorly represented in pollen spectra, this possibly indicates a growth local to Profile I and perhaps on drier areas at this point.

iii) Herbaceous pollen is markedly more important being dominated by Gramineae and Cyperaceae throughout the sequence but also with peaks of *Chenopodium* type, *Filipendula* and Compositae types.

Since the monolith from which pollen counts are based was taken from the same stratigraphy their contemporaneity is likely. Given this, it is apparent therefore that the autochthonous pollen components highlight two slightly differing environments. Profile I has been shown to be a predominantly damp wooded environment dominated by *Quercus*, *Corylus* type and *Salix.* The quantities of *Corylus* pollen, *Hedera* and a little

Ilex support the contention that this was of open canopy character to allow flowering. Profile II is apparently more open as indicated by the herbaceous pollen percentages. It can be noted here that this might have resulted in the slightly greater input of long distance transported elements such as the *Pinus*. If this openness had been due to human or animal factors, secondary woodland regeneration might have been expected to have occurred considering the long temporal span of this peat accumulation. The increase in herbaceous diversity noted can, however, be seen to be to some extent from aquatic, semi-aquatic/marginals and Gramineae and Cyperaceae. There is no evidence of large areas of open water, but it seems likely that the area of Profile II was one of wetter ground conditions with a community of *Salix* carr (Salicetum) and ground flora of wetland taxa. This contrasts with the midden peats of Profile I, the pollen of which indicates slightly drier conditions of *Quercus-Corylus* woodland with some *Salix*. The quite significant differences between these two profiles illustrates the value of the analysis of multiple pollen profiles from apparently identical stratigraphical profiles, especially in wooded environments where pollen transfer is more restricted.

THE INNER PEAT
(Profile III)

Palynologicaly this profile is somewhat different from those from the outer peats, as might be expected from its apparently substantialy later date. The principal differences are the higher values of *Alnus* pollen and of herbs indicative of open ground (*Plantago lanceolata* and cereal pollen are notable, Fig. 14). Arboreal and shrub taxa are dominated by *Quercus*, *Alnus* and *Corylus* with only a minimal presence of *Betula*, *Pinus*, *Tilia* and *Fraxinus*. The latter two - *Fraxinus* and *Tilia* - may be somewhat under-represented in the spectrum due to their poor pollen production and dispersal characteristics. Gramineae is the dominant herb throughout with values of 25-45% TP (total pollen) Cyperaceae are also important (20-35%). *Chenopodium* type (includes *Atriplex* spp.) is present in substantial quantity in the base of the diagram between 20cm and 34cm. Of the remaining herbs, *Plantago lanceolata* is evident in the middle of the diagram (12-24cm).

Discussion

As in the two sequences from the outer peat area (Profiles I and II), it is apparent that the local arboreal vegetation, where it remained, comprised *Quercus* and *Corylus*. Although slight, pollen representation of *Tilia*, *Fraxinus* and *Ilex* may indicate that these genera may have been of importance at some distance from the sample site and on better soils inland. *Alnus* appears as one of the dominant arboreal taxa, but in view of its high pollen productivity and over-representation in pollen spectra, the percentages of up to 15% of total pollen here are unlikely to represent its growth close by. It is likely that an Alnetum existed in a similar habitat but at some other point in the Westward Ho! area. Turner in Kerney (1976b) has also noted high values of *Alnus* and the presence of carr woodland at Blue Anchor Bay but which has been dated to the Atlantic (6730$\pm$150 BP, Q-1343) and therefore pre-dates this profile but which is contemporaneous with the outer peat profiles I and II. The peat macro-stratigraphy and the pollen record suggest that the autochthonous vegetation comprised a Cyperaceae and possibly Gramineae fen community. *Chenopodium* type at the base of the profile is of interest. It seems likely that this taxon, which includes *Chenopodium* and *Atriplex*, (the latter being dominant in maritime halophytic plant communities) may have resulted from the influence of such nearby ecology. The relative abundance of the pollen of these plants (which are anemophilous) is a phenomenon encountered in pollen spectra from coastal and estuarine peat and sediment sequences. The decline in its importance from the base of the profile here may tentatively be suggested as being due to eustatic changes and declining sea levels with the consequent distancing of such halophytic communities from the sample site. The relative increases in herbaceous diversity and the increases of *Plantago lanceolata* may perhaps be interpreted as evidence of anthropogenic activity during this period. This is also evidenced by the presence of cereal pollen. Alternatively, many of the herbaceous types might reflect a coastal dune, short-turf graminaceous plant community.

II.v - THE PLANT MACROFOSSILS

D. Vaughan

INTRODUCTION

The plant remains were recovered from six columns cut from the blocks removed from the site. The columns were divided at vertical intervals into samples. The samples from the first four columns (Tables 12-15) were from Block 26 and they were sorted after the insect remains had been extracted. The remaining samples (Tables 16 and 17) were from Block 29. They were examined after the molluscan remains had been extracted; these two columns were sorted dry and the condition of the seeds was not as good as that of the seeds from the insect samples, which had been kept wet. The full results are given in Tables 12-17, the nomenclature following Clapham *et al* (1981). Almost all the botanical remains had been preserved by waterlogging, the few charred remains are indicated separately in the tables.

The assemblages from the six columns were similar in terms of their species composition and relative abundance of taxa. However, they differed quantitatively, sample column 8311031 (Table 12) showing the greatest concentration of seeds. In all the columns, the peat samples were the richest in the concentration of macroscopic plant remains and showed the greatest range of species. The number of seeds diminished down through the midden levels to the clay, which contained hardly any plant fragments. The few plant remains from the clay had probably made their way down through root channels and crevices.

All the plant taxa identified from the midden were also present in the peat while the occurrence of species exclusive to the peat need only have been due to the greater concentration of remains in that layer. The flora suggested by the seeds and other macroscopic plant remains from the six columns is fen woodland throughout the period of deposition.

THE GENERAL ENVIRONMENT

The majority of the plant remains are suggestive of damp conditions; a significant element of the flora is unable to tolerate permanent waterlogging. Taxa requiring flowing water, or water more than a few centimetres deep, were absent but the occurrence of *Callitriche* (starwort) fruits, especially from the peat, is indicative of muddy patches or seasonal pools of water. In contrast, *Fragaria vesca* (wild strawberry), also identified from the peat, is probably only capable of growing on the driest of fen peats. Remains of saltmarsh plants were absent. Both the midden sediments and the peat seem to have accumulated under freshwater conditions. The abundant *Cornus sanguinea* (dogwood) stones throughout the sequence suggest a somewhat calcareous substrate.

Woodland or scrub is indicated by the comparatively large number of remains from woody taxa and only slight evidence for light-demanding herbaceous plants. The following large shrubs were represented by stones or nuts: *Prunus spinosa* (sloe), *Crataegus* cf. *monogyna* (hawthorn), *Cornus sanguinea* (dogwood) and *Corylus avellana* (hazel). Buds and catkins of *Salix* sp. occurred throughout the midden and peat sequence. Leaf fragments from some of the samples suggested the species concerned to be *Salix caprea*, *S. cinerea* or one of their hybrids (sallows). The buds also seem to include *Quercus* sp. (oak) and a single bud of *Populus* sp. (poplar) was discovered from the peat.

The tree canopy is likely to have been quite low and somewhat open, probably enabling *Cornus sanguinea* to contribute to the canopy. This vegetation may have been in part a fen woodland or damp woodland of *Corylus avellana* and *Quercus* sp. in which *C. sanguinea* comprised a major part of the shrub layer. *Prunus spinosa* and *Crataegus monogyna* are also likely to have been growing in the drier areas. They are thorny shrubs with relatively high requirements for illumination, *P. spinosa* being intolerant of dense shade. Their occurrence would have been favoured by breaks in the tree canopy and by the activities of browsing animals.

Table 12. Macroscopic plant remains, Block 26, 8311031.

depth (cm)	0-5	5-10	10-15	15-20	20-25	25-30	30-35
weight (kg)	0.70	0.40	0.75	0.90	1.10	0.83	0.86
sediment	peat	peat	midden	midden	midden	midden	clay
Ranunculus acris/repens/bulbosus	2	3	0	0	0	0	0
cf. *Ranunculus* sp.	0	0	0	0	0	1	0
Viola sp.	0	3	0	0	1	0	0
Filipendula ulmaria	0	0	2	0	0	0	0
Rubus fruticosus agg.	*c*38	*c*25	2	0	0	0	0
Potentilla palustris	1	0	0	0	0	0	0
Fragaria vesca	2	0	0	0	0	0	0
Prunus spinosa	2	3	1	0	0	0	0
Crataegus cf. *monogyna*	1	0	0	2	0	0	0
charred	0	0	0	1	0	0	0
Callitriche sp.	4	0	0	0	0	0	0
Cornus sanguinea	12	11	0	4	2	0	0
Rumex sanguineus/conglomeratus	1	0	0	0	0	0	0
Rumex sp.	15	5	0	3	0	1	0
Corylus avellana							
large nut	7	1	1	2	1	0	0
small nut	4	6	2	0	0	1	0
withered nut	16	7	1	2	5	0	0
fragments	+++	+++	++	+	+	+	0
charred fragments	0	0	0	+	+	0	0
Solanum dulcamara	15	11	3	6	3	0	0
Ajuga reptans	5	4	1	3	0	0	0
Eupatorium cannabinum	0	5	4	0	0	0	0
fragment	3	0	0	0	0	0	0
Carex spp. (see Note, below)	*c*300	*c*150	41	25	58	6	2
Indeterminate seeds	0	1	0	0	0	0	0
Total no. of seeds	425	237	58	47	70	9	2
Buds (excluding *Salix*)	0	5	0	5	4	1	0
Bud scales	0	8	0	13	3	0	0
Leaf abscission pads	2	1	0	0	0	0	0
Salix buds	0	0	0	0	2	0	0
Salix catkins	0	1	0	0	0	0	0
moss	x	x	x	x	x	x	x
Seeds/100g of sediment	61	60	8	5	6	1	<1

x = present, not quantified

Note.

In this and subsequent tables the totals for *Carex* spp. are made up of the following taxa in the proportions indicated:

C. lepidocarpa/demissa	+++
C. disticha	++
C. cf *hirta*	+
cf *C. strigosa* or *palescens*	+++

Table 13. Macroscopic plant remains, Block 26, 8311032.

depth (cm)	0-5	5-10	10-15	15-20	20-25	25-30	30-35
weight (kg)	0.55	0.50	0.58	0.65	0.78	0.67	0.61
sediment	peat	peat	peat	midden	midden	midden	clay
Ranunculus acris/repens/bulbosus	4	1	1	1	0	0	0
Viola sp.	0	0	0	1	0	0	0
Filipendula ulmaria	1	0	0	0	0	0	0
fragment	0	0	1	0	0	0	0
Rubus fruticosus agg.	28	3	1	0	0	1	0
Potentilla palustris	0	1	0	0	0	0	0
Rosa sp.	0	0	1	0	0	0	0
Prunus spinosa	0	0	0	1	1	0	0
Crataegus cf. *monogyna*	0	3	0	0	1	0	0
Callitriche sp.	0	10	1	1	0	0	0
Cornus sanguinea	12	6	1	1	1	0	0
Rumex sp.	4	0	0	3	0	1	0
Corylus avellana							
large nut	4	5	1	0	0	0	0
withered nut	8	3	2	2	0	2	0
fragments	++	++	++	++	+	0	0
Solanum dulcamara	4	8	4	4	1	1	0
Ajuga reptans	0	1	1	0	0	0	0
fragment	1	0	0	0	0	0	0
Eupatorium cannabinum	7	8	13	1	2	3	0
Carex spp.	150-200	*c*120	*c*120	*c*70	*c*50	12	0
fragment	0	0	0	0	0	0	1
Indeterminate seeds	1	0	0	1	0	0	0
Total no. of seeds	250	169	147	86	56	20	0
Buds (excluding *Salix*)	*c*20	11	7	0	0	0	0
Bud scales	28	15	0	10	2	6	0
Leaf abscission pads	0	0	3	0	0	0	0
Rubus prickles	5	0	0	0	0	0	0
Salix buds	4	2	2	0	0	0	0
Salix catkins	8	0	0	0	2	1	0
moss	x	x	x	x	x	x	x
Seeds/100g of sediment	46	34	25	13	7	3	<1

There may have been a tall field layer of *Filipendula ulmaria* (meadowsweet), *Solanum dulcamara* (woody nightshade), *Eupatorium cannabinum* (hemp agrimony) and *Carex* spp. (sedges) growing on the wettest areas of the woodland floor. Four of the five or six species of *Carex* were identified: *C. strigosa* or *C. pallescens* are characteristic of woodland or glades in woodland, *C. lepidocarpa/demissa* grows on base-rich fens and *C. disticha* in herb-sedge fen and wet meadows while *C. hirta* grows on hedge-banks, meadows or grassland (Jermy *et al*, 1982).

The seeds from the herbaceous plants suggest a lower growing ground flora of *Ranunculus* sp., *Fragaria vesca* (wild strawberry) and *Ajuga reptans* (bugle) growing in the drier spots in the woodland. *Rubus fruticosus* agg. (blackberry) is ubiquitous and its presence supports the idea that the woodland was fairly open. Although it occurs in almost all woods whatever the degree of shade, it flowers and fruits readily only where there is plenty of light.

A curious aspect of the Westward Ho! assemblages was that many of the numerous *Corylus* nuts were much smaller than normal and were very markedly ribbed and mis-shapen, as if pressed together. Several interpretations are possible to explain their peculiar appearance. They may have been immature nuts whose growth had aborted because of the browsing of young low-growing branches by animals such as deer or they may have come from branches cut by people in the summer before the nuts were ripe. However, as the

Table 14. Macroscopic plant remains, Block 26, Column 8311014.

depth (cm)	0-6	6-12	12-18	18-24	24-30	30-34
weight (kg)	0.76	1.14	1.07	0.93	1.14	0.60
sediment	peat	midden	midden	midden	clay	clay
Ranunculus acris/repens/bulbosus	3	0	0	0	0	0
Rubus fruticosus agg.	22	0	0	0	0	0
Fragaria vesca	1	0	0	0	0	0
Prunus spinosa	1	0	0	0	0	0
Callitriche sp.	1	0	0	0	0	0
Cornus sanguinea	8	0	0	2	0	1
Corylus avellana						
large nut	2	0	0	0	0	0
fragments	++	0	0	+	0	0
charred fragments	0	+	0	0	0	0
Solanum dulcamara	3	2	0	1	0	0
Ajuga reptans	1	1	0	0	0	0
Eupatorium cannabinum	3	2	0	0	0	0
Carex spp.	59	2	0	1	0	0
Total no. of seeds	104	7	0	4	0	0
Buds	13	1	0	0	0	0
charred	0	0	0	1	0	0
Bud scales	4	0	0	0	2	0
Rubus prickles	2	0	0	0	0	0
Salix catkins	2	0	0	0	0	0
Seeds/100g of sediment	14	<1	<1	<1	<1	<1

nuts are present throughout the peat layer (*ie* after the midden was abandoned) this latter explanation seems unlikely. A further possibility is that the *Corylus* were diseased, but this too seems an improbable explanation as the community would probably not have survived for such a long time if their nuts had been continually immature. It has been suggested (Mark Robinson, pers comm) that the condition of the nuts was due to environmental stress which resulted from the *Corylus* growing at the extreme of its tolerance to waterlogging. Although this explanation seems plausible, it is doubtful if the woodland was damp enough to cause such an extreme effect.

COMPARISON BETWEEN THE MIDDEN AND THE PEAT FLORAS

The waterlogged plant remains from the midden do not offer any evidence of human activities in the area. *Corylus* nut shells and remains of various edible fruit were identified from the midden deposits, but they were more abundant from the overlying peat, with its greater concentration of seeds. Charred fragments of *Corylus* nut shell, however, occurred in the midden samples from four of the six columns and there was a single charred stone of *Crataegus* cf *monogyna* from Sample Column 8311031 (Table 12). The macroscopic plant remains do not provide any information on the environmental changes consequent on peat formation.

SUMMARY

The macroscopic plant remains serve to confirm and, in some aspects, add to the evidence from the pollen for the character of the woodland on the site. They emphasise the importance of some entomophilous (insect pollinated) shrubs that are either poorly represented or absent from the pollen record.

The number of taxa represented by the macroscopic plant remains from the Westward Ho! deposits is not very great. They indicate that the environment was that of a fairly damp and relatively open woodland. The waterlogged plant remains show that a variety of edible fruit and nuts were available in the woodland around the site while the few carbonised remains show that the Mesolithic occupants may have exploited at least some of them.

The macroscopic plant remains recorded from this investigation correspond well with those listed by Churchill (1965) from the peat of the Westward Ho! submerged forest bed. However, he thought that the woodland was a "fairly dry fen". The brackish element to the flora noted by Reid (in Rogers, 1908) was absent but the location of the peat sampled by Rogers is likely to have been different (see above).

ACKNOWLEDGEMENTS

My thanks to James Greig for identifications of the sedges; to Sue Colledge for some of the identifications and for reading the first draft of this report; and to Mark Robinson for his identification of buds and leaves and for his revision and correction of both the identification of the seeds and the text; and to Nick Balaam for his support.

Table 15. Macroscopic plant remains, Block 26.

	8311023	8311030
depth (cm)	0-10	20-35
weight (kg)	0.95	2.03
sediment	peat	midden & clay
Ranunculus acris/repens/bulbosus	5	0
Filipendula ulmaria	2	0
Rubus fruticosus agg.	*c*60	1
Fragaria vesca	1	0
Prunus spinosa	1	0
Crataegus cf. *monogyna*	1	0
Cornus sanguinea	11	0
Rumex sp.	7	0
Corylus avellana		
large nut	9	0
small nut	3	0
withered nut	18	0
fragments	+	+
Solanum dulcamara	19	0
Mentha sp.	1	0
Ajuga reptans	4	1
Eupatorium cannabinum	9	0
Carex spp.	*c*250	3
Total no. of seeds	400	5
Buds (excluding *Salix*)	*c*40	1
Bud scales	*c*30	4
Leaf abscission pads	1	0
Populus buds	1	0
Rubus prickles	11	0
Salix buds	5	0
Salix catkins	9	0

Table 16. Macroscopic plant remains, Block 29, 8311219.

depth (cm)	0-5	5-10	10-15	15-20	20-25	25-30
sample number	8311220	8311221	8311222	8311223	8311224	8311225
sediment	peat	midden	midden	midden	clay	clay
Ranunculus acris/repens/bulbosus	2	0	0	0	0	0
Rubus fruticosus agg.	5	3	0	0	0	0
fragments	0	2	0	0	0	0
Prunus spinosa	0	1	0	0	0	0
Cornus sanguinea	3	2	0	0	0	0
Corylus avellana						
small nut	0	2	0	0	0	0
fragments	+	+	0	0	+	0
charred fragments	+	0	+	++	+	0
Solanum dulcamara	3	0	0	0	0	0
Eupatorium cannabinum	4	0	0	0	0	0
Carex spp.	*c*40	21	0	1	1	0
Total no. of seeds	*c*57	31	+	1	1	0
Buds	24	0	0	0	0	0
Bud scales	31	0	0	0	0	0
moss	x	x	x	x	x	x

Table 17. Macroscopic plant remains, Block 29, 8311201.

depth (cm)	0-5	5-10	10-15	15-20	20-25	25-30	30-35
sample number			8311212	8311213	8311214	8311215	8311216
sediment	peat	midden	midden	midden	clay	clay	clay
Ranunculus acris/repens/bulbosus	1	0	0	0	0	0	0
cf. *R. acris/repens/bulbosus*	1	0	0	0	0	0	0
Cornus sanguinea	0	1	0	0	0	0	0
cf. *Cornus sanguinea*	1	0	0	0	0	0	0
Corylus avellana							
fragments	+	+	+	0	0	+	0
charred fragments	+	+	+	+	+	+	+
Ajuga reptans	0	1	0	1	0	0	0
Eupatorium cannabinum	12	0	0	0	0	0	0
Carex spp.	11	1	1	0	0	0	0
Indeterminate	1	0	0	0	0	0	0
Total no. of seeds	27	3	1	1	+	+	1
Bud scales	0	0	2	1	0	0	0

II.vi - THE INSECT FAUNA

M. Girling, and M. Robinson, with a contribution by B.J. Wilkinson

INTRODUCTION

The waterlogged deposits at Westward Ho! provided the rare opportunity to investigate insect assemblages from a Mesolithic site. Numerous sub-samples from Block 26 were subjected to paraffin flotation onto a 0.3mm mesh and the insect remains extracted. Insects were also recovered from some of the molluscan samples from Block 29. Coleoptera were identified by Dr Girling and Trichoptera larval remains from some of the sub-samples were identified by Miss Wilkinson. This report was written by M. Robinson following the death of Dr Girling.

The insect assemblages from the sub-samples of Blocks 26 and 29 are too small for individual analysis. The results have therefore been combined into single stratigraphic sequences for each of the two blocks. Sample correlation is given in Tables 18 and 19. The coleopteran results are listed in Tables 20 and 21, which give the minimum number of individuals present at each level. Nomenclature follows Kloet and Hincks (1977). The sizes of some of the sub-samples had not been measured, but the total volumes of midden material and peat investigated from Block 26 were similar.

The concentration of insect remains from the underlying clay (samples g and D) was very low. It is possible that they had all been introduced by root disturbance. The midden and peat layers (samples a - f and A - C) produced rich insect faunas. Preservation of insect remains in the midden sub-samples was variable, but most sub-samples showed good preservation. While it is possible that some mixing occurred as the midden deposits, followed by the peat, accumulated, there was no evidence of subsequent mixing. The insect assemblage comprised a fauna of fen woodland. The insect remains seem to have accumulated in small pools on the woodland floor. There was no evidence for the transport of insect remains from other habitats by flooding.

The concentration of insect remains was about twice as great in the overlying peat as in the midden. The fauna from the peat was also more varied. The differences between the faunas, however, seem to have resulted from very local environmental changes, so general conditions which persisted throughout deposition will be considered first.

THE GENERAL ENVIRONMENT

The evidence from the water beetles suggests the presence of peaty pools of stagnant water. The fully aquatic species included *Hydroporus rufifrons*, which tends to occur in fen pools, *Agabus affinis*, a beetle which is restricted to peaty pools and acid water, and *A. melanarius*, which inhabits springs and pools with dead leaves on the bottom in forests (Balfour-Browne, 1940: 318-9; Balfour-Browne, 1950: 73 and 139). *A. melanarius* is now a rare species which no longer occurs in North Devon. Various other beetles which tolerate a range of aquatic conditions were present, but none is restricted to flowing water. Amphibious beetles and beetles of waterside habitats were particularly well represented. They included *Ochthebius minimus*, one of the most numerous beetles from the samples, which readily leaves water to crawl on mud around the edge and *Anacaena globulus*, which is as often found under wet leaf litter as in water. Scirtidae were consistently abundant throughout the sequence. They have aquatic larvae but the adults congregate on tall waterside vegetation. Another beetle that was consistently well represented was *Lesteva heeri*, which occurs by ponds, in willow swamps, in marshy meadows and amongst decaying reeds (Donisthorpe, 1939: 48; Horion, 1963: 125).

A few beetles which feed on aquatic and marsh plants were present (Table 22). Cyperaceae would have been capable of growing amongst the ground flora of the fen woodland. However, the floating aquatic plant *Lemna* sp., which is indicated by the beetle *Tanysphyrus lemnae*, requires a relatively high level of illumination. This would suggest that there were some gaps in the tree canopy. A single beetle which feeds on *Phragmites australis* was present. It is possible that there were some areas of developing fen that had not been colonised by trees.

Table 18. Westward Ho! Block 26: Sample correlation.

a) Top of peat	8311331, 8311343, 8311351, 8311362, 8311372, 8311380, 8311032 0-5cm
b) Middle of peat	8311033, 8311330, 8311342, 8311349, 8311350, 8311361, 8311371, 8311379, 8311031 0-5cm, 8311032 5-10cm, 8311614 0-4cm
c) Bottom of peat	8311329, 8311341, 8311348, 8311360, 8311369, 8311370, 8311378, 8311031 5-10cm, 8311032 10-15cm, 8311614 4-8cm
d) Top of midden	8311340, 8311359, 8311368, 8311377, 8311031 10-15cm, 8311032 15-20cm, 8311614 4-8cm
e) Middle of midden	8311339, 8311347, 8311358, 8311367, 8311376, 8311031 15-20cm, 8311032 20-25cm, 8311614 14-20cm
f) Bottom of midden	8311030, 8311332, 8311335, 8311338, 8311346, 8311357, 8311365, 8311366, 8311375, 8311031 20-30cm, 8311032 25-30cm, 8311614 20-28cm
g) Clay	8311333, 8311336, 8311337, 8311031 30-35cm 8311032 30-35cm, 8311614 28-34cm

None of the aquatic and waterside Coleoptera suggest saline conditions, indeed some are intolerant of brackish water. Despite its location well below modern high water level, the site seems to have been a freshwater fen.

The Carabidae (ground beetles) from the site, not surprisingly, included many hygrophilous species, such as members of the genera *Bembidion* and *Agonum*. The five species of *Agonum* identified (Table 20) occur in such habitats as "marshy deciduous forests amongst leaves", "damp soil with vegetation of *Phragmites* etc." and "on damp shaded peat" (Lindroth, 1974: 82-86). A few of the Carabidae suggest that some areas of the woodland floor were slightly less waterlogged although still damp, for example, *Cychrus caraboides* and *Stomis pumicatus* (Lindroth, 1974: 21 & 69).

Very different ground conditions, however, are suggested by single specimens each of the carabid *Harpalus sabulicola*, and the tenibrionids *Phylan gibbus* and *Opatrum sabulosum*. *H. sabulicola* is a very rare beetle of dry sandy ground near the coast (Lindroth, 1974: 98). *P. gibbus* occurs on sandy coastal areas, including dunes, as does *Opatrum sabulosum* (Brendell, 1975: 8-10). It is probable that there were exposed sand dunes beyond the area of fen woodland.

A full fauna of tree dependent Coleoptera was present, ranging from leaf feeders through beetles which feed on wood in all stages of decay. One of these species, *Tropideres sepicola*, which occurs in decaying branches, particularly of senile oak trees, is now very rare in Britain (Fowler, 1891: 112-3). Table 22 lists some of the more host-specific species. They indicate that the trees and shrubs included *Quercus* sp., *Salix* sp. and *Corylus avellana*, with the strong likelihood that rosaceous shrubs and *Ulmus* sp. were also present. The wood boring weevil *Acalles ptinoides* was identified from throughout the sequence. The continental European literature on this species suggests that it develops in dead twigs of *Calluna vulgaris* (eg Freude *et al*, 1983: 167; Hoffman, 1958: 1387), but in Britain it has also been recorded from dead twigs in woodland (Walker, 1906: 96), so its presence at Westward Ho! is not out of place. Other less host-specific, tree dependent species included *Cerylon fagi* and *C. ferrugineum*, predators which feed in rotten wood and in scolytid burrows under bark (Beaver 1966; Freude *et al*, 1967: 214-15), various wood tunnelling Cerambycidae, such as *Grammoptera homelina* and *Pogonocherus hispidus*, and Anobiidae of dry dead wood such as *Grynobius excavatus* and *Anobium punctatum*.

The range of trees and shrubs indicated confirms the other evidence from the insects for contrasting conditions over the woodland floor. *Salix* sp. could be quite capable of growing in the swampy carr with pools of water on the peat surface suggested by the water beetles and the caddis larvae. *Corylus avellana*, however, cannot tolerate permanent waterlogging so there must have been areas of dry fen. The phytophagous Coleoptera do not provide much information on the ground flora of the woodland, but in addition to the Cyperaceae-feeding species the weevil *Sciaphilus asperatus* was identified, which has larvae that normally feed on *Primula vulgaris*, a plant requiring rather drier conditions than most Cyperaceae.

Table 19. Westward Ho! Block 29: Sample correlation.

A) Peat	8311203, 8311220
B) Top of midden	8311207, 8311221
C) Bottom of midden	8311213
D) Clay	8311226

Whilst woodland Coleoptera predominated, the three dune species were not the only open country beetles. A *Trifolium*-feeding *Apion* and the chrysomelid *Chrysolina banksi* were each represented by single individuals. *C. banksi* has a somewhat coastal distribution in Britain and feeds on Labiatae. *Ballota nigra*, *Marrubium vulgare* and *Teucrium* sp. being given as host plants (Bedel, 1901: 251-2; Fowler, 1890: 304). These two beetles could have lived amongst dunes. Elaterid beetles which have larvae that tend to feed on roots in grassland were present. One of these, *Agrypnus murinus*, does seem to be restricted to well drained grassland while the other, *Althous* sp., including an individual identified as *A. haemorrhoidalis* has been recorded from soil under pastures, meadows, leaf litter and hedges (Horion, 1953: 258-9) but the adults are readily attracted to *Crataegus* blossoms and sallows (Donisthorpe, 1939: 79). Dung beetles of the genera *Geotrupes* and *Aphodius* were present, suggesting the occurrence of large grazing or browsing herbivorous mammals, but the numbers of these beetles were so low that they do not imply the presence of tracts of pasture land.

Beetles of decaying plant material were abundant throughout the sequence. They are predominantly species which favour the type of debris which would be accumulating on the floor of fen woodland, such as *Oxytelus fulvipes*, which occurs in leaf litter in marshy woodland and *Corylophus cassidioides*, which lives in rotting reeds or sedges. Some of the beetles of decaying organic materials can inhabit manure heaps and refuse of the sort that accumulates around human settlements but none is so restricted. It is significant that *Cercyon sternalis*, a rare beetle of marshes, was the only occurrence at this site of a genus which includes many dung and compost inhabiting species that are frequently found in archaeological deposits. Beetles undoubtedly indicative of human activity were absent from the sequence.

COMPARISON BETWEEN THE MIDDEN AND THE PEAT ASSEMBLAGES

Some of the more abundant taxa of Coleoptera from the site, such as *Lesteva heeri*, *Stenus* spp. and Scirtidae (all beetles of marshes and waterside habitats) appeared in similar proportions of the totals from both the midden and the overlying peat. However, phytophagous beetles, species of recently dead wood and water beetles occurred at a higher percentage frequency in the peat. The reason for this was probably not that the availability of their habitats increased during the period of peat accumulation but that the abundance of beetles of decaying plant material in the midden caused them to be underrepresented in the first part of the sequence. The beetles that were much more abundant in the midden than the peat included members of the Ptiliidae, which occur in various types of habitats involving compost, rotten wood or dead plant material in marshy places. *Carpelimus* sp., which lives in such habitats as vegetable refuse on mud, and *Bryaxis* sp., which occurs in grass tussocks, reed refuse and in disintegrated rotten wood (Donisthorpe, 1939: 48, 54, 56). *Abraeus granulum*, a rare beetle of very rotten wood (Donisthorpe, 1939: 62) was present, as was the large and brightly coloured beetle

Table 20. Coleoptera from Block 26.

	minimum number of individuals						
	Peat			Midden			Clay
COLEOPTERA	a	b	c	d	e	f	g
Cychrus caraboides (L.)	1	1	-	-	-	-	-
Notiophilus palustris (Duft.)	1	-	-	-	-	-	-
Elaphrus cupreus Duft. or *uliginosus* F.	-	1	-	-	-	-	-
Elaphrus sp.	1	-	-	-	-	-	-
Trechus obtusus Er. or *quadristriatus* (Schr.)	-	-	-	-	-	1	-
Bembidion decorum (Zenk.)	-	-	-	-	-	-	1
B. harpaloides Serv.	2	-	-	-	-	-	-
B. guttula (F.)	1	2	-	-	3	2	-
B. biguttatum (F.) or *unicolor* Chaud.	-	-	-	1	-	-	-
Bembidion sp.	-	-	1	-	1	-	-
Stomis pumicatus (Pz.)	1	1	-	-	-	-	-
Pterostichus niger (Schal.)	1	1	-	-	-	1	-
P. nigrita (Pk.)	-	1	-	-	-	-	-
P. strenuus (Pz.)	1	-	-	-	-	-	-
Agonum fuliginosum (Pz.)	-	1	-	-	-	-	-
A. gracile Sturm	2	1	-	-	-	-	-
A. livens (Gyl.)	1	-	-	-	-	-	-
A. obscurum (Hbst.)	-	1	-	-	-	-	-
A. thoreyi Dej.	1	1	-	-	1	-	-
Harpalus sabulicola (Pz.)	-	-	-	1	-	-	-
Badister dilatatus Chaud.	1	-	-	1	-	-	-
B. sodalis (Duft.)	-	1	-	-	-	-	-
Hydroporus rufifrons (Mull.)	3	5	-	1	-	1	-
Agabus affinis (Pk.)	1	1	-	-	-	-	-
A. bipustulatus (L.)	1	1	-	-	1	-	-
A. melanarius Aubè	-	1	-	-	-	-	-
A. sturmi (Gyl.)	-	1	-	-	-	-	-
Agabus sp.	1	1	-	-	-	-	-
Colymbetes fuscus (L.)	-	2	1	-	-	-	-
Cercyon sternalis Sharp	4	-	1	1	3	-	1
Megasternum obscurum (Marsh.)	-	-	1	-	-	-	-
Hydrobius fuscipes (L.)	3	2	1	1	1	-	-
Anacaena globulus (Pk.)	-	2	-	-	-	1	-
Anacaena sp.	4	1	1	-	-	-	-
Chaetarthria seminulum (Hbst.)	-	-	-	-	1	1	-
Abraeus granulum Er.	-	-	-	-	1	2	-
Ochthebius bicolon Germ.	6	6	1	-	-	2	-
O. minimus (F.)	16	15	6	4	5	3	-
Hydraena riparia Kug.	4	3	-	-	-	-	-
H. testacea Curt.	2	2	1	-	-	-	-
Limnebius sp.	1	-	2	-	-	-	-
Ptiliidae gen. et sp. indet.	-	5	3	2	5	11	1
Agathidium cf. *convexum* Sharp	-	1	-	-	-	-	-
Catops nigricans (Spence)	1	-	-	-	-	-	-
Catops sp.	-	-	-	-	-	1	1
Choleva or *Catops* sp.	-	-	-	-	1	-	-
Stenichnus collaris (Mull.)	1	2	-	1	2	2	-
Scydmaenidae gen. et sp. indet.	1	-	-	-	-	1	-
Micropeplus staphylinoides (Marsh.)	-	1	-	-	-	-	-

Table 20. Coleoptera from Block 26.(continued)

	Peat			Midden			Clay
COLEOPTERA	a	b	c	d	e	f	g
Megarthrus affinis Mill.	-	-	-	-	1	1	-
Olophrum piceum (Gyl.)	2	1	-	-	1	1	-
Olophrum sp.	1	-	-	-	-	1	-
Lesteva heeri Fauv.	7	13	4	2	6	2	-
Omalium sp.	1	-	1	-	-	-	-
Carpelimus sp.	1	1	2	-	5	6	1
Anotylus rugosus (F.)	-	-	-	-	-	1	-
Oxytelus fulvipes Er.	2	1	1	1	1	1	-
Stenus spp.	8	10	4	2	5	5	-
Lathrobium terminatum Grav.	-	1	-	1	-	-	-
Lathrobium sp.	1	1	1	-	-	-	-
Astenus immaculatus Step.	-	-	-	1	1	1	-
Rugilus sp.	-	-	-	-	1	-	-
Xantholinus sp.	-	-	1	-	-	2	-
Erichsonius sp.	2	-	-	-	2	-	-
Philonthus sp.	1	4	1	-	-	1	-
Quedius sp.	1	-	-	-	-	-	-
Tachyporinae gen. et sp. indet.	-	1	1	-	-	2	2
Drusilla canaliculata (F.)	-	-	-	-	2	1	-
Aleocharinae gen. et sp. indet.	13	16	7	7	12	15	-
Bryaxis sp.	2	1	2	1	4	11	1
Pselaphus heisei (Hbst.)	-	-	-	-	1	-	-
Geotrupes spiniger (Marsh.)	-	1	-	-	-	-	-
Geotrupes sp.	1	-	-	-	-	-	-
Aphodius depressus (Kug.)	1	-	-	-	-	-	-
Aphodius sp.	-	-	1	-	-	-	-
Melolontha melolontha (L.)	1	1	1	-	-	-	-
Cetonia aurata (L.)	1	1	-	-	1	1	-
Scirtidae gen. et sp. indet.	22	27	9	7	10	9	1
Dryops sp.	1	1	-	-	-	-	-
Agrypnus murinus (L.)	1	-	-	-	-	-	-
Athous haemorrhoidalis (F.)	-	1	-	-	-	-	-
Athous sp.	3	-	1	-	-	1	1
Selatosomus sp.	1	-	-	-	-	-	-
Dalopius marginatus (L.)	-	1	-	1	-	-	-
Melasius buprestoides (L.)	2	2	1	-	1	-	-
Cantharis sp.	-	2	-	-	-	1	-
Rhagonycha sp.	-	-	-	1	-	-	-
Grynobius planus (F.)	1	-	-	-	1	1	-
Hemicoelus fulvicornis (Sturm)	3	2	2	-	1	-	-
Anobium punctatum (Deg.)	-	2	-	-	-	-	-
Malachius sp.	1	-	-	-	-	-	-
Brachypterolus sp.	1	-	-	-	-	-	-
Rhizophagus dispar (Pk.)	-	-	1	-	-	-	-
Cryptophagus sp.	1	1	1	-	-	-	-
Cerylon fagi Bris.	1	1	-	-	-	-	-
C. ferrugineum Step.	3	-	-	-	-	1	-
Corylophus cassidoides (Marsh.)	1	2	1	1	4	-	-
Orthoperus sp.	-	-	-	-	-	1	-

Table 20. Coleoptera from Block 26 (continued).

	Peat			Midden			Clay
COLEOPTERA	a	b	c	d	e	f	g
Enicmus cf. *fungicola* Thom.	-	-	-	1	-	-	-
E. rugosus (Hbst.)	1	1	-	-	-	-	-
Dienerella separanda (Reit.)	-	-	-	-	-	4	1
Corticarina similata (Gyl.)	1	2	-	-	-	-	-
Corticarina sp.	2	2	1	1	-	-	-
Synchita humeralis (F.)	-	1	-	-	-	-	-
Synchita sp.	1	-	-	-	-	-	-
Opatrum sabulosum (L.)	1	-	-	-	-	-	-
Gonodera luperus (Hbst.)	1	-	-	-	-	-	-
Mordellistena variegata (F.)	-	-	-	1	-	-	-
Grammoptera homelina Pool	1	1	-	1	1	1	-
G. ruficornis (F.)	-	1	-	-	-	-	-
Pogonocherus hispidus (L.)	1	1	1	-	-	-	-
Plateumaris braccata (Scop.)	-	1	-	-	-	-	-
P. discolor (Pz.)	2	-	-	-	-	-	-
P. sericea (L.)	-	-	1	-	-	-	-
Chrysolina banksi (F.)	-	-	1	-	-	-	-
Chalcoides fulvicornis (F.)	-	1	-	-	-	-	-
Tropideres sepicola (F.)	-	1	-	-	-	-	-
Rhynchites caeruleus (Deg.)	-	1	-	-	-	-	-
Apion dissimile Germ.	-	1	-	-	-	-	-
Apion sp.	3	3	1	2	2	1	-
Otiorhynchus ovatus (L.)	1	-	-	-	-	-	-
Phyllobius sp.	-	1	-	-	-	-	-
Polydrusus pterygomalis Boh.	4	3	2	-	1	1	-
Sciaphilus asperatus (Bons.)	-	-	1	-	-	1	-
Brachysomus echinatus (Bons.)	1	-	-	-	-	-	-
Strophosomus melanogrammus (Forst.)	-	2	-	-	-	-	-
Tanysphyrus lemnae (Pk.)	-	1	1	-	-	-	-
Eremotes ater (L.)	1	-	-	-	-	-	-
Rhyncolus lignarius (Marsh.)	-	1	-	1	-	-	-
Acalles ptinoides (Marsh.)	3	9	1	1	1	1	-
Bagous sp.	-	1	-	-	-	-	-
Dorytomus sp.	-	-	1	-	-	-	-
Ceutorhynchus sp.	-	1	-	-	-	-	-
Phytobius comari (Hbst.)	-	2	-	1	-	-	-
Curculio nucum L.	2	1	1	-	1	-	-
Rhynchaenus quercus (L.)	1	3	-	-	-	-	-
Ramphus pulicarius (Hbst.)	-	1	-	-	1	-	-
Acrantus vittatus (F.)	2	1	1	-	-	-	-
Xyleborus dryophagus (Ratz.)	1	1	-	-	-	-	-
Total	181	204	75	47	92	104	11

Cetonia aurata, whose larvae develop in very rotten wood or compost (Britton, 1956: 26; Harde, 1984: 240). Presumably there was a greater concentration of decaying vegetation or rotting wood on the woodland floor in the vicinity of the site when the midden deposits were accumulating than when the peat covered the site.

Two other differences between the faunas are possibly significant. An increase in the diversity of water beetles from the peat compared to the midden does not seem to have been due to the increase in the number of individuals alone. The arrival of such species as *Hydraena riparia* was possibly a reflection of the changing conditions on the woodland floor shown by the caddis larvae, which suggest an onset of water trickling over the site at the start of peat formation. It is also noticeable that the beetles which tend to be indicative of grassland. *Geotrupes*, *Aphodius*, *Agrypnus* and *Athous* spp. were almost confined to the peat, reaching their highest concentration in Sample a, the uppermost sample of Block 26. It is possible that some opening up of the woodland canopy was taking place.

DISCUSSION

The results from the insect analysis support the other lines of environmental evidence that deciduous woodland with a freshwater swamp or fen substrate grew on the site throughout the period of accumulation of the midden sediments and peat. The beetles comprised an "old woodland fauna". Some of the species of decaying trees are now associated with relatively undisturbed unmanaged woods in which elderly trees have been left to die and rot. Such a fauna would be expected from a site of this date, although the Westward Ho! assemblage did not include any of the species which are now extinct in Britain but have been identified from other old woodland deposits (Girling, 1982: 140). The Coleoptera largely support the evidence from the macroscopic plant remains and pollen for the composition of the woodland.

Table 21, Coleoptera from Block 29.

	minimum number of individuals			
	Peat	Midden		Clay
COLEOPTERA	A	B	C	D
Bembidion decorum (Zenk.)	1	-	-	-
B. doris (Pz.)	1	-	-	-
B. guttula (F.)	1	-	-	-
B. unicolor Chaud.	-	1	-	-
Cercyon sternalis Sharp	-	1	-	-
Megasternum obscurum (Marsh.)	-	-	1	-
Hydrobius fuscipes (L.)	1	-	-	-
Anacaena sp.	1	-	-	-
Ochthebius minimus (F.)	2	-	-	-
Hydraena riparia Kug.	2	-	-	-
Catops nigricans (Spence)	1	1	-	-
Catops sp.	-	-	-	1
Lesteva heeri Fauv.	1	-	-	1
Oxytelus fulvipes Er.	1	-	-	-
Stenus spp.	2	-	-	1
Lathrobium terminatum Grav.	1	-	-	-
Quedius sp.	-	-	-	1
Scirtidae gen. et sp. indet.	3	-	-	2
Athous sp.	1	-	-	-
Melasis buprestoides (L.)	1	-	-	-
Hemicoelus fulvicornis (Sturm)	1	-	-	-
Phylan gibbus (F.)	-	1	-	-
Otiorhynchus ovatus (L.)	1	-	-	-
Polydrusus pterygomalis Boh.	1	-	-	-
Acalles ptinoides (Marsh.)	-	-	1	-
Total	23	4	2	6

The evidence of the caddis larvae, water beetles and freshwater bivalve molluscs suggests that the Mesolithic midden sediments accumulated in a shallow water-filled depression, although the depth of the water need only have been a few centimetres. The area remained very wet after the midden had silted up. The overlying peat seems to have had shallower pools and water trickling across its surface. In the immediate vicinity, however, during both the midden and peat phases, there was somewhat less waterlogged ground.

The caddis larvae from the peat phase comprised a fauna appropriate to a series of wet flushes along a spring line. The underlying clay, which slopes gently away from the land, might have formed sufficiently impermeable a barrier that water draining through the peat was forced to emerge at the surface in the vicinity of the midden. Dry fen peat, which supported the element of the flora intolerant of permanent waterlogging, would perhaps have existed above such a spring line whereas below it would have been swampy fen. It is uncertain how far the Mesolithic shore line was to the west of the site, but the insect evidence is for freshwater conditions. However, the flow of water through the peat would probably have reduced any trend towards brackish conditions. There is evidence from the Coleoptera for sand dunes beyond the woodland. It is uncertain where they were located. It is possible that there were dunes on the seaward side of the midden and that they have been entirely eroded with the rise in sea level. A series of coastal dunes would probably have impeded drainage and it is possible that the site was on the landward side of an extensive area of fen and swamp behind dunes.

Unlike the macroscopic plant remains and the molluscs, the insects give no certain evidence of human activity. The high proportion of beetles of decaying plant remains from the midden samples need not be due to decaying occupation refuse cast into the midden. It is possible that the decaying plant remains were in the form of a platform of cut branches and vegetation. However, debris of this sort could also have accumulated naturally. The increase in the number of grassland related beetles at the top of the peat sequence was not sufficiently great to be definitely attributable to clearance activity and it would have long post-dated the use of the midden. The uppermost pollen spectra from the peat possibly show the *Ulmus* decline so it is possible that this slight change reflected the onset of the Neolithic.

Table 22. Host plant requirements of the phytophagous beetles.

AQUATIC AND WATERSIDE PLANTS	
Plateumaris braccata	*Phragmites australis* (Cav.)
P. discolor	Cyperaceae
P. sericea	*Iris pseudacorus* L. & *Carex* spp.
Tanysphyrus lemnae	*Lemna* sp.
HERBACEOUS PLANTS	
Chrysolina banksi	Labiatae esp. *Ballota nigra* L.
Apion dissimile	*Trifolium* spp.
Sciaphilus asperatus	larvae on woodland herbs esp. *Primula vulgaris* Huds.
SHRUBS, TREES AND ROTTING WOOD	
Chalcoides fulvicornis	leaves of *Salix* spp.
Tropideres sepicola	decaying branches esp. *Quercus*
Rhynchites caeruleus	Rosaceae, mostly trees and shrubs
Polydrusus pterygomalis	leaves of young *Quercus*, *Corylus* and other deciduous trees
Eremotes ater	decaying *Pinus*, *Quercus* and other deciduous trees
Rhyncolus lignaruius	decaying *Quercus*, *Ulmus*, *Hedera* etc.
Dorytomus sp.	catkins of *Salix* and *Populus* spp.
Curcculio nucum	nuts of *Corylus avellana* L.
Rhynchaenus quercus	*Quercus* leaves
Ramphus pulicainus	leaves of *Salix* spp., also *Populus* and *Betula* spp.
Acrantus vittatus	under bark of dead *Ulmus*, occasionally *Fraxinus excelsior* L.
Xyleborus dryophagus (*dryographus*)	moribund wood, often *Quercus*

CADDIS FLY (TRICHOPTERA) REMAINS

B.J. Wilkinson

Some of the samples investigated for Coleoptera were analysed quantitatively for Trichoptera remains. Species presence was recorded for the remaining samples. Trichoptera remains were quite common and consisted entirely of larval remains. These comprised pronotal, parietal and frontoclypeal sclerites and cases, one case containing numerous fragments of the larval exuviae. Few species were present and most were limnephilids belonging to the Limnephilinae. The results are given in Table 23.

Overall the most common genus from the site was *Limnephilus*. All specimens were compared with modern specimens or photographs of modern specimens for species identifications. Species of *Limnephilus* are characteristic of slow flowing or still water (Wallace, 1980). The species from this site can be split into three main groups. The majority of the *Limnephilus* found are of a small type with a small brown frontoclypeus, brown parietals and pale brown pronota and mesonota. The frontoclypeus has a pale hind angle and length is not much greater than breadth; the constriction is about half way between anterior and posterior ends. These specimens compare well with *L. affinis* Curtis, *L. auricula* Curtis and *L. incisus* Curtis. Hiley (1976) does not consider himself capable of reliably separating larvae of *L. affinis* and *L. incisus*, hence only tentative identifications of them have been given in Table 23.

All of these species live in small, often grassy, still water bodies; *L. auricula* lives in temporary water bodies. *Limnephilus* group "a" has a brown pronotum similar to that of the *L. affinis* group, these individuals may belong to this group. The subfossil called *Limnephilus* cf. *rhombicus* is of an individual represented by a frontoclypeus with a broad dark stripe on a pale background spreading towards the anterior margin where the length is more or less equal to the width at the anterior margin. Modern species similar to the subfossil are *L. marmoratus* Curtis, *L. politus* McL. and *L. rhombicus* (L.). These species all live in a wide variety of lenitic waters. One pronotum of a *Limnephilus* similar to *L. subcentralis* Brauer was recovered. Other species showing comparable features are *L. borealis* (Zett.) and pale forms of *L. lunatus* Curtis. These species are found in most types of still and slowly flowing water.

Numerous pieces of a limnephilid more or less identical with *Glyphotaelius pellucidus* (Retz.) were recovered from all units of this site. Those identified as *G. pellucidus* are subfossil frontoclypeal apotomes. The frontoclypeus has a length/breadth ratio which is substantially greater than any British *Limnephilus*. The pronotum of *Glyphotaelius* is very similar to certain forms of *Limnephilus*, represented in part here by the *Limnephilus rhombicus* group. These are referred to in the table as ?*Limnephilus* group "b". *G. pellucidus* is usually found in association with the dead leaves of trees, eg, in forest pools. It builds its case from fragments of tree leaf. It is not found in brackish waters.

Two specimens were recovered of the frontoclypeus of a member of the Limnephilinae with the area of the anterior part in front of the central constriction too great for *Limnephilus*. It is, however, very similar to allied genera. These subfossils possessed a dark uniform brown frontoclypeus with a pale fleck either side of the hind angle. This pattern is characteristic of the *Phacopteryx* subgenus of *Anabolia*. The only member of this subgenus in Europe is *Anabolia brevipennis* (Curtis), which these specimens closely resemble. *Anabolia brevipennis* inhabits "small, stagnant often swampy water bodies" (Lepneva, 1971) and "ponds or lakes with the bed covered with dead leaves" (Lestage, 1921), for example in peat bogs. It is not tolerant of brackish water.

The only non limnephilid from the samples analysed quantitatively was a species of leptocerid. This was represented by a pupal case containing some larval mandible and other head capsule fragments, and possibly by another empty case. Leptocerids habitually throw out the larval skin before pupation, so the discovery of even these meagre fragments was the greatest good luck. From these fragments and the case itself it is possible to identify this species as *Athripsodes aterrimus* (Steph.). Wallace (1981) states that *A. aterrimus* is found in "ponds, lakes and slow flowing waters, also faster flowing water of lake exits and below weirs; usually among plants and on muddy sand; widespread and abundant". It is also slightly tolerant of brackish conditions. I have frequently encountered this species as a subfossil in the past, the only one on this site of which I can say this (Elias and Wilkinson, 1983; Wilkinson, 1984).

site of which I can say this (Elias and Wilkinson, 1983; Wilkinson, 1984).

Remains from three other families of Trichoptera in addition to Limnephilidae were identified from the samples for which species presence alone was recorded. The Lepidostomatidae was represented by *Crunoecia irrorata* (Curtis). It inhabits very small, fast flowing forest streams, often being found clinging to submerged roots. A species of Beraeaidae from the genus *Beraea* was also present. It most closely resembled *B. pullata* (Curtis). *Beraea* larvae inhabit very small streams and springs, both British species sometimes being semi-terrestrial. The final species was *Psychomyia pusilla* (F.), a member of the Psychomyidae. It occurs in the head waters and main parts of streams and rivers.

Table 23. Trichoptera from Block 26.

	minimum number of individuals						
	Peat			Midden			Clay
TRICHOPTERA	a	b	c	d	e	f	g
Psychomyia pusilla (F.)	-	+	-	-	-	-	-
Psychomyia sp.	+	-	-	-	-	-	-
Limnephilus cf. *auricularia* Curtis	-	-	2	-	-	1	-
L. cf. *incisus* Curtis	-	1	1	-	-	-	-
L. cf. *rhombicus* (L.)	1	-	-	-	-	-	-
L. subcentralis group	-	1	-	-	-	-	-
Limnephilus group "a"	-	6	-	-	-	-	-
?*Limnephilus* group "b"	-	3	-	-	-	-	-
Limnephilus sp.	-	1	4	3	1	3	-
Glyphotaelius pellucidus (Retzius)	1	5	2	-	1	3	-
Anabolia brevipennis (Curtis)	-	1	1	-	-	1	-
Limnephilidae gen. et sp. indet.	4	-	-	2	1	5	-
Crunoecia irrorata (Curtis)	+	+	+	-	-	-	-
Beraea cf. *pullata* (Curtis)	+	+	-	-	-	-	-
Anthripsodes aterrimus (Steph.)	-	-	-	-	-	1	-
?Leptoceridae (case)	-	1	-	-	-	-	-
Trichoptera (case)	-	-	-	-	-	1	-
Total	6	19	10	5	3	15	0

+ = *only present in other samples*

Samples used for quantitative analysis: 8311030, 8311031 0-35cm, 8311032 0-35cm, 8311033 and 8311614 0-34cm.
For correlation see Table 18.

The overall impression given by the larval ecology of the caddis identified is of sediments and peat accumulating in small pools with dead leaves in the bottom under woodland conditions. *Glyphotaelius pellucidus*, one of the species which suggests such a habitat and occurred almost throughout the sequence is intolerant of brackish conditions. Some change, however, can be seen within the succession. *Athripsodes aterrimus*, only found from the bottom of the midden, is characteristic of rather more open water than the *Limnephilus affinus* group, species of small, sometimes temporary bodies of water, from the overlying peat. Flowing water species were absent from the midden, which seems to have accumulated in a still woodland pool. However, three species which are restricted to flowing water occurred in the peat, their remains being most abundant in the uppermost samples. *Crunoecia irrorata*, the most stenotropic of these species is restricted to very small woodland streams, sometimes no more than trickles. Another of these species, from the genus *Beraea*, is sometimes semi-terrestrial. It is probable that at this stage, the environment of the site comprised fen woodland with wet flushes on the peat surface dribbling freshwater. There is no need to postulate discrete streams flowing across the peat.

II.vii - MAGNETIC MEASUREMENTS

A.E.U. David and A.J. Clark

Inspection of the "blue clays" exposed on the beach at Westward Ho! showed that they were likely to exhibit depositional remanent magnetisation (DRM) suitable for the measurement of former magnetic directions. When a silt contains magnetised particles, these tend to align with the prevailing direction of the Earth's magnetic field while suspended in water. On settlement, this direction becomes fixed in the sediment as a DRM and is retained effectively for ever. Measurement of the angles of dip (inclination) and declination in samples of the deposit can be matched against reference curves of the secular variation of the Earth's field to obtain the date of the deposit.

Altogether, five separate groups of samples were taken, the first three during the exposures of January 1983 and the remaining two in March 1984. On each occasion, owing to weather and tides, the sampling had to be hasty and somewhat curtailed: ideally, more leisurely and detailed sampling would be required and, on the basis of the results presented here, would seem to be worthwhile.

In addition to palaeomagnetic dating, each sample was measured for magnetic susceptibility and intensity, and these results are also discussed.

DRM MEASUREMENTS

In all sampling operations, a number of samples must be taken to compensate for random variations in the remanent direction recorded in the material. Groups 1, 2, and 3 each come from a single layer at each sampling point and are intended only to date this horizon of sedimentation in each case. Groups 4 and 5 are made up of pairs of samples taken one above the other through the deposits in an attempt to identify any sequential change in DRM directions through time. To obtain oriented samples, a roughly level surface was prepared and PVC tubes 5cm in diameter and length were pressed downward into the clay by means of a plate of perspex, which was kept level during the process by observing a bullseye spirit level attached to it. The direction of magnetic north was marked on the top of the tube using a trough compass, and the tube filled with sample was then carved out of the deposit. Table 24 identifies each group of samples and their contexts.

Table 24. Magnetic dating, sampling details.

Group 1: 10 samples top 20cm of silt underlying peat in Area 1

Group 2: 6 samples top 20cm of "lower clay" in Area 3

Group 3: 10 samples top 20cm of "upper clay" in Area 3

Group 4: 16 samples 8 horizontal pairs of samples spaced with 2cm gaps through 0-54cm depth of the "upper clay", in Area 3

Group 5: 8 samples 4 horizontal pairs of samples spaced with 2cm gaps through 0-26cm depth of the "lower clay", in Area 3

In the laboratory, the natural remanent magnetisation (NRM) of the samples was measured in a Digico spinner magnetometer. For Groups 1-3 a pilot sample was then subjected to increasing steps of alternating field partial demagnetisation followed by remeasurement, to test its magnetic stability and to remove any low coercivity viscous magnetisation acquired since deposition. The material was found to be very stable, on average falling to only half its original intensity at 30 microtesla peak-to-peak field. There was no systematic change in the inclinations, but the declination tended to fall fairly steadily but slightly with increasing field by about 2° at this field level, and although it seemed that some viscous magnetisation remained, further demagnetisation would probably have tended to destabilise the basic remanence, and the remaining samples were therefore demagnetised at 30 mT. In the case of Group 4, each individual sample was tested for its optimum demagnetisation level, and as a result the samples were demagnetised at levels of 50, 100, or 150 mT. A pilot sample from Group 5 showed no significant change with demagnetisation, and therefore only the NRM values were used. The results are shown in Table 25.

These results were compared with the curves of Turner and Thompson (1982), based upon measurements on the silts of three British lakes and extending back to 10,000 years cal. BP. The rates of deposition of these silts had themselves been calibrated against radiocarbon measurements of the organic content of the silt cores, and the marker point provided by the elm decline pollen horizon, assumed to be at 5300-5200 cal. BP. Recently obtained archaeomagnetic data indicate that the declination values used in the Turner and Thompson curve are rather too far west, and accordingly it is assumed here that the directions are 7-10° further east than those published. Also, inclinations are reduced by about 1.4° to compensate for the difference in latitude between Westward Ho! and the more northerly mean latitude of the lakes.

A difficulty with magnetic dating is that similar, and sometimes identical, field directions occur at different times, and ancillary evidence is often needed to make the correct choice from alternative possible dates. Such evidence is important in the present case, where the horizontal groups of samples produced remarkably similar readings, and is provided by the radiocarbon dates for the midden material and peat, which constitute an "exclusion zone" from about 9750 cal. BP to 7550 cal. BP.

The dating significance of the results given in Table 25 is discussed below, roughly in chronological order.

"Lower Clay"

Consideration of this deposit must necessarily assume that it is of post-glacial origin, as the relevant calibration does not extend further than 10,000 cal. BP. An earlier origin may be a possibilty, and has been discussed above.

Group 2: Both the NRM and demagnetisation values fit the curves between 8400 and 7800 cal. BP. A distinctive indicator for this period is the inclination, which falls rapidly by 18° from a peak of about 70° before 8400 cal. BP. It does not rise anywhere near this level back to 10,000 cal. BP where the curve ends. A date before this, where the calibration is unknown, cannot be ruled out.

Group 5: This short column of samples fits the trend of the curves between about 8500 cal. BP and 7800 cal. BP. NRM only was measured, because tests on a pilot sample showed that demagnetisation made no significant difference to the results.

The evidence from both these groups, from separate sampling points, combined with radiocarbon dating, indicates that the midden was established soon after the clay ceased to accumulate. This is suggestive of a drop in sea-level at the time.

"Upper Clay"

Group 3: NRM and demagnetisation values suggest alternative date spans of 4200-3800 cal. BP, and 2100-1200 cal. BP.

Group 4: It is tentatively suggested that this column of samples might span the period 3500-1200 cal. BP. This would involve making allowance for slight shallowing of the inclination, as occurs in many sediments. This interpretation depends heavily on the equation of the swing to the east in the middle part of the column with the very substantial swing of the reference curve, which reaches its maximum of about 46° E at, or close to, 2700 cal. BP. A coincident peak in magnetic susceptibility and intensity of the samples suggests a period of major inland ground clearance which might perhaps be associated with the beginning of the Iron Age period. If this interpretation of the age span is correct, then the later of the two alternatives for Group 3, which coincides with the top of the column, would fit in just after this peak on the basis of susceptibility, although the intensity seems too high.

In summary, this deposit was rather unsatisfactory for palaeomagetic work. Considerable discrepancies existed between some of the samples at the same level, and even the easterly swing, upon which the above interpretation relies heavily, showed strongly in

Table 25. Directional magnetic measurements.

GROUP	NRM		AFTER DEMAG	
	Dec	Inc	Dec	Inc
1. Single level	4.0W	70.2	6.4W	69.0
	5.4W	67.1	9.5W	66.4
	3.1W	72.1	7.8W	71.8
	0.4E	68.7	2.3W	67.3
	4.0W	68.5	7.2W	68.9
	0.2W	64.3	0.1E	64.3
	9.2W	67.2	12.5W	62.0
	12.0W	66.5	11.1W	63.5
	0.1W	71.2	0.7E	70.2
	3.2E	66.5	1.7E	64.6
Mean	3.5W	68.2	5.6W	66.8
	α-95=1.7		α-95=2.2	
2. Single level	3.5w	65.2	6.4W	64.6
	6.9W	65.7	1.3W	65.5
	1.7W	67.9	5.8W	66.5
	3.2E	57.3	5.0E	55.6
	8.0W	61.6	14.4W	61.8
	2.7W	64.5	7.0W	63.4
Mean	3.1W	63.7	4.7W	63.0
	α-95=3.4		α-95=4.1	
3. Single level	11.0E	69.6	4.7E	70.9
	6.4E	63.3	4.4E	63.2
	1.9E	64.0	0.7W	62.1
	3.4W	64.2	4.1W	62.7
	1.9W	62.6	2.9W	60.7
	0.7E	74.8	3.4E	75.3
	4.3E	66.4	1.0E	66.0
	5.5W	68.6	9.1W	67.5
	10.3W	70.1	11.9W	67.8
	7.3E	66.3	3.8E	64.4
Mean	1.1E	67.1	1.3W	66.1
	α-95=2.7		α-95=3.0	
4. Column	4.1W	65.6	2.0W	59.5
	0.3W	66.6	10.9E	63.3
	2.7W	69.0	2.1W	67.1
	14.7E	68.8	18.3E	66.2
	3.9E	66.2	13.8E	65.0
	8.2E	68.0	14.4E	67.6
	4.0E	66.1	8.6E	67.4
	0.6E	67.8	4.6E	68.8
5. Column	17.5E	67.5		
	0.4E	63.5		
	0.9W	64.8		
	3.4E	62.2		

α-95 is the angle from the direction given within which there is 95% confidence that the true mean direction of a group of samples lies.

only one of the pair of samples at that level. There were also substantial lateral variations in magnetic susceptibility, which ranged from 6 to 46 x 10^{-8} SI/kg. These problems might be explained by bioturbation or unstable conditions during deposition.

The susceptibility variation could perhaps be due to the samples being taken at the disturbed interface of high and low susceptibility silts. Alternatively, if disturbance was slight, minor height differences in sampling positions relative to the bedding of the silt could have caused samples to include varying proportions of high and low susceptibility material. The implication of both these possible explanations is that the interface between high and low susceptibility silt was narrow and the change therefore quite sudden.

AREA 1 CLAY

Group 1: This clay deposit was separate from those above, and it was hoped that magnetic measurements might enable it to be linked with one or the other. However, the single set of samples, from the upper few centimetres of the layer, produced a directional result closely similar to both of them. A distinctive feature of this deposit is the ratio of the figures for magnetic susceptibility and intensity (see below), approximately 1:13. The only comparable relationship is found in Group 3, from the "upper clay" of Area 3, where the ratio is 1:12.5, although the absolute values are four times as large. On balance, the evidence cannot be taken as confirming a correlation between this and either of the other clay deposits.

MAGNETIC SUSCEPTIBILITY MEASUREMENTS

In addition to the measurement of remanent magnetic directions, each sample was also measured for magnetic susceptibility. Variations in susceptibility reflect both the natural magnetic content of the soil derived from the parent rock, and also the extent to which this magnetism has been enhanced by subsequent burning and, possibly, decomposition - both of which processes are often associated with human occupation. Such magnetic enhancement of soil is a well-known phenomenon, and if anthropogenic soils are destabilized through erosion and introduced into local depositional systems, specific horizons of magnetically characteristic sediment can be recognized. The measurement of magnetic susceptibility and related magnetic parameters alongside other analyses of pollen, trace elements, charcoal and radiocarbon can then be correlated with possible episodes of anthropogenic activity. The use of mineral magnetic measurements in

Table 26. Magnetic susceptibility values.

Group	*n*	Mag. susc. (means of *n* samples)	Intensity (means of *n* samples)
1 (Clay, Area 1)	10	5.88	77.6
2 (Lower Clay, Area 3)	6	5.74	22.1
3 (Upper Clay, Area 3)	10	21.60	268.5
4 (Upper Clay, Area 3)	2	4.2	3.7
depth	2	4.4	4.4
	2	22.7	62.2
	2	66.3	205.0
	2	59.1	178.2
	2	12.8	50.1
	2	9.0	28.8
	2	9.0	22.3
5 (Lower Clay, Area 3)	2	8.5	16.9
depth	2	9.9	31.0
	2	9.6	31.4
	2	11.5	25.5

Units: Magnetic susceptibility - 10^{-8} SI/kg
Intensity - mA/m

study at Westward Ho! is very tentative we feel that this site and others like it can provide considerable potential for future research in this field.

Each of the samples taken for measurement in the spinner magnetometer was measured for magnetic susceptibility in the Bartington Magnetic Susceptibility Meter Type MS1, and the resulting values are tabulated above, alongside those for magnetic intensity derived from the magnetometer.

What is immediately striking about the values in Table 26 is their wide range and their variation in magnitude between different stratigraphic units. Both groups of samples from the "lower clay" in Area 3 have very weak values with little or no magnetic enhancement, which is surprising in that the midden directly overlies this clay and intermingles with its surface at the two sampling points. Burnt material from the midden might be expected to have been introduced into the "lower clay", at least in its upper horizon, and to elevate its magnetic intensity. However, no such enhancement is detectable, and it is therefore a possibility that there were no hearths or burning episodes in this immediate area of the midden. Certainly, the presence of charcoal and a few burnt flints (approximately 10% of those seen) is equivocal evidence of *in situ* burning.

Although there are no direct archaeological associations, the increased magnetic values for the "upper clay" in Area 3 and the clay in Area 1 are suggestive of a link with local human activity contributing magnetically enhanced material to the contemporary sediment load (see above).

The magnetic susceptibility measurements discussed above were made rather as an afterthought to the DRM study, and we feel that both approaches to the deposits at Westward Ho! could well be taken a great deal further if future opportunities arise. Sites such as Westward Ho!, and similar post-glacial catchments elsewhere, invite the application of environmental magnetic techniques alongside more conventional dating and environmental studies: an exercise in interdisciplinary correlation which would surely benefit our understanding of mans' impact on his surroundings.

II.viii - THE FLINTS

A.E.U. David

The flint collection described here was assembled during the field visits in 1983 and 1984, and also includes a number of flints collected by Dr Jacobi during a visit in 1980. A large part of the assemblage was recovered from samples of midden and adjacent deposits removed to the laboratory for the extraction of environmental evidence. The remainder was collected as circumstances allowed from Areas 2 and 3 during the survey and sampling exercise. No flints were found in Area 1 during the recent fieldwork.

Earlier collections

Jacobi has traced a total of 1859 struck flints (and cherts) amongst several former collections, and has tabulated these in detail (1979: 80). Their character is generally later Mesolithic, but provenances are not fully reliable and the number of diagnostic pieces is too small to allow a more satisfactory attribution (Jacobi, pers comm). As we have seen earlier in this paper, there are at least three and probably more locations from which flints may have been collected in the past, from both peat and clay layers.

The present collection

The great majority (1138 pieces) of the collection to be discussed was retrieved from Area 3, mostly from the vertical edges of the erosion gulleys and runnels which are incised into the remaining outcrop of peat and clay here. Flints were common in the exposed midden itself and in the uppermost level of the "lower clay" and the zone of transition between this and the overlying peat. A few pieces were also collected from similar stratigraphic positions from each of two smaller outcrops of peat and clay to the north and north-east, respectively, of the main outcrop in Area 3. Although the latter were not visibly connected to the main site it is a fair assumption that they belong to it and that the artefacts are from the same horizons. An indication of the provenance of most of the lithic material is shown in Figure 3.

In addition to the flints found in and around Area 3, twelve pieces were recovered inshore at Area 2 where another clay-peat sequence was noted and seen to be penetrated by wooden stakes, one of which has been radiocarbon dated to 4840±70 BP (HAR-5642). Two of the flints were found in the peat, the remainder in the underlying clay.

Raw material

Apart from 1.4% of the material, which is of Greensand chert, the remainder is composed of beach pebble flint derived either from contemporary beaches or from exposures of raised beach deposits. A wide variety of small flint pebbles are common on the present-day Pebble Ridge from which it is possible to select raw material very similar to that in use on the prehistoric site. Greensand chert pebbles can also be found on the beach in a proportion roughly equivalent to their occurrence on the site. Portland chert is apparently absent.

The flint is mostly unpatinated (66%), dark grey to black, with a thin pebble cortex and frequent large opaque grey inclusions. The texture is generally fine to very slightly granular and appears to have responded well to knapping. Patination varies widely, from a very light blue-grey tinting of the surface to an opaque white.

THE FLAKED ASSEMBLAGES

Of a total of 1167 pieces from all the find-spots combined, about 40% are less than 10mm in size and are of little dignostic value. The large number of very small pieces is due to their recovery during the fine sieving, through a 5mm mesh, of sediment samples for environmental analysis. Much of this very small material must be the spalls and debris from knapping and retouch, and also includes microlith fragments and microburins. Retouched pieces amongst the remainder of the assemblage are scarce and fragmentary. 10% of the material is burnt. Most of the retouched pieces are illustrated in Figure 16, whilst the composition of the entire assemblage is outlined in Table 27.

Area 3

By far the greater part of the assemblage is composed of the debitage and fragments of core reduction and tool manufacture, the remains of which, along with at least one instance of conjoining primary flakes, suggest that all stages of knapping were practised on the site.

Extrapolating from the abandoned flaked flint pebbles present in the assemblage, core reduction has been initiated on pebbles varying in weight between at least 41g and 320g, resulting in a range of platformed bladelet cores of which there are 22 here. 187 cores are known from earlier collections, and measurement of the maximum dimension (core "height") from a sample of 136 of these shows an average height of 31mm with the majority within a size range (less than 36mm) believed to be typical of later Mesolithic technologies present on the coast of the south-west peninsula and west Wales (Johnson and David, 1982: 80-84, Fig. 10). All the cores (25) collected recently from Wesward Ho! conform with this data, having an average height of 32mm, with extremes at 17 and 48mm. A selection of these is illustrated in Figure 17.

Considerable attention seems to have been given to the production of bladelets, and the very high representation of microburins (40) demonstrates that many of these were intended for microlith manufacture, with the strong implication that an important preoccupation at the site was with the preparation and repair of hunting or fishing equipment. The finished microliths are scarce, however, being heavily outnumbered by microburins (1:13), and may be presumed to have been lost elsewhere. Only 15 microliths are known amongst all the material available from Westward Ho!, and only three of these come from the present collection: two lanceolate forms and a portion of a scalene triangle (Fig. 16, 30-32). Both types occur commonly within the typological range of the later Mesolithic in southern England, but the microliths from Westward Ho! do not form a sufficiently large or representative sample for inclusion within the suggested seriation for the later Mesolithic as part of a "Southern English Territory" (Jacobi, 1979; Jacobi and Tebbutt, 1981; Smith, 1982: 50, Fig. 20).

Contrasting with the abundant evidence for microlith manufacture is the poor representation of processing tools such as burins and scrapers. Only three burins are recorded, all from the earlier collections, but scrapers are rather more frequent, numbering twelve, of which five are from this collection (Fig. 16, 45-49). Three of these are steeply retouched primary flakes with a convex and somewhat denticulated outline (45, 47,and 48), whilst the remaining two are developed on secondary flakes. Number 50 is an example of a piece intermediate between scraper and core, common on many sites but of uncertain status (but see below). All these scraper types are entirely consistent with a later Mesolithic attribution, and indeed all of the remaining retouched and utilized items (ie, nos. 38-44) could be accommodated within any of the more prolific Mesolithic artefact scatters along West Country and Welsh coasts.

Microscopic examination of seven of the tools (a retouched primary flake from Jacobi's collection, and Fig. 16, nos. 45 - 50) by Dr E.H. Moss did not result in any reliable indications of usage. Although the flint was suitable for the preservation of use-wear traces, only very slight traces were seen on two of the scrapers (46 and 45) and the core/scraper (50), and in no case was it possible to identify the worked material. Interestingly, the core/scraper showed definite indications of usage which "match those produced by planing red deer antler which has been soaked in water" (E. Moss, pers comm), suggesting that at least one of these ambiguous artefacts may have served as a tool.

A notable absence from the assemblage so far recovered from Area 3 is that of the heavier duty tools such as picks, choppers, and bevelled or ground stone tools. Bevelled pebbles are a frequent, if enigmatic, component of later Mesolithic lithic inventories on western coastal sites. If their absence from Westward Ho! is real, it would suggest a corresponding absence, eastwards from Hartland Point, of some activity relevant to rock-shore environments.

Dating

The earliest evidence for human activity at Westward Ho! comes from a single feature observed in the "lower clay" (see above) containing at least fourteen flints and charcoal. The latter (identified as *Quercus* and *Corylus*) has given a radiocarbon date of 8180+150 BP (HAR-5643), and this precedes by a full millennium the sequence of dates obtained for the occupation material in the top of the "lower clay" and for the midden deposit itself. The flints found within this feature are unfortunately composed of undiagnostic debitage indistinguishable in character from those seen amongst the midden material, and discussion of this putative earliest evidence of a Mesolithic presence cannot, on this evidence, be taken any further.

Radiocarbon dates on material from the top of the "lower clay" and from the midden span the seventh millennium BP (uncalibrated) (see Table 1), a dating in broad agreement with the typological evidence for a later Mesolithic attribution suggested by the flints found intermingled amongst these deposits. The sample of lithic equipment recovered is insufficient to make any independant contribution to the chronological record for the locality. It has to be admitted that the database and interpretational framework currently available for the Mesolithic of southern England is so thin that, even were the lithics data from Westward Ho! more distinctive, these could not, with any real confidence, be given an independant age (Jacobi, pers comm).

Within the peat in Area 3 there is some slight evidence for mesolithic activity that might post-date the midden: a few flints including a microburin were found here in 1983 and a small and discrete group of apparently Mesolithic flints were reported by Rogers (1946: 129, Plate XXI). No specifically Neolithic or Bronze Age type flints have been

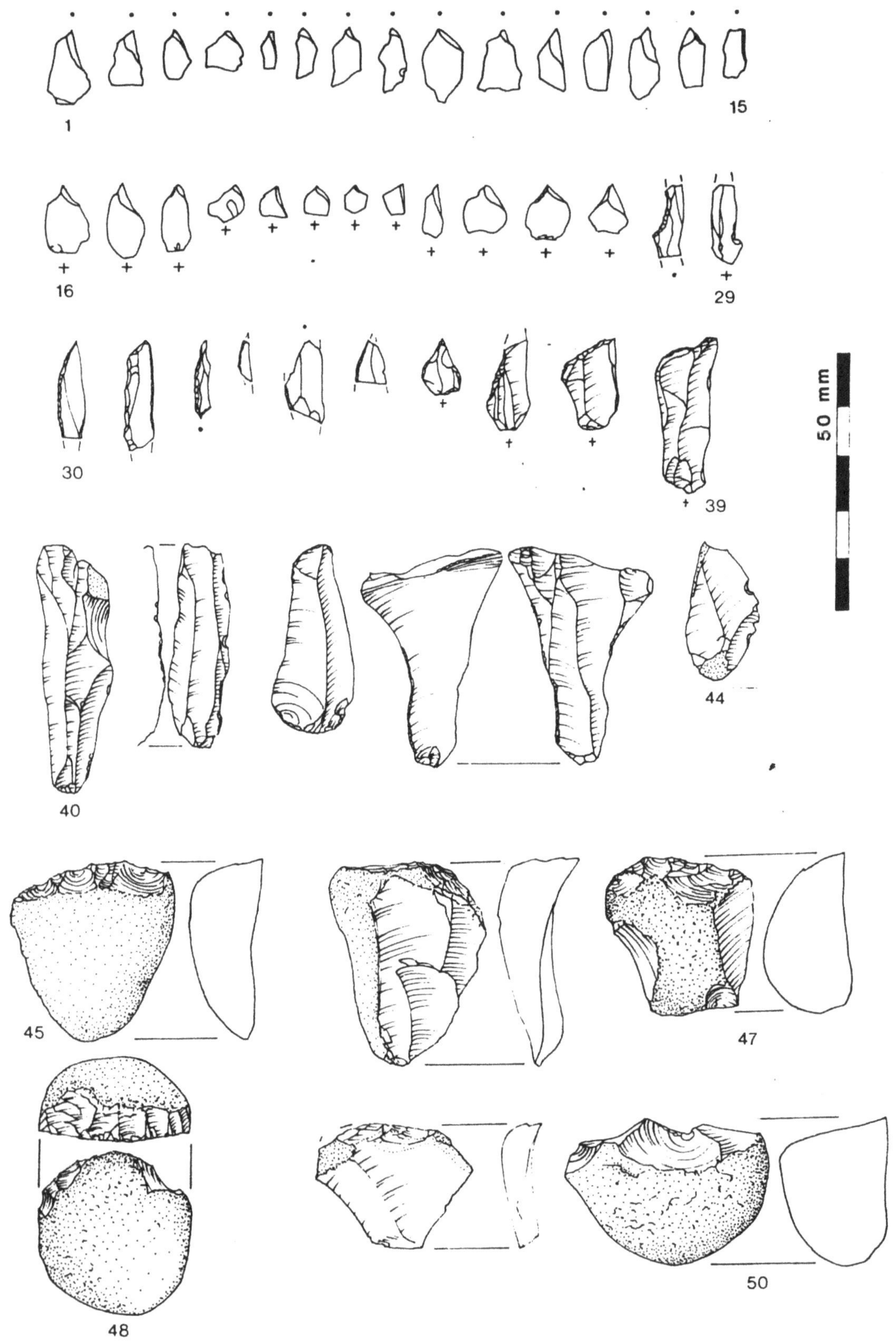

Figure 16. The flint flakes.

recorded from the peat.

Area 2

The affiliation of this small group of flints (see Table 27) must remain in doubt as there are no fully diagnostic items, and no independant dating evidence. The presence of Neolithic stakes in this exposure may be irrelevant to the flints whose occurrence near the surface of a clay, overlain by a peat, strongly suggests a Mesolithic attribution similar to that proposed for Area 3.

An indication of the degree of erosion on the beach, and also that later prehistoric activity may have been present in the general vicinity of Area 2 is suggested in a report by Ellis (1867) in which he noted flints occurring between 1 and 8 inches (*c* 2 to 20cm) from the top of a 4 feet (*c* 1.2m) peat deposit at a point mid-way between high and low water mark.

Area 1

Although no flints were found in this area during the survey work, it would not be surprising if they were present, either as residual deposits or *in situ*. Mr D. Trapnell who walks the beach regularly has noted flints hereabouts, and of those collected by him and seen by the writer at least one is a struck flake.

CONCLUSIONS

The only flint assemblage of any consequence at Westward Ho! comes from the exposures in Area 3. Here, one can detect an emphasis on bladelet production, with the presence of a large number of microburins which may be taken as indicative of "gearing up", or the making use of locally available pebble flints to provide armatures for equipment to be used elsewhere. There is no direct evidence for the purpose of such armatures but there is at least the possibility, based on the find of similar microlith fragments associated with a pig skeleton found in peat at Lydstep Haven, Dyfed (Leach, 1918: 50-51), that they were used in the hunt for land mammals.

Although there are a few items of processing equipment alongside the products of microlith manufacture, there can again be little more than speculation on their usage or their relationship to the faunal and floral remains detected in the midden. Bevelled pebbles, notable by their absence, may therefore belong with an activity such as seal hunting which is unrepresented at Westward Ho! If there is to be anything more than a continuing guesswork as to the character of later Mesolithic sites such as this, a more concerted approach to the functional analysis of the relevant lithic tool types is one requirement. This, alongside considerations of their raw material, context and taphonomy, might provide important interpretative information where detailed organic preservation is lacking.

The main outcrop of submerged exposures at Area 3 is contained within some 1600m2, a size well within that of many dry-land Mesolithic flint scatters. If this site is or was contiguous with the finds of flint in the subsidiary exposures to the north and east, and even with Area 2, then a considerably more substantial area of Mesolithic activity, covering a hectare or more, has to be considered. Judging from earlier accounts such as that of Hall (1879), the greater part of an extremely rich and extensive later Mesolithic site (or sites) has gone. The significance of the surviving deposits and their contained lithic and organic material can only be considered with great caution, representing, as they do, only a very small and arbitrary selection dictated by the whim of tidal erosion and exposure. Nevertheless we must be thankful here in catching a glimpse of a type of site located at or near contemporary sea level, which elsewhere may have been removed or concealed by subsequent marine encroachment. Apart from Blashenwell (Preece, 1980), Westward Ho! remains the only site in southern England where a later Mesolithic technology has been set into a convincing environmental setting

Table 27. Composition of the lithic assemblages collected during 1980 (by R.M. Jacobi) and during 1983-4 (CEU sampling and survey).

	R M Jacobi collection 1980, Area 3	Area 3	Exposure N of Area 3	Area 2	Total	% of total assemblage
Tools						
Classifiable microliths	-	3	-	-	3	
Microlith fragments	1	8	-	-	9	
Scrapers	-	5	-	-	5	
Truncated pieces	-	2	1	-	3	
Awls	-	1	-	-	1	
Denticulated pieces	1	-	-	1	2	
Retouched pieces	4	2	-	-	6	
Notched	-	1	-	-	1	
? Utilised pieces	-	6	-	-	6	
TOTAL	6	28	1	1	36	3.1
By-Products						
Microburins (bulbar)	7	12	-	-	19	
Microburins (distal)	3	15	1	-	19	
Microburins (bulbar/distal)	-	3	-	-	3	
Microintermediate	-	1	-	-	1	
Mis-hits	2	2	-	-	4	
TOTAL	12	33	-	-	46	3.9
Debitage						
Pieces >10mm	23	436	-	2	461	39.5
Primary flakes/blades/frags.	23	101	4	1	129	11.1
Secondary flakes/blades/frags.	116	290	8	7	421	36.1
Cores	6	16	2	1	25	2.1
Core rejuvination flakes	-	2	-	-	2	0.2
Flaked nodules	4	8	-	-	12	1.0
Unflaked nodules	-	1	-	-	1	0.1
Unclassified fragments (>10mm)	1	32	1	-	34	2.9
TOTAL	173	886	15	11	1085	93.0
ABSOLUTE TOTAL	191	947	17	12	1167	100.0
Burnt pieces (>10mm)	18	91	5	3	117	10.0
Patinated pieces (>10mm)	108	272	9	4	393	33.7
Greensand chert pieces (>10mm)	3	13	-	-	16	1.4

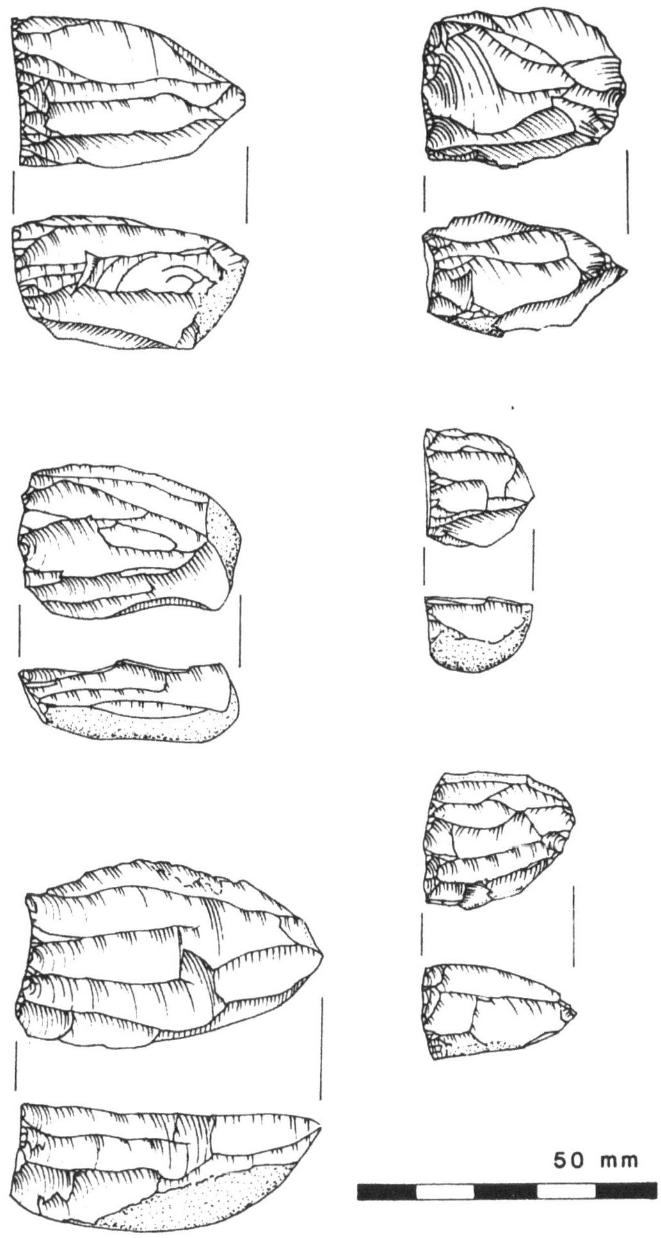

Figure 17. The flint cores.

REFERENCES

Andersen, S.Th. 1970. The relative pollen productivity and pollen representation of North European trees and correction factors for tree pollen spectra *Danm. geol. Unders.* Ser. II, **96**.

Arber, M.A. 1982. The Bideford Museum geological collection of Inkerman Rogers (1866-1959) *The Geological Curator* **3.4**, 199-208.

Arnold, E.N. and Burton, J.A. 1978. *A field guide to the reptiles and amphibians of Britain and Europe* London: Collins.

Balfour-Browne, F. 1940. *British water beetles 1* London: Ray Society.

Balfour-Browne, F. 1950. *British water beetles 2* London: Ray Society.

Bate, C.S. 1866. Attempt to approximate the date of the flint flakes of Devon and Cornwall *Rep. and Trans. Devon. Assoc.* **1**, 128-136.

Beaver, R.A. 1966. Notes on the fauna associated with elm bark beetles in Wytham Wood, Berks., 1 Coleoptera *Entomologist's Monthly Magazine* **102**, 163-170.

Bedel, L. 1901. *Faune des Coleoptes du Bassin de la Seine V* Paris: Societe Entomologique de France.

Birks, H.J.B., Deacon, J. and Peglar, S. 1975. Pollen maps for the British Isles 5000 years ago *Proc. Roy. Soc. London B* **189**, 87-105.

Brendell, M.J.D. 1975. *Coleoptera: Tenebrionidae* (Handbook for the identification of British insects 5, pt 10). London: Royal Entomological Society.

Britton, E.B. 1956. *Coleoptera: Scarabaeoidea* (Handbook for the identification of British insects 5, pt 11). London: Royal Entomological Society.

Bullock, P., Federoff, N., Jongerius, A., Stoops, G., and Tursina, T. 1985. *Handbook for soil thin section description* Wolverhampton: Waine Research Publications.

Churchill, D.M. 1965. The kitchen midden at Westward Ho!, Devon, England: ecology, age and relation to changes in land and sea level *Proc. Prehist. Soc.* **31**, 74-84.

Clapham, A.R., Tutin, T.G. and Warburg, E.F. 1981. *Excursion flora of the British Isles* Cambridge: University Press.

Clason, A.T. 1967. *Animals and man in Holland's past* (2 vols) Groningen: J.B. Walters.

Clutton-Brock, J. 1985. *A biometrical analysis of neolithic antler picks from Grimes Graves, Norfolk and Durrington Walls, Wiltshire* London: British Museum.

Coles, J.M. 1971. The early settlement of Scotland: excavation at Morton, Fife *Proc. Prehist. Soc.* **37ii**, 284-366.

Coles, B. and Coles, J.M. 1986. *Sweet Track to Glastonbury* London: Thames and Hudson.

Dimbleby, G.W. 1976. The history and archaeology of heaths *in* J.H.P. Sankey and H.W. Mackworth-Praed (eds) *The Southern Heathlands* Guildford: Surrey Naturalists Trust.

Donisthorpe, H.St J.K. 1939. *A preliminary list of the Coleoptera of Windsor Forest* London: N. Lloyd.

Duchaufour, P. 1982. *Pedology* London: George Allen & Unwin.

Elias, S.A. and Wilkinson, B.J. 1983. Lateglacial insect fossil assemblages from Lobsigensee (Swiss Plateau) *Revue de Paléobiologie* **2**, 189-204.

Ellis, H.S. 1866. On a flint-find in a submerged forest of Barnstaple Bay, near Westward Ho! *Rep. and Trans. Devon. Assoc.* **I(v)**, 80-1.

Ellis, H.S. 1867. On some mammalian bones and teeth, found in the submerged forest at Northam *Rep. and Trans. Devon. Assoc.* **2**, 162-3.

Evans, J.G. 1972. *Land Snails in Archaeology* London: Seminar Press.

Evans, J.G. and Jones, H. 1973. Subfossil and modern land-snail faunas from rock-rubble habitats *J. Conchology* **28**, 103-29.

Findlay, D.C., Colborne, G.J.N., Cope, D.W., Harrod, T.R., Hogan, D.V. and Staines, S.J. 1983. *Soils of England and Wales. Sheet 5, South West England (1:250,000)* Harpenden: Soil Survey of England and Wales

Fowler, W.W. 1890. *The Coleoptera of the British Islands 4* London: L. Reeve.

Fowler, W.W. 1891. *The Coleoptera of the British Islands 5* London: L. Reeve.

Freude, H., Harde, K.W. and Lohse, G.A. 1967. *Die Käfer Mitteleuropas 7* Krefeld: Goecke and Evers.

Freude, H., Harde, K.W. and Lohse, G.A. 1983. *Die Käfer Mitteleuropas 11* Krefeld: Goecke and Evers.

Geological Survey, 1977. *Bideford Sheet* Southampton: Ordnance Survey.

Girling, M.A., 1982. Fossil insect faunas from forest sites *in* M.G. Bell and S. Limbrey (eds) *Archaeological aspects of woodland ecology* (Symposia of the Association for Environmental Archaeology No 2.) Oxford: British Archaeological Reports (International Series **146**), 129-146.

Godwin, H. 1940. A Boreal transgression of the sea in Swansea Bay. Data for the study of post-glacial history VI *New Phytologist* **39**, 308-332.

Godwin, H. 1975a. History of the natural forests of Britain: establishment, dominance and destruction *Phil. Trans. Roy. Soc. B* **271**, 47-67.

Godwin, H. 1975b. *The History of the British Flora* Cambridge: University Press.

Hall, T.M. 1870. The raised beaches and submerged forests of Barnstaple Bay *The Student and Intellectual Observer (of Science Literature and Art)* **4**, 338-349.

Hall, T.M. 1879. The submerged forest of Barnstaple Bay *Quart. J. Geol. Soc. Lond.* **35**, 106.

Hanak, V. 1977. *A colour guide to familiar mammals* London:Octopus.

Harcourt, R. 1974. The dog in prehistoric and early historic Britain *J. Archaeol. Sci.* **1**, 151-175.

Harde, K.W., 1984 *A field guide in colour to beetles* London: Octopus.

Hawkins, A.B. 1971. The late Weichselian and Flandrian transgression of South West Britain *Quaternaria* **14**, 115-130.

Hawkins, A.B. 1973. Sea level changes around South West England *in* D.J. Blackman (ed) *Marine Archaeology* London: Butterworths.

Heyworth, A. 1978. Submerged forests around the British Isles: their dating and relevance as indicators of post-glacial land and sea level changes *in* J. Fletcher (ed) *Dendrochronolgy in Europe* Oxford: British Archaeological Reports (International Series **51**).

Heyworth, A. and Kidson, C. 1982. Sea Level changes in southwest England and Wales *Proc. Geol. Assoc.* **93 (1)**, 91-111.

Hiley, P.D. 1976. The identification of British limnephilid larvae (Trichoptera) *Systematic Entomology* **1**, 147-167.

Hoffman, A. 1958. *Coleopteres Curculionides Faune de France 62* Paris: Lechevalier.

Horion, A.D. 1953. *Faunistik der Mittleuropäischen Käfer 3* München: G. Frey.

Horion, A.D. 1963. *Faunistik der Mittleuropäischen Käfer 9* Überlingen-Bodnsee: Aug. Feyel.

Jacobi, R.M. 1973. Aspects of the "Mesolithic Age" in Britain. *in* G.K. Kozlowski (ed) *The Mesolithic in Europe* (Papers read to the International Symposium on the Mesolithic in Europe) Warsaw: University Press, 237-266.

Jacobi, R.M. 1975. Aspects of the Postglacial Archaeology of England and Wales. Unpublished PhD thesis. University of Cambridge.

Jacobi, R.M. 1978. Population and landscape in Mesolithic lowland Britain *in* S. Limbrey and J.G. Evans (eds) *The effect of man on the Landscape: the lowland zone* London: Council for British Archaeology (Research Report 21), 75-85.

Jacobi, R.M. 1979. Early Flandrian hunters in the South-West *Proc. Devon Archaeol. Soc.* **37**, 48-93.

Jacobi, R.M. 1980. The early Holocene settlement of Wales. *in* J.A. Taylor (ed) *Culture and Environment in Prehistoric Wales* Oxford: British Archaeological Reports (British Series, **76**, 15-100).

Jacobi, R.M. and Tebbutt C.F. 1981. A late Mesolithic rock shelter at High Hurstwood, Sussex *Sussex Archaeol. Collect.* **119**, 1-36.

Johnson, N. and David, A.E.U. 1982. A Mesolithic site at Trevose Head and contemporary geography *Cornish Archaeol.* **21**, 67-103.

Jones, R.T., Wall, S.M., Locker, A.M., Coy, J. and Maltby, M. 1981. Computer based osteometry data capture user manual (1): Ancient Monuments Laboratory Report No. 3342.

Kerney, M.P. 1976a. *Atlas of non-marine Mollusca of the British Isles* Monks Wood: Institute of Terrestrial Ecology.

Kerney, M.P. 1976b. Two Postglacial faunas from South-West England *J. Conchology* **29**, 71-3.

Kerney, M.P. 1977. A proposed zonation scheme for late-glacial and post-glacial deposits using land Mollusca *J. Archaeol. Sci.* **4**, 387-90.

Kerney, M.P., Preece, R.C. and Turner, C. 1980. Molluscan and plant biostratigraphy of some late Devensian and Flandrian deposits in Kent *Phil. Trans. Roy. Soc. London B* **291**, 1-43.

Kidson, C. 1982. Sea level changes in the Holocene *Quaternary Science Reviews* **1**, 121-151.

Kidson, C. and Heyworth, A. 1973. The Flandrian sea level rise in the Bristol Channel *Proc. Ussher Soc.* **2**, 565-584.

Kidson, C. and Heyworth, A. 1976. The Quaternary deposits of the Somerset Levels *Quart. J. Eng. Geol.* **9**, 217-235.

Kidson, C. and Heyworth, A. 1979. Sea "level" *in* K. Suguio, T.R. Fairchild, L. Martin and J.-M. Flexor (eds) *Proceedings of the 1978 International Symposium on Coastal Evolution in the Quaternary* Sao Paulo, Brazil, 1-28.

Kloet, G.S. and Hincks, W.D. 1977. *A check list of British insects, 2nd edition (revised): Coleoptera and Strepsiptera* (Handbook for the identification of British insects 11, pt. 3). London: Royal Entomological Society.

Lacaille, A.D. and Grimes, W.F. 1955. The prehistory of Caldey *Archaeologia Cambrensis* **104**, 85.

Leach, A.L. 1918. Flint working sites on the submerged land (submerged forest) bordering the Pembrokeshire coast *Proc. Geol. Assoc.* **29 (2)**, 46-64.

Legge, A.J. 1981. The agricultural economy *in* R.J. Mercer (ed) *Grimes Graves, Norfolk excavations 1971-72 (1)* London: HMSO.

Lepneva, S.G. 1971. *Fauna of the U.S.S.R. Trichoptera* Vol 2, No 2 *Larvae and pupae of Integripalpia.* Jerusalem: Israel Program for Scientific Translations.

Lestage, J.A. 1921. *in* E. Rousseau *Les larves et nymphes aquatiques des insectes d'Europe 1* Brussels.

Levitan, B. *forthcoming.* The vertebrate remains from Hazleton North barrow. *in* A. Saville *Hazleton North 1979-1982: the excavation of a Cotswold-Severn tomb in Gloucestershire.*

Lindroth, C.H. 1974. *Coleoptera: Carabidae* (Handbook for the identification of British insects 4, pt. 2) London: Royal Entomological Society.

McAndrews, J.H. 1966. Post-glacial history of Prairie Savanna and forest in North Western Minnesota *Mem. Torrey Bot. Club* **22**, 1-72.

Mellars, P.A. 1978. Excavation and economic analysis of Mesolithic shell middens on the island of Oronsay (Inner Hebrides) *in* P.A. Mellars (ed) *The Early Postglacial Settlement of Northern Europe* London: Duckworth, 371-96.

Miedema, R., Jongmans, A.G. and Glager, S. 1974. Micromorphological observations on pyrite and its oxidation products in four Holocene soils in the Netherlands *in* G.K. Rutherford (ed) *Soil Microscopy: 4th International meeting on Soil Microscopy* Kingston, Ontario: Limestone Press.

Moore, P.D. and Webb, J.A. 1978. *An Illustrated Guide to Pollen Analysis* London: Hodder and Stoughton.

Pearson, G.W., Pilcher, J.R., Baillie, M.G., Corbett, D.M. and Qua, F. 1986. High precision 14C measurement of Irish Oaks to show the natural 14C variations from AD 1840 to 5210 BC *Radiocarbon* **28(2b)**, 911-934.

Pengelly, W. 1868. The submerged forest and the pebble ridge of Barnstaple Bay *Rep. and Trans. Devon. Assoc.* **2**, 415-422.

Preece, R.C. 1980. The biostratigraphy and dating of the tufa deposit at the Mesolithic site at Blashenwell, Dorset, England *J. Archaeol. Sci.* **7**, 345-62.

Reid, C. 1896. An early Neolithic kitchen-midden and tufaceous deposit at Blashenwell near Corfe Castle *Proc. Dorset Nat. Hist. and Antiq. Fld. Club* **17**, 67-75.

Risdon, T. 1811. *The Chorographical Description or Survey of Devon* London.

Rogers, E.H. 1946. The raised beach, submerged forest and kitchen midden of Westward Ho! and the submerged stone row of Yelland *Proc. Devon Archaeol. Expl. Soc.* **3**, 109-135.

Rogers, I. 1908. On the submerged forest at Westward Ho!, Bideford Bay *Rep. and Trans. Devon. Assoc.* **40**, 249-259.

Scaife, R.G. 1980. Late-Devensian and Flandrian palaeoecological studies in the Isle of Wight. Unpublished PhD thesis. University of London, Kings College.

Scaife, R.G. 1982. Late-Devensian and early Flandrian vegetational changes in southern England *in* S. Limbrey and M.G. Bell (eds) *Archaeological Aspects of Woodland Ecology* (Symposia of the Association for Environmental Archaeology No 2). Oxford: British Archaeological Reports (International Series **146**), 57-74.

Scaife, R.G. 1983. Pollen analysis of a monolith from Westward Ho!: Ancient Monuments Laboratory Report 4239.

Scaife, R.G. 1984. A history of Flandrian vegetation in the Isles of Scilly: Palynological investigations of Higher Moors and Lower Moors peat mires, St. Mary's. *Cornish Studies* **11**, 33-47.

Scaife, R.G. 1986. Further pollen analysis of the Westward Ho! submerged peat deposits: Ancient Monuments Laboratory Report 4884.

Simmons, I.G. and Dimbleby, G.W. 1974. The possible role of ivy (*Hedera helix* L.) in the Mesolithic economy of western Europe *J. Archaeol. Sci.* **2**, 291-296.

Smith, M. 1969. *The British amphibians and reptiles* (4th edn) London: Collins.

Smith, A.G. 1970. The influence of Mesolithic and Neolithic man on the British vegetation: a discussion *in* D. Walker and R.G. West (eds) *Studies in the Vegetational History of the British Isles* Cambridge: University Press, 81-96.

Smith, A.G. 1981. The Neolithic *in* I.G. Simmons and M. Tooley (eds) *The Environment in British Prehistory* London: Duckworth, 125-209.

Smith, A.G. and Pilcher, J.R. 1973. Radiocarbon dates and the vegetational history of the British Isles *New Phytologist* **72**, 903-914.

Smith, G.H. 1982. The excavation of Mesolithic, Neolithic and Bronze Age settlements at Poldowrian, St. Keverne, 1980. *Cornish Archaeol.* **21**, 23-62.

Stephens, N. 1970. The west country and southern Ireland *in* C.A. Lewis (ed) *The Glaciations of Wales and Adjoining Regions* London: Longman, 267-314.

Stephens, N. 1973. South-west England *in* G.F. Mitchell, L.F. Penny, F.W. Shotton and R.G. West (eds) *A Correlation of Quaternary Deposits in the British Isles* London: Geological Society of London (Special Report No 4), 36-45.

Stuart, A. and Hookway, R.J.S. 1954 "Coastal erosion at Westward Ho! Devon": Report to the Coast Protection Committee (Special) of the Devon County Council.

Tansley, A.G. 1939. *The British Isles and their Vegetation* Cambridge: University Press.

Tooley, M.J. 1978. *Sea level changes in North West England during the Flandrian stage* Oxford: University Press.

Tooley, M.J. 1982. Sea-level changes in northern England *Proc. Geol. Assoc.* **93(1)**, 43-51.

Turner, G.M. and Thompson, R. 1982. Detransformation of the British geomagnetic secular variation record for Holocene times *Geophys. J. Roy. Astr. Soc.* **70**, 789-792.

Thompson, R. and Oldfield, F. 1986. *Environmental Magnetism* London: Allen and Unwin.

Thorley, A. 1981. Pollen analytical evidence relating to the vegetational history of the chalk *J. Biogeography* **8**, 93-106.

von den Dreisch, A. 1976. *A guide to the measurement of animal bones from archaeological sites* Harvard: Peabody Museum (Bulletin No. 1).

Walker, J.J. 1906. Preliminary list of Coleoptera observed in the neighbourhood of Oxford from 1819 to 1907 *Ashmolean Natural History Society Report for 1906*, 49-100.

Wallace, I.D. 1980. The identification of British limnephilid larvae (Trichoptera): Limnephilidae) which have single-filament gills *Freshwater Biology* **10**, 171-189.

Wallace, I.D. 1981. A key to larvae of the family Leptoceridae (Trichoptera) in Great Britain and Ireland *Freshwater Biology* **11**, 273-297.

Welin, E., Engstrand, L. and Vaczy, S. 1971. Institute of Geological Sciences Radiocarbon dates III *Radiocarbon* **14**, 331-335.
Went, A.E.J. 1966. Some ancient Irish salmon fishing weirs *Industrial Archaeol.* **3**
Went, A.E.J. 1969. Ancient sprat fishing weirs in the South of Ireland *Industrial Archaeol.* **6**
Wilkinson, B.J. 1984. Interpretation of past environments from sub-fossil Caddis larvae *in* J.C. Morse (ed) *Proceedings of the Fourth International Symposium on Trichoptera*
Woodman, P.C. 1981. A Mesolithic camp in Ireland *Scientific American* **245**, 92-100.
Woodman, P.C. 1986. *Excavations at Mount Sandel, Ireland* Belfast: HMSO.
Woolner, D. and Woolner, A. 1966. A midden at Borough, Stoke Gabriel, Devon *Proc. Devon Archaeol. Soc.* **23**, 31.

www.ingramcontent.com/pod-product-compliance
Lightning Source LLC
LaVergne TN
LVHW072250300926
849160LV00018B/1636
* 9 7 8 0 8 6 0 5 4 5 0 1 9 *